Preventing Industrial Acc

Herbert William Heinrich has been one of the most influential safety pioneers. His work from the 1930s/1940s affects much of what is done in safety today – for better and worse. Heinrich's work is debated and heavily critiqued by some, while others defend it with zeal. Interestingly, few people who discuss the ideas have ever read his work or looked into its backgrounds; most do so based on hearsay, secondary sources, or mere opinion. One reason for this is that Heinrich's work has been out of print for decades: it is notoriously hard to find, and quality biographical information is hard to get.

Based on some serious "safety archaeology," which provided access to many of Heinrich's original papers, books, and rather rich biographical information, this book aims to fill this gap. It deals with the life and work of Heinrich, the context he worked in, and his influences and legacy. The book defines the main themes in Heinrich's work and discusses them, paying attention to their origins, the developments that came from them, interpretations and attributions, and the critiques that they may have attracted over the years. This includes such well-known ideas and metaphor as the accident triangle, the accident sequence (dominoes), the hidden cost of accidents, the human element, and management responsibility.

This book is the first to deal with the work and legacy of Heinrich as a whole, based on a unique richness of material and approaching the matter from several (new) angles. It also reflects on Heinrich's relevance for today's safety science and practice.

Carsten Busch has studied Mechanical Engineering, Safety, and Human Factors. He has over 25 years of experience in Safety and Quality Management at various levels in organisations ranging from railway to oil & gas to police in The Netherlands, the United Kingdom, and Norway. He is professionally active on various forums, owner of mindtherisk.com, tutor at Lund University Human Factors and System Safety programme, and author of *Safety Myth 101*, *Veiligheidsfabels 1–2–3*, and *If You Can't Measure It... Maybe You Shouldn't*. His main research interests include the history of knowledge development and discourse in safety.

Preventing Industrial Accidents

Reappraising H. W. Heinrich – More than Triangles and Dominoes

Carsten Busch

Routledge
Taylor & Francis Group

LONDON AND NEW YORK

First published 2021
by Routledge
2 Park Square, Milton Park, Abingdon, Oxon OX14 4RN

and by Routledge
52 Vanderbilt Avenue, New York, NY 10017

Routledge is an imprint of the Taylor & Francis Group, an informa business

British Library Cataloguing-in-Publication Data
A catalogue record for this book is available from the British Library

Library of Congress Cataloging-in-Publication Data
A catalog record has been requested for this book

ISBN: 978-0-367-34380-4 (hbk)
ISBN: 978-0-429-32539-7 (ebk)

Typeset in Times New Roman
by codeMantra

Contents

List of figures xi
Preface xiii
Acknowledgements xv
List of abbreviations xvii

1 Introductions 1
 1.1 Why this book? 1
 1.2 Isn't the past, the past? 2
 1.3 Citing Heinrich in safety literature 3
 1.4 Limitations 5
  1.4.1 The lenses of today (1) 5
  1.4.2 Conflicting aims 5
  1.4.3 Was Heinrich right? 6
  1.4.4 Developments and critique 6
  1.4.5 Standing in Heinrich's shoes 6
  1.4.6 The lenses of today (2) 7
  1.4.7 100,000 Words 7
 1.5 About references 8

2 A biography 11
 2.1 Early years 11
 2.2 The Travelers 13
 2.3 First steps in safety 14
 2.4 Emerging foundations 16
 2.5 The book 20
 2.6 Dominos and cause code 22
 2.7 Meanwhile at Travelers 22
 2.8 Teaching, speaking, and another book 24
 2.9 War! 25
 2.10 Post-war work 27

2.11 Heinrich, the person 28

2.12 Retirement 29

3 Heinrich's work   **34**

3.1 General overview 34

    3.1.1 Phases 35

    3.1.2 Developments and changes 35

    3.1.3 Industrial Accident Prevention – an overview 37

3.2 Main themes 39

    3.2.1 Scientific approach 40

    3.2.2 The economics of safety 40

    3.2.3 Causation 40

    3.2.4 The human element 40

    3.2.5 The role of management 41

    3.2.6 The triangle and reacting on weak signals 41

    3.2.7 Axioms of safety 42

    3.2.8 Professionalisation of safety 42

    3.2.9 Practical accident prevention 42

    3.2.10 Safety management 42

    3.2.11 Social engagement 43

    3.2.12 What about risk? 43

3.3 Influences 44

    3.3.1 "Best of" or original work? 44

    3.3.2 The Travelers 45

    3.3.3 Bibliography 47

    3.3.4 Associates and acquaintances 54

    3.3.5 Others 55

3.4 Success and longevity 56

3.5 Heinrich's influence and developments 58

    3.5.1 Lateiner 59

    3.5.2 Bird and loss control 59

    3.5.3 Petersen and the fifth edition 60

    3.5.4 Behaviour Based Safety 62

    3.5.5 Manuele 65

    3.5.6 Serious injuries and fatalities 66

    3.5.7 "New view" safety 67

    3.5.8 Heinrich or not Heinrich? 68

3.6 Conclusion 69

4 A scientific approach   **73**

4.1 Introduction 73

    4.1.1 Critique 74

4.2 Why did "scientific" make sense? 75
    4.2.1 Understanding the term 75
    4.2.2 Structured method, facts, and principles 77
    4.2.3 Practicable and useful 80
    4.2.4 Marketing 83
    4.2.5 Scientific management and Heinrich 84
4.3 What about the missing data? 88

5 The economics of safety   **94**
5.1 Early safety and economics 94
5.2 Efficiency 96
5.3 Hidden costs 99
    5.3.1 The idea 99
    5.3.2 Research and the 1:4 ratio 100
    5.3.3 The study's importance 101
    5.3.4 After Heinrich 102
5.4 Heinrich and the Great Depression 104
5.5 A continuous theme 105

6 Accidents are caused   **109**
6.1 The importance of causes 109
    6.1.1 Early 1900s causation 110
    6.1.2 Real causes? 112
    6.1.3 Why did "real cause" make sense? 113
6.2 Dualism of causation 115
    6.2.1 Origin of accidents 115
    6.2.2 Developments 118
    6.2.3 Cause code 120
    6.2.4 Critique 122
    6.2.5 Dismantling the "grand statement" 124
    6.2.6 Another reflection 127
6.3 Accidents as processes 128
    6.3.1 The dominos 129
    6.3.2 A powerful metaphor 131
    6.3.3 Critique 1: linearity 131
    6.3.4 Why did linearity and simplicity make sense? 133
    6.3.5 Critique 2: direct causes 134
    6.3.6 Why did focus on direct causes make sense? 135
    6.3.7 Evolution 138
6.4 A more complete accident model 140
    6.4.1 Multiple causes 140
    6.4.2 Reasons and subcauses 141

6.4.3 Underlying causes 143
6.4.4 Enhancing Heinrich's accident model 145
6.5 Final reflections 146

7 The human element   **150**
7.1 Humans as an element to control 150
7.1.1 Why did this make sense? 151
7.1.2 Dealing with humans 152
7.1.3 Causes and actions 156
7.2 Various concepts 158
7.2.1 Psychology 158
7.2.2 Accident proneness 161
7.2.3 Ancestry 163
7.2.4 Human error? 165
7.2.5 Carelessness 166
7.2.6 Shame and blame 168
7.3 Humans as an asset 169
7.3.1 The reward of merit 171

8 The role of management   **176**
8.1 Foremen and supervisors 177
8.1.1 Contemporary thinking 177
8.1.2 Heinrich and foremen 179
8.1.3 Professionalisation 180
8.1.4 Supervisors and investigation 183
8.1.5 Post-Heinrich 185
8.2 Top management 186
8.2.1 Planning 188
8.3 Responsibility 189
8.4 Integrating safety 191
8.5 Safety management 193
8.5.1 Early developments 193
8.5.2 Axioms 194
8.5.3 Structured management tools 197
8.5.4 Another metaphor 198
8.5.5 Misconceptions 201

9 The triangle   **205**
9.1 Origin and development 205
9.1.1 Conception 205
9.1.2 A ziggurat 206
9.1.3 Researching and building 209

9.1.4 Earlier ideas 210
9.1.5 1931 211
9.1.6 The triangle 212
9.1.7 Variations 212
9.1.8 1950 214
9.1.9 The triangle expanded 216
9.1.10 Post-Heinrich triangles 217
9.2 *Studying the triangle 221*
9.2.1 Ways to read the triangle 221
9.2.2 Underlying principles 224
9.3 *Interpretations and critique 230*
9.3.1 Behaviour 231
9.3.2 Fixed ratios 232
9.3.3 Proportional reduction 234
9.3.4 Prediction 234
9.3.5 Metric 237
9.3.6 Zero 238
9.4 *Some reflections 240*
9.4.1 Opportunity 240
9.4.2 Challenges and limitations 241
9.4.3 Conclusion 244

10 Other main themes   **248**
10.1 *Professionalisation of safety 248*
10.2 *Social engagement 250*
10.3 *Practical remedy 252*

11 Heinrich in the 21st century   **254**
11.1 *Famous last words 258*

Appendix 1: The Heinrich bibliography   259
References   263
Index   281

Figures

1.1	Citations of *Industrial Accident Prevention* per year	4
1.2	An illustration of the use of primary Heinrich sources	7
6.1	*Origin of Accidents* chart	117
6.2	The 1941 *Origin of Accidents* chart	119
6.3	The 1954 *Origin of Accidents* chart	121
6.4	The accident sequence	129
6.5	Removing the middle domino "prevents the accident"	134
6.6	A more complete model of Heinrich's causation	145
8.1	The chain of supervision	182
8.2	Heinrich's safety ladder	199
9.1	Foundations chart	208
9.2	1941 version of the triangle	213
9.3	A novel depiction of the 1:29:300 ratio	214
9.4	The graphic representation from *Basics of Supervision*	215
9.5	The expanded 1959 triangle	217
9.6	A simple event tree scenario – reducing the input might theoretically reduce outcomes proportionally	235
9.7	A counter-intuitive effect	235

Preface

Like many other safety professionals, I was raised and educated with Heinrich's metaphors and ideas. During my early years, I even conducted some incident investigations in which I placed causes and events in various stages of the dominos. I never believed that it was a purely linear five-step sequence; it was just a very practical way of thinking about events and ordering them in a logical and branching sequence.

Fast forward, almost a decade. I decided that I should work on a proper professional library and quickly acquired books by people like James Reason, Charles Perrow, and a few others. I also decided that I had to own one of the founding texts of the profession and dashed out a decent amount of cash for two different versions (1941 and 1980) of *Industrial Accident Prevention*.

While I read the other books more or less as soon as they arrived, it would take several years for me to finally turn to Heinrich's work. Annoyed by the fact that many safety professionals discussed his work without ever having read it, I decided that I was not going to fall into that trap. In a way, this marked the beginning of my journey as a "safety mythologist," even though I was not aware of it at the time. It also sparked my curiosity. Why did Heinrich look at certain things the way he did when such ideas did not make sense to *me*? His "fixation" on "direct causes" was one of the things I would have loved to ask him about.

Working on my thesis and this book provided me with an opportunity to do exactly that. Not ask him in person, of course, but attempt to find answers in his writings and the historical context. This proved to be an at times fascinating archaeological quest that evoked a "Yes!!" feeling whenever another small piece of the Heinrich puzzle(s) came to light, or I learned something surprising. Who would have thought that Heinrich has written management books besides his safety classic? Also, there was the fascinating archaeology into Heinrich's influences – providing greater insight into the foundations of my profession.

Sometimes, the research also evoked mixed feelings – shifting from enthusiasm about great insights a long time ago to shock at things like the original dominos with ancestry. Also, it makes me sad how much of a negative impact the 88% ratio and wrong readings of the triangle have had. This leaves me conflicted in my study of Heinrich. He fascinates me and has done great things for the profession, but he has also written some stuff that leaves me sad...

Anyway, this was an inspiring intellectual challenge that made me think and rethink concepts and ideas several times. I hope that others also learn something from reading this – and enjoy it with at least a fraction of the fun I had writing it.

Acknowledgements

It is not one person who writes a book but an entire network around him that contributes, knowingly and unknowingly. So, let me thank a small selection of this network, even if it runs the risk of forgetting some.

Guy Loft and Matthew Ranscombe at Taylor & Francis for helping in the making of this book.

My mentors at Lund, Johan Bergström and Anthony "Smokes" Smoker, for their guidance and wisdom.

Tusen Takk to Bill Lischeid and Mary-Beth Davidson from Travelers Insurance. Without their help, I would never even have come close to the depth and width in this study and discussion of Heinrich's work. Both were extremely generous with their time and resources, providing all the information they had. Mary-Beth, especially, made several tours into the Travelers Archives to check out yet another possible clue. Thanks so much. Many thanks to Travelers Insurance for their kind permission to use some of their material. If all companies showed the level of cooperation that Travelers did during my research, I think researchers will have a much easier life.

I think we should all thank Jesse Bird, who took the effort to painstakingly document Heinrich's work and life, and create a rich – but regrettably unpublished – source of information on one of safety's pioneers. It is sad that Jesse, who passed away in February 2010, cannot witness this, but I am glad that I can present some of his work to a wider audience.

Paul Swuste and Coen van Gulijk for helping me to get hold of some of Heinrich's and Lateiner's papers, and doing some of the groundwork with their series of historical papers, on which I could build.

Donald Lateiner for providing information about his father's life and work. Roger Brauer of the historical society for enabling access to some hard-to-get sources and doing some digging into the backgrounds of early authors, which helped me a lot. I must also mention Alaina Kolosh of the National Safety Council, Lucas Clawson of Hagley Museum and Library, and Sue Trebswether of the ASSP for helping me to get access to a couple of hard-to-find articles.

Of course, I also need to thank countless safety professionals that have engaged in discussions on Heinrich, which helped to shape and improve my thinking on the subject. I am sure that I will forget some, but let me at least mention, in no particular order, Phil La Duke, Todd Conklin, Alan Quilley, Cary Usrey, Robert Long, Scott Gesinger, Craig Marriott, Nick Gardener, Martijn Flinterman, Frank Guldenmund, David van Valkenburg, Bart Vanraes, Peter Booster, Linda Wright (our work together kinda sparked this), Jos Villevoye (great closing quote), Erik Hollnagel (Heinrich and quality), Ron Gantt (discussing prediction), Jean-Christophe Le Coze (models, visuals, and much more), Richard Cook (no reports!), Roel van Winsen (discourse and stuff), David van Valkenburg (FRAM and H), Johan Roels and Wim Top (for sharing some stories about Frank E. Bird Jr. with me), Dave Rebbitt (pyramids!), Jean Pariès (more pyramids), Andrew Hale (even more pyramids), Linda Bellamy (yup, pyramids), Jim Nyce (for introducing me to some other Hartford intellectuals and a great quote), Walter Zwaard (I am so sorry that our awesome book on risk is delayed; hopefully soon), and IPS and Mats Lindgren for the encouragement by awarding my thesis.

Thanks to Martijn Flinterman, Richard Abbott, and Rosa Carrillo for reading some chapters and giving some valuable feedback. Special thanks to Tristan for doing an awesome job on the proofreading. It is good practice to state that all the mistakes are mine, and they are. Tell me if you find them.

Obviously, the main dedication of this book is to my lovely wife, Annemarije, and my family, who have been super supportive.

Abbreviations

AEC	American Engineering Council
ASME	American Society of Mechanical Engineers
ASSE	American Society of Safety Engineers
ASSP	American Society of Safety Professionals
BBS	Behaviour Based Safety
BLS	Bureau of Labor Statistics
CWA	Civil Works Administration
DNV	Det Norske Veritas
ETTO	Efficiency Thoroughness Trade-Off
HOP	Human & Organizational Performance
ILCI	International Loss Control Institute
IOSH	Institution of Occupational Safety and Health
LTI	Lost Time Injury
NSC	National Safety Council
NVVK	Nederlandse Vereniging voor Veiligheidskunde (Dutch Society for Safety Science)
OHS	Occupational Health and Safety
PDCA	Plan Do Check Act
SCAT	Systematic Cause Analysis Technique
SIF	Serious Injuries and Fatalities
SOAT	Systematisch Oorzaken en Analyse Technieken

Chapter 1

Introductions

1.1 Why this book?

There is this quote, commonly attributed to Dan Ariely, which I have paraphrased on several occasions. It says, "Big data is like teenage sex: everyone talks about it, nobody really knows how to do it, everyone thinks everyone else is doing it, so everyone claims they are doing it."[1] It serves as an icebreaker in some settings, often leading to nervous laughter. At the same time, there is truth in it.

I think it also applies to the work of Herbert William Heinrich. Pop into a random professional safety forum, and there is a high chance of finding a discussion about one of Heinrich's concepts.[2] If you hang around for a while, you may notice that little of that discussion is fuelled by actual knowledge. Many, instead of researching and trying to understand, "fall into the Heinrich was a god versus Heinrich was a monster argument" (La Duke, 2019).

This is not restricted to practitioners. In recent years, Heinrich's legacy received increasing critique from contemporary safety authors, from those counted among the "new view," as well as some rooted in more traditional approaches. However, few authors seem to engage with Heinrich's work properly and systematically. Instead, many authors and safety professionals revert to an extreme position by either unquestioningly accepting and echoing Heinrich's ideas (or a contemporary derivate) or dismissing them entirely – with rather little middle ground (Busch, 2019b).

This book aims to rectify some of that, by diving deep into Heinrich's work and (re)viewing it in its context and in combination with older and contemporary thinking in safety. This book offers safety professionals and students (and others interested in backgrounds and history of safety science and practice) to acquire good second-hand knowledge of Heinrich's work and some of its backgrounds.

Heinrich's work is out of print for decades and most of it is hard to find. One of the aims of this book is to make Heinrich's work accessible for others. Still as a secondary source, but more complete and hopefully more nuanced than alternative sources and with some new insights. On a few occasions the book will also try to figure out why it made sense for Heinrich to write what he wrote at the time.

The text will do so by extracting cues and looking for patterns (Weick, 1995), and by exploring questions that inform us about what could have been his *local rationality*.[3] What was his knowledge at the time? What were his objectives? What outside factors affected him? What was the context he acted (wrote) in? Applying the local rationality perspective to some topics in Heinrich's writings that *we* may perceive as strange, bizarre, or erroneous, but possibly were entirely sensible to *him*, may give us better understanding.

1.2 Isn't the past, the past?

One may wonder whether it makes any sense to dive this deep into theories and ideas that go back eight decades and more. Should time and resources be spent on theories and ideas that have been used and abused so much that most of their contemporary use is rather counterproductive? Should we instead follow Marriott's suggestion when he finishes his discussion of the "triangular fallacy," stating, "Unfortunately, it has been so tainted now, that it would be better to remove it altogether and consign it to history" (Marriott, 2018, p.30)? He is not alone, also others advocate leaving the past behind and look forward, "…it is time for the old ideas to be swept into history…" (Gesinger, 2018, p.2).

My immediate reply to the question whether we should bother at all is a clear "Yes!" Heinrich's work has been of great importance for safety practice and theory. Dwyer pinpoints *Industrial Accident Prevention*, as the start of safety as "an academic and a practical discipline" (Dwyer, 1992, p.266). As we will see later, this assessment is open for discussion. More correct and reasonable is Manuele's statement from his seminal book *On the Practice of Safety*, "H.W. Heinrich has had more influence on the practice of safety than any other author. Heinrich's premises have been adopted by many as certainty. They permeate the safety literature" (Manuele, 2013, p.57).

Although Heinrich's legacy has proven to be a subject of discussion among practitioners and academics for years, many safety professionals (and scholars) have not even read his work. However, lack of information and knowledge has never stopped humans of having an opinion. The analysis in this book, and the discussion of numerous misconceptions and attributions, shows us that there is some need for clarification and better understanding. It is important to have an *informed* opinion, which means one must engage and study. Study critically, taking in view both how things looked back then and from the benefit of today's knowledge.

Knowing about history and where one comes from is important, also as a profession, and for science, "…safety should be examining its history for trends and at least trying to use that study to identify the challenges it may have to face in the future" (Townsend, 2013).

To study, one needs to engage with an open mind. Instead of flatly reject-
ing an "old" approach as some seem to suggest, it may be wise to first take
a look at the baby and only throw out the dirty water from the bath – even
though there may be the possibility that there is only dirty water to be found
in the tub. However, I think that much of Heinrich's work is still relevant
today, at least within certain applications.

Manuele's *On the Practice of Safety* quotes Dan Petersen from the 1998
edition of *Techniques of Safety Management* saying,

> In the safety profession, we started with certain principles that were
> well explained in Heinrich's early works. We have built a profession
> around them, and we have succeeded in progressing tremendously with
> them. And yet in recent years we find that we have come almost to a
> standstill. Some believe that this is because the principles on which our
> profession is built no longer offer us a solid foundation. Others believe
> they remain solid but that some additions may be needed. Anyone in
> safety today at least ought to look at that foundation – and question it.
> Perhaps the principles discussed here can lead to further improvements
> in our approach and further reductions in our record.
>
> (Petersen, 1998, p.27 quoted in Manuele, 2013, p.13)

I think there is some truth in this. More and critical research and reflection
with a non-binary view is advisable, and I hope this book kickstarts some.

1.3 Citing Heinrich in safety literature

Before we turn to Heinrich's life and work, let us have a quick look at how
safety science publications reference Heinrich. Heinrich's work has been influential for safety theory and practice for many decades. Still, reviewing
the academic attention for his work shows some surprising results. A simple
search in Scopus, showing the number of publications citing *Industrial Accident Prevention*,[4] indicates a continuously increasing number of publications citing Heinrich's book. Almost nine decades after the publication of
the first edition, four decades after the publication of the last version, and
with the book being out of print for a long time, "common sense" might
suggest a picture of declining interest as new theories and new authors enter
the field. Figure 1.1 seems to paint another picture, however.

From the first citation in 1970, the diagram shows a slow increase until the
early 2000s. The average number of citations per year from 1970 to 1999 is
below five. Some years see no citation at all. In 2002, the number of citations
suddenly starts a steep climb, peaking in 2015 (119 times referenced) and 2019
(170). Even the half year shown for 2020 is higher than any year before 2010.

The below picture only shows academic quotations. However, the vast
amount of safety literature is rather *practical* than *academic*. Much of safety

Industrial Accident Prevention references (source: Scopus, 3 June 2020)

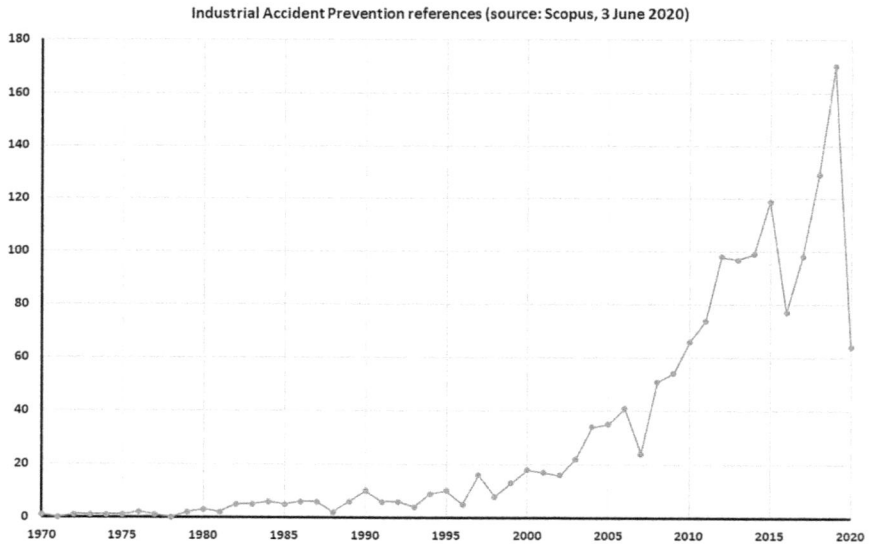

Figure 1.1 Citations of *Industrial Accident Prevention* per year.
Source: Scopus.

literature aimed at practitioners can be found in professional magazines (e.g. from associations as ASSE, NSC, IOSH, or NVVK), books, course materials, guidance documents, and websites.[5] These sources are rarely found in academic databases like Scopus, and attempting some form of systematic bibliographic research in "practical" literature encounters serious difficulties.

Another limitation of the search is that it returns mainly papers published in academic journals and hardly any books. A cursory check shows that books by for example Dan Petersen (multiple works from the 1970s to early 2000s), Dekker, and two of Fred Manuele's notable works, dealing (partly) with Heinrich's work, *On the Practice of Safety* and *Heinrich Revisited*, are missing. Interestingly, several of Hollnagel's books *are* included – but not all of them. Neither is Frank E. Bird's work, strongly influenced by, and continuing many of Heinrich's core concepts, included.

The search and diagram above are thus a rather crude way of looking at the matter, but it may nevertheless suggest a direction and indicate a growing number of academic papers citing Heinrich. Whether this is due to the general increase of safety publications in the 2000s (Busch, 2019b), an increased interest in his work, or the fact that authors cite him because they feel they should do so as part of the "academic name dropping" (Hopkins, 2014, p.8),[6] one cannot say without diving deeply into the material.[7]

1.4 Limitations

Literature-based research, and especially studying "ancient" literature, brings a number of implicit limitations that one must acknowledge.[8] In order to keep the scope and material, depth and width within what fits within this book, there have to be some additional limitations, which means that there must be efficiency thoroughness trade-offs (ETTO) (Hollnagel, 2009) on the way.

1.4.1 The lenses of today (1)

Jean-Christophe Le Coze remarked,[9] "history is always seen through the lenses of today." This creates challenges when reading literature from past generations. One encounters archaic language. One does not fully understand the implication of certain words and expressions. Words may look familiar, but they may have had somewhat different meanings at the time. Texts may describe processes and conventions that are unknown to us, and in general, we will lack much of the context (social, cultural, political, economic, technical, etc.).[10] Also, we risk assigning contemporary meaning and understanding to the words and phrases we read.

This means that this book sets out for an almost paradoxical quest of evaluating safety-related texts critically against current knowledge and state of the art while at the same time trying not to judge against today's standards, both scientifically and regarding the language chosen. Additionally, it tries to identify elements that may be relevant even today, or in the future.

Reading with an open mind can be difficult, because author and readers alike bear with them much professional (and cultural, social, and other) bias. Many have been in pro/contra Heinrich discussions before and that should have left an effect. It is important to try to read *what he actually did write*, not what we have heard about it, or what we imagine he wrote.

1.4.2 Conflicting aims

This book aims to present Heinrich and his work, put it in context, review critique and developments, and add some analysis to this. This means the book has to strike a difficult balance between standing back and observing on one side and interacting and providing critique on the other side. And in-between the book tries to systematise, structure, analyse, and make sense.

Most likely, the balance will tip towards presenting information, but it is important to try to understand Heinrich as well, and much of the critique he has received will be seen in that light. And yet, the book will not shy away from critical reflections, and even proper critique of his work where this is thought to be due.

1.4.3 Was Heinrich right?

Many people would surely like a conclusive answer to that question. This book is not primarily intended as an appraisal of Heinrich's work. The aim of this book is not to look into whether certain ideas or theories (like the much-criticised triangle) are right or wrong. Neither is there any intent to defend Heinrich or his work. One of the prime objectives is to look at why things may have made sense to him – based upon the available material – and present a *second story*[11] (Woods et al., 2010).

1.4.4 Developments and critique

This book discusses developments of Heinrich's ideas by people such as Bird, Petersen, or Krause. However, the space to do so in-depth is limited. Neither will there be an in-depth study how unnamed educators and practitioners may have twisted and altered concepts in their communication and practice. It is likely that these developments have affected how we view Heinrich today, but it is hard to assess in what way without a major study – if it is possible at all.

The same applies to the critique that Heinrich has received over the years. This book only allows a selection, which will be discussed with varying depth instead of doing a full review of *all* critique and developments. Some essential and important elements, as the triangle and the 88:10:2 ratio, get more attention, because of their impact on safety theory and practice.

1.4.5 Standing in Heinrich's shoes

We must also ask ourselves if we actually can determine why things made sense to Heinrich. By trying to get into his shoes, studying his work closely and trying to suspend judgement we may come some way. We must not forget, however, that just as we cannot interview a pilot who became one of the fatal victims of a plane crash, we cannot ask Heinrich in person what he thought when he wrote what he wrote.[12] So, as with the plane, we must rely on the "black boxes" left behind and what we can learn from the context – fragmentary and incomplete as this may be, decades after.

I have tried to compensate for this limitation as much as possible in an attempt to get a better – or at least less superficial – picture of the man and his work, by reading as much of Heinrich's writings as possible (several times over) and systematically annotating them. Also, by researching newspapers and studying and expanding his biography, and many of his (possible) influences and contemporaries in the hope that this will give a better picture of him as a person and professional.

Figure 1.2 illustrates the richness of primary Heinrich-sources in this book compared to previous studies of Heinrich's work.

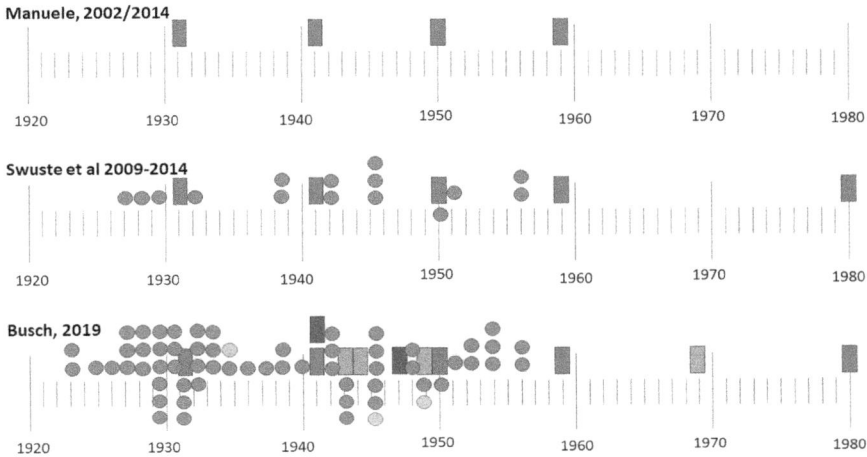

Figure 1.2 An illustration of the use of primary Heinrich sources.

1.4.6 The lenses of today (2)

Another pitfall is to attribute today's insights to Heinrich. There is a possibility that with our 2020 knowledge and filters, we may construct or read things into Heinrich's writing that Heinrich did not intend or understand in that way at all. For example, when one reads "Individuals have hobbies and pet aversions which often lead them (conscientiously enough, but mistakenly) to emphasize unnecessarily the importance of specific hazards" (HWH1931a, p.258), one may interpret this as Heinrich talking about risk perception (Slovic, 1987). It seems that he understood the concept but did not know the term.[13]

Another example is, when Heinrich notes "…the use of severity as a measure can accomplish no more than frequency and has the disadvantage of overemphasizing the injury instead of the real objective…" (HWH1931a, p.260) that he talks about outcome bias (Baron & Hersey, 1988), a term that was not even coined at the time.

Several other of these "insights" can be found throughout Heinrich's work, and we have to be cautious concluding to what this means to us or him.

1.4.7 100,000 Words

That may seem a lot, but it is not if you desire to give a rich account. Consequently, this book has to manage within its limits and as a proper ETTO try to provide the basic information regarding Heinrich's life and work, the context he operated in and touch upon some developments and critique.

To write means to leave out, to edit, to select. Research time is inevitably limited, and some sources are hard or impossible to access. Even if the author manages to use primary sources and wants to present things as they were said and written as much as possible, full texts cannot be provided, and quotes can only give so much information. Leaving parts out and presenting some without their context, may colour the message in some ways, or prevent the reader from making up his/her own mind. Besides, the selection of quotes may be biased – despite the attempt to give a complete and balanced picture. This means that there will be many nuances lost underway and only a limited picture can be provided.[14]

Sometimes this text may be just scratching the surface, but hopefully at the same time nudge the reader into further exploration. The book intends to give the reader a glimpse and definitely lead towards thinking critically, exploring and investigating. Do follow sources and endnotes, and feel free to contact the author.

1.5 About references

Emphasis in quotes is from the original source, unless indicated otherwise.

References are collected in the list of references, except for Heinrich's works. An overview of his (known) work is found in Appendix 1.

References follow APA-style with the exception of Heinrich's work where I have chosen to omit the author's name and instead indicate them by HWH, followed by year, and if relevant the page number of a quote.

Many of the referenced early safety books, including a few Heinrich and Travelers sources, are available online through either https://archive.org/, https://jstor.org; https://hathitrust.org/, or https://books.google.com/

References to newspaper articles from *The Hartford Courant* and other newspapers are included as endnotes. These can be accessed through https://courant.newspapers.com/

Endnotes provide additional depth and reflection about parts of the main text. Sometimes they also comment on sources. Endnotes are no essential reading but often fun or useful to follow-up.

Now, to quote the man,

...it is time to stop talking, roll up the sleeves, and *go to work*.

(HWH1931a, p.268)

Notes

1 Dan Ariely posted it on his Facebook page in 2013: https://www.facebook.com/dan.ariely/posts/904383595868
2 To count as a "Heinrichian theme" in this book, the subject has to be connected directly to Heinrich, for example by the mention of one of his metaphors

(e.g. domino, triangle) or ratios. A mere discussion of causation or human error in general does not qualify.

3 The concept of local rationality draws on the concept of "bounded rationality." Simon (1969) explained that information in situations exceeds by far human ability to process it. Humans have limited (sensory) means to gather information. Knowledge and memory are limited. Due to the limited resources to process information, humans have to make decisions within these limitations. Humans can never be fully rational; at best, their rationality is "bounded." In all variations on the concept since Simon, there is an assumption that "full rationality" exists. It just eludes humans. In his discussion of the evolution of bounded rationality, Dekker notes that at some point the term "local rationality" (Woods et al., 1994) was introduced to explicitly make clear that,

> The agent not only does not need perfect rationality; it can never achieve it. It locally senses its inputs and notes its outputs (actions), inducing regularities that hold in its environment. What these agents do is locally rational - assessments and actions make sense given the agent's active goals, current knowledge and focus of attention.
>
> (Dekker, 2017, p.556)

One of the best ways of describing local rationality may be a thought attributed to Jens Rasmussen:

> As Jens Rasmussen is purported to have said "If you don't understand why it made sense for people to do what they did, it is not because they were behaving really strangely, bizarrely, or erroneously, it is because your perspective is wrong." We need to understand what was going on in the world of those involved and see through their eyes at the time of the incident.
>
> (Long, 2018, p.4)

4 Conducted 3 June 2018. Note that the search returns all five editions of the book.
5 Another factor is that much professional literature is written in the language of the country it is published in, which may limit its inclusion in academic databases and its international use.
6 Hopkins discusses this in the context of Perrow's Normal Accident Theory. The wonderful full quote:

> I suspect the fact is that while people continue to make reference to the theory, this is no more than lip service. We are dealing here with one of the more unfortunate aspects of academic practice. People refer to the works of others not necessarily because that work supports their arguments or are in any other way relevant to what is being said, but simply to establish that they are aware of the relevant literature. Such citations amount to little more than academic name dropping. I have myself been cited by people who seem unaware that my point is quite the reverse of theirs and that my work undermines their own conclusion, rather than supporting it. I suspect that this process of catch-all citation is part of the reason the theory of normal accidents continues to be cited.
>
> (Hopkins, 2014, p.8)

7 Also, note that this is an increase in *absolute numbers* of references to Heinrich in recent years. Looking at a *relative* measure, for example the number of publications mentioning Heinrich divided by the total number of papers published that year, instead of an increase, one would rather see a somewhat randomly fluctuating line.

8 The following section is an updated version of the limitations chapter from my thesis (Busch, 2019b).

9 1 May 2018, Learning Lab *A Critical View on Safety Culture* at Utrecht.

10 One good example, relevant here, is how hard it is to understand for us what role insurance companies played in safety at the time, and how and why they did so. Most have experiences with insurance companies and seeing them in a (partly) humanitarian role may just be hard to imagine.

Another example of the use of language differing strongly from today is the "sexist," or rather gender-biased, language in early safety literature. The sources speak mostly of "safety men," "foremen," or "men" when speaking of workers. Some authors did acknowledge female workers, but the texts were written in a time when many workplaces (especially industrial ones) and the safety profession were mostly man-dominated (despite Crystal Eastman's outstanding contribution as early as 1910). To a certain degree the safety profession still is today...

11 Talking about first and second stories is a way to view accidents and other events. "First stories, biased by knowledge of outcome, are overly simplified accounts of the apparent 'cause' of the undesired outcome" (Woods et al., 2010, p.24). This is not about understanding organisational events, but about analysing safety texts. We will think of multiple accounts as one guide in our search to understand the texts because "When you pursue second stories, the system starts to look very different," and "Through these deeper insights learning occurs and the process of improvement begins" (Woods et al., 2010, p.242).

12 Even that would be assuming that he was able to answer that question. Maybe all we would get is a mere hindsight rationalisation...

13 More about Heinrich and Risk in Section 3.2.

14 Some of the edited material can be found on www.mindtherisk.com

Chapter 2

A biography

Although Heinrich's influence on safety has been enormous (Petersen, 1971; Manuele, 2013), surprisingly little information about him is around. A first source to check for anyone with a cursory interest might be Wikipedia. There are only a few Wikipedia pages[1] – all with limited information. Only a few safety scholars have written about him, notably Manuele (2002, 2013) and a Dutch team around Swuste and Van Gulijk (2009, 2010, 2013, 2014, 2019), even though not extensively so. Much has been written about his concepts. However, without paying attention to the backgrounds.

This lack of information[2] has led some to speculate sometimes wildly. De Groot (2005) makes him a psychologist, La Duke (2014) a statistician, an occupational safety researcher according to Marsden (2017), and Long (2012) calls him repeatedly an "insurance salesman." Marriott comes close by "insurance industry researcher, who reviewed thousands of workplace accidents in order to determine their likelihood of occurrence to support actuarial calculations" (Marriott, 2018, p.23). Even the well-researched Dutch team of scholars around Van Gulijk and Swuste somewhat miss the point by calling him a "former military working in insurance" (Van Gulijk et al., 2009, p.80).

So, who was Heinrich, and what did he do?[3]

2.1 Early years

Herbert William Heinrich was born 6 October 1886[4] in Pownal, a little town in Bennington County, Vermont, United States. Pownal is located in the utter south-west of the state, bordering both Massachusetts in the south and New York state in the west. Herbert William was the fifth child to wood-carver Rudolph Carl Heinrich and his wife Minna Rosamond Kortum.

Little is known about Heinrich's family.[5] They had emigrated from Germany to the United States in 1879. His father, Rudolph, originated from Rosen, where he was born on 4 March 1848. Rosen is most likely nowadays the village of Rożnów in the Lower Silesian Province of Poland. Heinrich's mother, Minna, was born on 19 January 1853 in Sagan, currently the capital of Żagań County in the region of Silesia, Poland. Rudolph and Minna married in 1876 or 1877. Of the children,[6] only two older brothers to Herbert William are known, Carl R. (born August 1877) and Oscar R. (born December 1883). Rudolph worked as a woodcarver and pipefitter, while his sons Carl and Oscar started out working as druggist and engraver.

Like all children, young Herbert William went to grammar school, but after completion of the sixth grade, he started working in local quarries and woodworking shops. Heinrich's unofficial biographer Jesse Bird comments on this sparse education:

> Looking back now from 1976, the six years in school appears quite scant. However, a review of the biographies of the founders of the industrial empires of the 1800s in the United States will reveal that a fourth-grade education was adequate for a frugal, risk taking, self-controlled person to become a business tycoon or inventor. Heinrich was burdened with two years of excess education if all he desired was to be a commercial success.
>
> (Lischeid & Bird, 2008, p.4)

Heinrich started an apprenticeship in the machinist trade at American Tool & Machine Company in South Boston. By 1901, he was a machinist, learning elements of tool making and design of electrical appliances.[7] In Boston, he became a "Polar Bear." To earn this nickname, one was required to cut a hole in the ice in Boston Harbour and swim for at least one minute. Heinrich was a member of the Polar Bear Club for several years.

In 1903, he went to sea. During his early sailing years, he worked in the engineering department of a steamship sailing on the Far East. During this time, he passed the third Assistant Engineer's written and oral examination for an Unlimited License – Ocean Steamers in 1904. Most likely, Heinrich acquired during his time at sea knowledge of mechanical engineering and thermodynamic principles, which would be beneficial in his later work. Through his apprenticeship and additional studies, he must have gained knowledge of mathematics. Apparently, his lack of formal education even brought an advantage:

> The maritime regulations favoured the apprentice in the machinist trade by requiring only three years of machinist apprenticeship time and one year's experience in the engine department of a steamer. By contrast, a graduate of M.I.T. with a degree in mechanical engineering at the time was required to have four years of instruction plus one year at sea.
>
> (Lischeid & Bird, 2008, p.4)

In 1906, he worked for the US Navy as a civilian employee as a machinist in the repair, maintenance and inspection of ships and onshore equipment on Mare Island, California. He would stick to this Navy job for a couple of years with assignments in Manila on the Philippines and in Honolulu, Hawaii, where he spent a couple of months in 1908.

After this, Heinrich worked briefly for General Electric Co., in West Lynn, MA with experimental and model work. Then he moved to Hartford, CT to work as a machinist for Arrow Electric Co. Here he met George E. Peterson, who was the assistant superintendent of the plant. In 1912, Heinrich moved briefly to Jersey City for a job at the Manhattan Electric Supply Co. only to return to Arrow Electric shortly after, working in the tool and die making and appliance design departments.

Peterson left Arrow in 1911 to start working for The Travelers Insurance in the Engineering and Inspection Division. He persuaded Heinrich to join The Travelers as well. Heinrich applied and started his first position there as a boiler and industrial plant inspector on 1 January 1913.

Later the same year another significant event for occupational safety happened. The United States Department of Labor[8] was established as a Cabinet-level Department by President William Howard Taft. Taft signed the bill on the last day of his presidency, 4 March 1913. William Bauchop Wilson, a former miner, union leader, congress representative and chairman of the 1911 congressional committee investigating Scientific Management, was appointed as the first Secretary of Labor on the next day by new President Woodrow Wilson. Taft had not been in favour of a new Cabinet-level Department but realised that he could not stop the development after decades of campaigning by organised labour and the early safety movement, promoting better working conditions.

2.2 The Travelers

Heinrich's employer for the next 43 years, the Travelers Insurance Co., was chartered in 1863. Before they actually started business, their President James G. Batterson verbally insured banker James Bolter in the sum of $5,000 against death if due to an accident occurring while Mr. Bolter walked from the Hartford Post Office to his residence in the same city, for the amount of two cents. This was the first ever accident insurance in the USA. One week later, on 1 April 1864, Travelers Insurance Co. formally opened for business (Malcolm-Smith, 1964).

In 1884 Germany implemented the first modern worker compensation act. Many countries followed. The USA was late to catch on with first worker compensation laws coming in 1911. It would take several decades before all States had adopted them. Compensation laws brought a change in how safety was perceived and practiced. Previously, under liability rules, when an employee was injured during work, he had to sue the employer and "prove" that the employer was at fault. The employer, however, would usually try to blame the accident on the worker. Prevention was not profitable, and employers felt little responsibility for safety. Compensation laws worked on the premise that an injured worker would receive compensation during the time that he was unable to work, and that medical treatment was paid for. The cost was to be borne by industry – adding to the cost of products. For this, companies insured themselves. Accidents increased the insurance premium and therefore under the new regime, it became profitable to try to prevent accidents, rather than fighting claims. This was an important stimulus to other initiatives to improve safety. Safe companies had low rates; unsafe companies got higher rates (Travelers, 1913a, b; Beyer, 1917; Aldrich, 1997).[9]

Insurance companies did not merely take a passive role and "gambled" on results. They took an active role in advancing safety through inspection, investigation, education, and giving recommendations. Insurance firms thus gained a broad knowledge base, serving many other companies, thus having the unique opportunity to share recommended practices and solutions (Blake, 1945). At some point it even gave them a competitive edge, "Competition between insurance carriers today is largely a competition between service departments, with safety service playing the most prominent part" (Williams, 1926, p.7).

The Travelers was one of the major insurance companies when Heinrich joined them. In 1904, they were the first company in the USA to organise a professional corps of safety engineers. By 1915 there were 339 people in the Engineering and Inspection Division, of which 203 were field inspectors and 90 were Home Office professional staff. The Travelers Insurance Co. had a history of publishing safety-related books, including the works *Grinding Wheels* (1912), *Accident Prevention on the Farm* (1914), *A Treatise on Safety Engineering as Applied to Scaffolds* (1915), *Safe Foundry Practice* (1920), *Safety in Building Construction* (1921 and 1927), *Coal Mining Hazards* (1916), *Airplanes and Safety* (1921), *Motor Vehicles and Safety* (1915), *Safety in Moving Picture Theatres* (1914), *Safety In the Machine Shop* (1920), *First Aid to the Injured* (1927), and booklets directed to foremen and employees as well as one about *Organization in Safety Work*. Besides these books and brochures, they issued regular publications, notably *The Travelers Protection* and *The Travelers Standard*. Especially the latter, a monthly publication of the Engineering & Inspection Division for the benefit of policyholders, contained many quality articles on safety.

2.3 First steps in safety

Heinrich was assigned to the home office staff for the first six months, mentored by Dr Allan Risteen who was a renowned safety scientist and chemist of the time who lectured at universities. He edited the chemistry sections of *Century Dictionary and Encyclopaedia Americana* and initiated the *Travelers Standard* in 1912. Risteen had previously worked at Hartford Steam Boiler, where he had been in charge of the publication *Locomotive*, which was started in 1867. In 1919 Risteen was transferred from the Engineering and Inspection Division to take charge of the Travelers Publicity Department.

In 1913, one of Risteen's roles was acting as the tutor of the inspectors being readied for field office assignments. During those first six months of training, Heinrich's task was to "scrutinize" reports. This meant reviewing reports and recommendations of inspectors, adding improvements and writing a letter to the insured. Scrutinisers were described as "...men of extended technical training, who examine all inspection reports that come in, and exercise a check over the field work of inspection, in addition to performing other services" (Travelers, 1913b, p.298). If the scrutiniser felt that other recommendations should have been made to the insured, they added them to the final report. The recommendations in the final report, according to the article, "represent the thought and experience of the entire engineering staff of the Company."

After several months, Heinrich was transferred to Boston to take the Massachusetts Boiler Inspector examination and to inspect for all lines of insurance. As a bachelor, Heinrich had to travel quite a lot between various offices in New England and New York, assisting the work there. At the time, most travelling was still done by horse and buggy, or by horse-drawn sleighs in winter, to visit insured locations that were not near railroad lines. In 1917, he was promoted to Senior Inspector and stationed at the Albany, NY office. Here he lived at 290 Hamilton Street.

Meanwhile, the First World War had arrived, and also Heinrich was drafted. First, he enrolled in the U.S. Naval Reserve as a Junior Lieutenant – a rank gained thanks to his previous engagement as Navy engineer. Despite the fact that his draft registration dates 5 July 1917,[10] it was not before the final months of the War that he was called into active duty. From 12 August 1918 on, he served as engineering officer aboard a merchant ship of the Naval Overseas Transport Service that transported troops and war supplies. Meanwhile, Heinrich passed the examination for chief engineer on ocean-going steam vessels, unlimited tonnage. Bird's biography tells an interesting anecdote from Heinrich's First World War experiences that adds a richer perspective to how his view of humans often is displayed:

> While on board, the main propulsion unit of the ship, a triple expansion steam reciprocating engine was due for complete overhaul and repair. Heinrich wrote over 200 pages of detailed instructions on the performance of the work to be done - instructions on disassembling, moving, temporarily storing, then reassembling all parts of the main engine and the major auxiliaries. He estimated the work would require three full days. He started writing the instructions at a European Port of Departure, three weeks before the ship was to arrive in a U.S. port. The ship was only several hours from the port of destination when he realized that the instructions would take longer to read and understand by his staff than the time to complete the entire job. He called his crew together and told them in ten minutes what to do. The crew completed the task in less than the allotted three days.
>
> (Lischeid & Bird, 2008, p.6)

Heinrich later told Jesse Bird that he learned that many situations exist where you have to depend on your staff to exercise common sense. Therefore, detailed instructions are apt to do more harm than good with experienced personnel.

Discharge from the Navy came in April 1919 with the rank of Senior Lieutenant. He spent his first month as a civilian in New York City. With several thousand dollars in his pocket, he tried himself on stock market speculation. Successfully, because he made a profit of over $100,000 in the first two weeks. Success did not last, however, and in the course of the next month all profits along with most of his original capital was lost.

Heinrich re-joined The Travelers in June 1919. He was promoted to take charge of the newly formed Indemnity Section of the Engineering & Inspection Division. As part of his work, Heinrich prepared and edited numerous articles for *Protection* and *The Standard*.[11] He also wrote speeches and started speaking at association and chamber of commerce meetings and courses.

In early 1923, Heinrich presented a speech at the Welding Conference in New York City. This speech appeared as a paper, *Welded Joints in Unfired Pressure Vessels* (HWH1923a), in *The Standard*. This paper is probably the first published work by Heinrich. He would return to these engineering roots after his retirement when he worked with the Uniform Boiler & Pressure Vessel Society.

Around this time, Heinrich married Viola V. King, born 29 June 1901 in Philadelphia.[12] On 27 January 1924, their first and only child, Virginia Ruth, was born. The couple lived in East Hartford. Hartford, founded in 1635 and thus among the oldest cities in the United States, is Connecticut's capital city. Hartford was the richest city in the States for several decades following the American Civil War. Today, it is one of the poorest cities in the nation, despite an over average economic production. Hartford is known for being the home of weapons manufacturer Colt. It is also the historic international centre of the insurance industry, with companies as Aetna, Conning & Company, Prudential Financial, The Travelers, and United Healthcare. Therefore, Hartford is nicknamed the "Insurance Capital of the World." Some of America's most famous authors lived in Hartford, including Mark Twain and Harriet Beecher Stowe. Twain wrote many of his classic novels during his 17 years in Hartford.

2.4 Emerging foundations

Later that year, in November, Heinrich presented a safety talk at the Bedford (Indiana) Stone Club (HWH1923b). This speech contained the first ever mention of one of Heinrich's main themes, namely the "indirect" costs of accidents, along with some basic guidelines for safety management.

Heinrich was promoted to Assistant Superintendent of the Engineering and Inspection Division in 1925. At the end of that year, during the New

York State Safety Congress at Syracuse, N.Y. on 2 December 1925, he mentioned research into hidden costs of accidents and an approximate 1:4 ratio. In a speech on *The Contractors' and Builders' Safety Problem*, Heinrich told the audience,

> In a list of one hundred minor accident cases selected at random, our records show that the average incidental accident cost – paid for by the contractor himself – was more than four times the loss covered by insurance. And yet some contractors, uninterested in accident prevention, say they have no time for safety – that they cannot afford to fool around with safety precautions. As Josh Billings so aptly put it, 'It ain't that there are so many ignorant people in the world, but that there are so many people who believe things that ain't so.
>
> (HWH1925, p.252–253)

The "hidden cost" principle was to become an important driver for safety, and a recurring theme in many of Heinrich's writings. The research continued and was presented more fully in a number of papers about the *Incidental Cost of Accidents* (HWH1926, 1927a, b, c). The idea – although not particularly new because others had written about it before – gained quite some interest. According to Aldrich (1997) the *Incidental Cost* paper was reprinted nine times in 1927 and 1928.[13] The *Travelers Yearbook for 1928* – the company's official report over the previous year to the World of Finance – spoke highly of the research, spending an entire page on the subject,

> The prevention of occurrences which would call for the payment of insurance benefits is a service to the Travelers Companies, of course, but it is an even more valuable service to those who are insured. In factories where safety first rule is the rule, dividend profits are greater to owners and wage profits are greater to workers. A careful study of the costs of accidents in industry by The Travelers indicates that capital and labor, jointly, lose four times as much from every accident as the amount of workmen's compensation paid to the man injured in the accident or to his family. Business has been slow to recognize this hidden cost of accidents because bookkeeping systems have not shown the losses as disbursements. But the costs, elusive as they may have been when sought by accountants, have existed just the same in the loss of time by workers near the scene of the accident, in loss of morale, in damage to machinery, and in spillage of goods. The ascertainment of this four-to-one ratio in the cost of industrial accidents has secured for the Travelers engineers more whole-hearted cooperation than ever from employers, plant superintendents and foremen.
> In engineering and inspection service the Travelers Companies have thus far expended nearly twenty million dollars. How many times that

amount has been saved to those insured by The Travelers can only be imagined. No satisfactory formula for estimating the profits to employers and employees has been evolved as yet. There is no way of counting the accidents which do not happen, nor measuring their seriousness.

(Travelers, 1928, p.20–21)

Meanwhile, Heinrich led another project by Travelers engineers to study the causes of 75,000 accidents.[14] The first results of this were presented in the *The Origin of Accidents* paper in March 1928, which for the first time presented the (in)famous 88:10:2 ratio, postulating that the ratio of direct causes of accidents are – in the words of the 1941 edition of *Industrial Accident Prevention* – 88% attributable to "unsafe acts," 10% "unsafe conditions," and 2% unpreventable. This group of numbers created many arguments. Jesse Bird noted in retrospect that they were "poorly understood."[15]

The ideas were not necessarily new, however, and while credited to Heinrich they most likely were the result of collaborative work. *Protection* magazine from March 1927 reported about a convention of Travelers senior engineers, held at the Home Office in February. It was attended by 82 delegates from Canada and the USA. Most likely many of the basic ideas that later appeared in Heinrich's work were already discussed and expressed by members of this group. The article told,

Special stress was laid on the fact that it is the human element on a risk that counts. If the man at the top can be convinced of the value of safety engineering and safety organization work, the men under him will quickly swing into line. It may take years for an idea to work up from the foreman or superintendent to the top; it only takes months for that same idea to permeate the entire organization if it starts at the top and works down.

(Travelers, 1927, p.8–9)

The new research gained quite some interest during 1928 and 1929. Heinrich presented his findings during the Seventeenth Annual Safety Congress of the National Safety Council in New York City in October. The magazine *Popular Science Monthly* had an item on it in their September issue,[16] and news items in newspapers over the whole country mentioned Heinrich and his work stressing the preventability of most accidents, or quoting some of the more "juicy" statements, like "...a man who is mentally disturbed is as much of a danger to himself and fellow workers as if he were physically disabled."[17]

Later in 1928, Heinrich presented another of his main ideas to the world. This built on some of the findings that already had been presented in *The Origin of Accidents*. Titled, *The Foundation of a Major Injury*, the 1:29:300 ratio (a.k.a. the triangle) was presented to the world in Heinrich's presentation before the 12th New York Industrial Safety Congress in Syracuse on 6

December 1928, and in several papers (HWH1929a, b) the month after. The research showed that there were many more no-injury accidents than accidents with minor injuries and major injuries. The research of the Travelers safety engineers found an average ratio of 300-to-29-to-1.

Inside the company, Heinrich's work continued to be highly regarded. The *Travelers Yearbook for 1929* praised the work done by Heinrich and his team.

> During the past year the Engineering and Inspection Division completed a study of the causes of 75,000 industrial accidents. This showed that 88 percent were due to human failings[18] that 10 percent were the result of mechanical defects – thus leaving only 2 percent as unavoidable. For several years Travelers engineers have known that more progress could be made in accident prevention by improving plant discipline, by more careful instruction of operatives and by better selection of workers according to type of work, than could be made by guarding machines, properly arranging machinery or improving plants, though such changes are important in reducing the accident toll.
>
> The results of this study, confirming their previous belief, have been of material assistance to the engineers in winning the cooperation of employers and employees in carrying out programs that have reduced accident ratios where The Travelers carried the Workers' Compensation and public liability insurance. Much money has thus been saved to employers and employees and much suffering and discomfort to workers and their families.
>
> (Travelers, 1929, p.25–26)

At a conference of Travelers managers from 8 to 11 January 1929, Travelers President Louis Fatio Butler gave highest praises to the Engineering Division in general, and to Heinrich in particular. This to some dismay of some other managers who got together and complained, pointing out that "...while safety is wonderful and it has its place, we (as managers) want to know how to fight the mutual competition" (Bird, 1976, p.9). Apparently, accident prevention could not be applied to accounts unless Travelers insured them. The mutual insurance companies were cutting rates and taking business away. Heinrich later remarked that he believed that this was the "kiss of death" of the research. However, President Butler had full confidence in Heinrich and backed Heinrich's work for full. According to Jesse Bird, Butler told one group of managers that if he were a younger man, he would start an insurance business with Heinrich as "idea man."

When the chairman of the subcommittee on classification of causes, L.L. Hall, formerly of the National Council on Compensation Insurance resigned because of work pressure due to a change of profession, Heinrich was appointed in Hall's place as the new chairman of this committee sometime in 1929 (BLS, 1930).

In late October 1929 the stock market crashed, which is generally considered as the start of the Great Depression, although a recession had started some months earlier. Despite efforts of the Roosevelt administration to restore economy, the recession would basically last until the approach of the Second World War, hitting its lowest point during the winter of 1932–1933 with sky-high unemployment numbers. Only in 1941 with war preparations in full motion, pre-Crash levels of employment and production returned (Petersen, 1990).

2.5 The book

The material from *Incidental Cost, Origin of Accidents* and *Foundation of a Major Injury* formed an important part of the core for his first book, *Industrial Accident Prevention, A Scientific Approach*, published through McGraw-Hill in 1931. Interestingly, there had been a book titled *Industrial Accident Prevention* before, written by David Beyer, published in 1916. It is one of the titles in Heinrich's bibliography.

Heinrich dedicated the book to the late Butler who had been extremely supportive, but never saw the book since he had passed away in October 1929. Jesse Bird suggests that while Heinrich is credited as the sole author, it was very much a collaboration with the academically educated Edward Granniss, who worked with Heinrich at The Travelers from 1925 to 1935. Granniss did supply, and is credited for, the first two appendices of the 1941 and 1950 editions of *Industrial Accident Prevention*, the chapter on personal protective devices (HWH1950a), and he would be credited as co-author for the fourth edition (HWH1959) – although in a smaller font than Heinrich.

It seems that there was some antipathy regarding Heinrich's book. Never before had anyone in The Travelers used the company's research and published this material under their own name. However, The Travelers' President Butler had given Heinrich permission to use the material, and of course Heinrich had led the personnel that had gathered the data and done the research for the various ratios.

There was a heavy speech schedule. Apparently, the book elevated Heinrich to a celebrity within safety circles. On 2 November 1932, he spoke at Maine's Fifth Annual Industrial Safety Conference in the House of Representatives, State Capitol Building in Augusta, Maine. He presented the *Mastery of the Machine* paper, which would be included in the second edition of *Industrial Accident Prevention*. The *Industrial Safety Bulletin* of the Maine Department of Labor and Industry spoke in glowing terms:

> The *famous* H.W. Heinrich, Asst. Supt., Travelers Insurance Company, next offered his paper 'Mastery of the Machine', an exposition many had travelled a hundred miles to hear.
>
> (Maine Department of Labor and Industry, 1932, p.4, emphasis added)

None of the other speakers, not even the governor of Maine who opened the conference, was spoken about in such superlatives. One wonders whether Heinrich's presence was one reason for a 30% increased attendance compared to the year before – despite depression times.

When J.L. Thompson, the Superintendent of the Engineering and Inspection Division retired in 1932, Vice President R.J. Sullivan took over direct control of the Division and all its activities. Heinrich requested to become Superintendent the next year, but Sullivan refused, finding Heinrich not suitable for the job. Instead, James A. Burbank, a graduate from MIT and Lieutenant Commander during the First World War, was appointed as Superintendent in 1934.

Heinrich was noted for his attention to details. This characteristic may have led to his not being advanced to Superintendent. Some characterised his fixation on details as being inflexible and complex in his approaches. Many years later, Heinrich told Jesse Bird that he had asked why he could not get the job. Sullivan told him that the engineering staff in the Home Office and field would not work for him, because he was too exacting and lacked flexibility.

Heinrich realised he was "frozen" in his position for the remainder of his work career at The Travelers. This sense of being "frozen" must have irked, especially given his "celebrity status" outside his own organisation. This may be one partial reason for Heinrich being even more prolific from this point on.

He was not the only one who was dissatisfied. Some personnel with top qualifications resigned, including Edward Granniss, who became an industrial staff engineer for the National Safety Council and in 1939 the Director of the Industrial Engineering Division of the Association of Casualty and Surety Companies, and Ralph Crosby, who went to Marsh & McLennan and became a lecturer at New York University. Especially Crosby must have been annoyed by Burbank's appointment. Like Burbank, he had been a Lieutenant Commander in the First World War, he was also a graduate of MIT, and in addition he had been with The Travelers for 12 years in charge of research.

In December 1933, Heinrich was appointed to direct the safety work in Connecticut State as the Safety Director of Civil Works Administration (CWA), the 20-week program to put unemployed to work. The Travelers had loaned out services to CWA.[19] The CWA's safety department consisted of Heinrich and 12 safety state inspectors, who worked with about 1,000 full-time or part-time inspectors, and some 2,000 foremen responsible for safety. Until the program stopped in April 1934, an average of almost 32,000 persons per week were employed in the various CWA-activities, working on, e.g., parks, playgrounds, streets, public lands, and pest control.[20]

Somewhere in this period, the Heinrich family moved from 827 Burnside Avenue in East Hartford to 24 Castlewood Road in West Hartford where Heinrich lived until his death.

2.6 Dominos and cause code

Later in 1934, The Travelers published the six-page accident sequence bro-chure, for the first time presenting one of Heinrich's most iconic ideas: the domino model. The brochure was based on a speech presented before the Down River Section of the Detroit Safety Council, 30 November 1934. Heinrich went on tour to present the domino accident sequence to the entire Engineering & Inspection Division field staff.

Heinrich was elected chairman of the subcommittee of The American Society of Mechanical Engineers (ASME) to study and prepare an improved accident cause code. The result, baptised the "Heinrich cause code," was published as a first draft in 1937:

> The underlying philosophy of this method of analysis is that industrial accidents are due to unsafe conditions and unsafe practices, which, if eliminated, would prevent a repetition of the same or similar type of accident. The effort therefore is to enable the statistician to identify and select the unsafe factors and then present the data to safety engineers for guidance in accident prevention.
>
> (Kossoris, 1939, p.526)

Aldrich notes that the cause code "proved cumbersome" (Aldrich, 1997, p.152[21]), but Max D. Kossoris from the Bureau of Labor Statistics was quite positive after a year of trial use: "In this way statistics and statisti-cal analysis are not merely compilations of questionable historical value, but vital tools towards a socially and economically desirable end-the pre-vention of accidents in industry" (Kossoris, 1939, p.532). In 1941, the new cause code was adopted as a national standard and three years later, the Bureau of Labor Statistics published a manual for accident records that was based on the code. Heinrich and Granniss contributed to this manual (Kossoris, 1944).

2.7 Meanwhile at Travelers

Much of Heinrich's time was filled visiting field offices ensuring the quality of inspectors and instructing them. Heinrich was responsible for interview-ing all new engineers who were assigned to all lines of work. The typical interview was arranged after the new employee had passed the examination to inspect boilers.

Jesse Bird reports about his own job interview by Heinrich in 1937. The 25-year-old man visited Heinrich, who explained how many accident re-ports received from the field were poorly written, lacking sketches and il-lustrations. The young Jesse Bird listened closely for an hour. Heinrich then abruptly swung around in his chair and inquired, "Any questions?"

BIRD: "Yes Sir, One!"

HEINRICH: "Shoot."

BIRD: "Suppose I have to investigate an accident on equipment I have never seen or heard of before and I am unable to make an important recommendation to prevent another serious or fatal accident. What is the procedure?"

HEINRICH: "That is very simple."

BIRD: "What is that?"

HEINRICH: "We will find a man, who can."

The young man was motivated "1937 Style." For the remainder of his career with The Travelers Jesse Bird was never without an appropriate important recommendation.

In 1938 Heinrich led a committee of 20 people made up of Home Office personnel and the New York City office staff. Their objective was producing a single page chart on how to service a risk. This document showed in a schematic way how the efforts of the Travelers safety engineers supported the objectives of both the customer and the company, both financially and in terms of accident prevention. The roles of the engineers were listed as "fact finders, advisors and consultants," which "recommend and assist, enthuse and inspire" (Travelers, 1938). The main product of their division was "oral or written recommendations or suggestions." In the document the author (most likely Heinrich himself) made an interesting critical remark that recommendations left much room for improvement regarding their helpfulness. The second part of the document was dedicated to helpful suggestions in writing good recommendations for customers.

Heinrich visited field offices during the year. He usually stayed for three days, quizzing each engineer separately on subjects as the engineer's objectives, principles of safeguarding and how to distinguish good and poor risks. This was followed by an interview and file review. They looked at how the major risk was serviced, what outstanding hazards there were and how these were handled and what recommendations were made to reduce future accidents. Failing to answer or explain the quiz or pass the file review correctly meant the engineer was terminated. Welfare and unemployment money were unheard of in 1938. According to Bird, the engineers who were fired found employment with the "easy companies." The Travelers was regarded as the toughest employer. However, they paid the best.

In September 1938, Virginia went to Northfield Seminary, Northfield, MA. Heinrich became member of Hartford Chamber of Commerce, chairman of the industrial safety committee,[22] and spent much of the next year on routine visits to selected offices, instructing new personnel in inspection and report writing techniques.

2.8 Teaching, speaking, and another book

Around this time, Heinrich joined his former colleagues Crosby and Granniss as lecturer at New York State University, teaching safety to post and undergraduate students for many years to come. He also taught at e.g. John Hopkins and spoke on occasions such as the Congress of the National Safety Council. They were joined by Bob Zawar of Liberty Mutual, Dr John Grimaldi, and other safety notables. Dr Herbert J. Stack and Dr Cutter were the guiding lights of this endeavour. Stack was director of the Center for Safety Education,[23] which was established at New York University in 1938.[24] Its first mission was the preparation of safety leaders and it provided training to thousands of teachers, engineers, supervisors, army officers with an ever-increasing number of courses per year (Stack, 1953).

By 1940, war preparations had started, and industrial activity increased. As an example, one can take Pennsylvania which was "...already grappling with the problem of accident control in industries humming with the new activity of defense production and the manufacture of munitions..." (Immel, 1942, p.127). During 1940, there was an increase of 250,000 employed workers with another increase of 235,000 workers the next year. The industrial injuries followed this trend, climbing from 109,475 in 1940 to 130,403 in 1941. Blake (NSC, 1942) reported that fatal incidents were on the rise – before the war about 15,000 to 16,000 rising to 20,000 in 1941 and a prognosis of even more in 1942, most of these in small plants.

The government realised the importance of safety to ensure efficient productivity and called upon "big names" in safety, including Heinrich, Granniss, and Crosby. The importance of safe and efficient work necessary for war preparations was reflected in Heinrich's publications from this period. Granniss, presiding the *Government Safety Service in Industry* session at the 1942 National Safety Congress, said,

> One of the most serious problems facing industry in its war effort is the control of waste. The manufacture of new and different products, the employment of new workers, including women, young and over-age employees, high-speed production and long hours, all of these have contributed to spoiled materials, broken equipment and most of all to lost work hours.
>
> (NSC, 1942, p.103)

Among the actions taken, was the development of safety education and safety training courses organised through the National Committee for the Conservation of Manpower in War Industries. This committee was formed during the summer of 1941 and in cooperation with the U.S. Office of Education. Thanks to the efforts of Captain Laurence Tipton of the U.S. Army working with the Department of Labor, they set out to organise safety training on a large scale throughout the entire nation.[25] The purpose of these courses was "…to assist defense industries by training key supervisory employees who have the responsibility for maintaining the safety, health and efficiency of war workers."[26]

All this was part of the Engineering Defense Program, authorised by Congress in October 1940, with the aim to organise short intensive courses designed to meet the shortage of engineers with specialised in fields essential to national defence. In July 1941 the program was expanded to include management and science, and after Pearl Harbor the program was baptised Engineering, Science, and Management War Training (ESMWT) – safety thus became as essential part of the war efforts.

The response was huge. H.H. Armsby, the U.S. Office of Education field coordinator for the ESMWT, informed that as of 1 October 1942, 216 colleges had been authorised for ESMWT courses with nearly 11,000 courses in more than 1,000 cities with around 600,000 trainees. Of these, over 15,000[27] had taken the safety courses (NSC, 1942). Heinrich and Robert E. Blake, senior safety engineer with the Department of Labor, were the ones to provide the texts for these ca. 90-hour courses. The war thus meant an enormous boost for Heinrich's work, and may have contributed to his continued popularity, also after the war.[28]

In 1941, the updated second edition of *Industrial Accident Prevention* was published, for the first time presenting the dominos, axioms of safety and the 1:29:300 presented in an actual triangle shape. The book became a textbook at various universities across the country[29] and abroad.

2.9 War!

When, after Pearl Harbor, the United States entered the war against Germany and Japan in December 1941, Heinrich came in even higher demand. Edward Granniss had been placed in charge of all safety activities of the U.S. Army with the rank of Colonel. He went directly to the President of Travelers Insurance and asked for Heinrich's services. This was granted. In a letter from November 1975 to Jesse Bird, Granniss commented that the Travelers gave Heinrich a premium-free insurance policy for covering him while he was in Granniss's service: "There were a couple of times while we were in the European theater I thought Virginia might appreciate it."

Indeed, Heinrich reported, "…we found out that the grandfather's clock in our apartment had been boobytrapped by the defeated Germans." And

> The author will be forever grateful to a sleeping bag which cushioned the shock when the German Opal [sic] in which he was riding with a party of Russians was struck head-on by a big oil truck. The two Russians in the front seat of our car were killed.
>
> (HWH1945e, p.7)

Foremen were a focal point for the Department of Labor. They were regarded as "key men" and thus important for the war efforts. Safety was elementary, and foremen were given courses. Heinrich did a series of speeches and papers on the role of the foreman. He had already paid much attention to the role of management in safety work, and from the early 1940s on, this became his main area of interest. In late 1942, he started working on a book dedicated to first line managers. This work was published the next year through the Alfred H. Best Company, first as a thin booklet, *The Supervisor's Safety Manual* (HWH1943d) and the year after as the book *Basics of Supervision* (HWH1944).

In December 1942 Superintendent Burbank presented his "Find, Name and Fix" theory. This competed with Heinrich's method of (1) identifying the Principal Risk Hazard, (2) the reason why the Principal Risk Hazard existed (3) giving recommendations to eliminate to the minimum level the Principal Risk Hazard. Heinrich more or less reconciled both approaches in two charts, the *Supervisory Thru Track* and *Hazard Thru Track*. These were used during the War for educational purposes and were later included in *Industrial Accident Prevention*.

The year 1943 brought much travel to various offices and U.S. Army locations for speeches and inspections. Another major event that year was that Heinrich's daughter Virginia married Charles Eugene Pressler.[30]

For diversion Heinrich conquered the violin. He did not particularly love music. His interest was aroused by an office conversation about the difficulty to learn to play a violin at an advanced age. Heinrich believed that he could learn and do and render a creditable performance in any field. He took up golf in the same manner.

During the Second World War, Heinrich took up several positions, including Chairman of the Safety Advisory Committee to the War Department. Heinrich was one of two civilians to be appointed by Chief of Staff General George C. Marshall to the Advisory Board on Fire and Accident Prevention of the Under Secretary of War. The other civilian of the Board was Percy Bugbee, general manager of the National Fire Protection Association of Boston. The military members included two brigadier generals, two colonels and a member of the engineer's corps. The Board's task was to analyse the War Department's organisation on fire and accident prevention, study accidents and incidents, and make recommendations on improvement along with providing a quarterly report on trends.[31] A third appointment followed in the summer, when Heinrich became Chairman of the War Department Safety Council.[32]

When the war came to an end, Heinrich served on a special committee formed by the Secretary of War to control accident fatalities in Europe. With Granniss, he travelled to Germany for several weeks to assist in the organisation of a program aimed to control risks of booby-traps and mines. Here he encountered difficult circumstances and several high-risk experiences.

Heinrich described his experiences in Europe in an article in the July/ August 1945 issue of *The Travelers Beacon*. After the War he did several presentations about his experiences in the Hartford area. His work during the war was awarded with a certificate for "patriotic service in a position of trust and responsibility," presented by Major General B.M. Bryan during a formal ceremony at The Pentagon.[33]

2.10 Post-war work

After the war, Heinrich stuck to the subject of the role of management and supervisory (safety) education. According to Jesse Bird, he would do this by means of long, detailed, and technical talks in which he tried to cover his two books. Heinrich and Granniss, who meanwhile had become manager of the Engineering Department of Eagle Royal, continued to work together. Both were on the Safety Committee of the American Society of Mechanical Engineers that had prepared an ASME standard for the self-evaluation of industrial plants which saw the light of day in 1947 (ASME, 1947). The next year, Heinrich presented a Boiler and Machinery Accident-Cause Code at the ASME Annual Conference in New York.

Since there was a large turnover in the Engineering and Loss Control Division after the war, Heinrich spent much time hiring and training new engineers. Meanwhile he worked on his second management book, *Formula for Supervision*, published by the National Foremen's Institute in 1949. The year after, in July 1950, he completed the third, reworked version of *Industrial Accident Prevention*, which saw an expansion on previous editions and included some of the work on supervision. In addition, Heinrich did much committee work for the American Standards Association (ASA – later renamed to ANSI).

Heinrich kept working on his personal development. In 1951, he took up a one-year night class in French, preparing himself for an assignment for a U.S. airport in French Morocco. He became quite fluent in the language. He also studied a successful preacher in the way he made sermons and presentations. Heinrich learned a great deal from this, and from that time on, he did short and concise talks with a limited number of ideas within a 20 minutes timeframe.

In 1952, Heinrich visited the construction work in Morocco. That year, he received the Arthur Williams Memorial Medal from the American Museum of Safety in New York, one of the nation's oldest safety organisations,[34] in recognition of "his many and valuable contributions to the cause of safety over a long period of years."[35] He also received the Medal of the Conservatoire Nationale des Arts et Métiers in Paris. In his speech, Dr Andre Samont mentioned that Heinrich received the award for his many basic contributions to the science of preventing loss of life and property, and because of the value of the material in his book, that was a standard text in France.

Heinrich led a project on *Effective Work in Accident Prevention*, intended to stimulate field engineers to find out how management talks about plant costs, and how engineers can relate accident costs to plant costs that management would understand. All the while, he kept up the routine of doing presentations and visiting field offices.

2.11 Heinrich, the person

Heinrich's work reveals little of his personal life, on the job or with his family. Only a few glimpses have survived through articles in Travelers publications, newspapers,[36] and the recollections of Jesse Bird.

Heinrich was about five feet eight inches tall, with light-blue eyes and light-brown hair.[37] Bird tells us that at the age of 55 he appeared red-faced, thin, and wiry. In mid-life he walked with the stride of an athlete. This he never lost, in fact his step appeared to be more rapid when around 70 than in the earlier years. His posture was erect, one could say quite military. He was partial to wearing his military overcoats of the Second World War.

Heinrich seemed to have a gift to make other persons feel that his interest and welfare was paramount. He listened intently to the other person, aided by gesture and posture. Heinrich was a good actor and an avid speaker. He used this also in other than job-related matters. Jesse Bird mentions that Heinrich loved to tell ghost stories at the employee club of The Travelers.

On a "negative" note,[38] it seems that Heinrich was a heavy consumer of cigarettes. Jesse Bird says:

> A hallmark of the Heinrich office was the lighted cigarette burning in the ashtray. This continued whether he was in or out of the office for short periods. He admitted to smoking one and a half packs of cigarettes a day and just burning another pack or two. Only two fires occurred in his office.[39] Both were in the old straw waste baskets.

Occasionally, Heinrich would find a puzzle which amused him. This he would use to "test" an engineer. Except at lunch, Heinrich spent practically all his time on accident prevention matters. During lunch hour he played ping-pong, chess, checkers, and cards, and in the 1920s, he played dominoes. He was quite good at all these games.

A typical eccentricity of Heinrich's consisted in posing a question, usually about a design, on the front cover of a publication. This took the form of a series of questions. Why was the design made as it was? Why was it not moved to the right, left, up, or down by a quarter of an inch? Why did the publisher choose that book size and colour? This kind of questioning was usually reserved for long-time acquaintances working directly or partially under his supervision. Jesse Bird reports that other than driving a few people to nearly climbing the walls, little harm appears to have been done.

The *Beacon* reported on Inspection Division parties, which provide a glimpse at the social life of the department. The first party, on 26 May 1925 at the Recreation Grounds, was called the "best party ever put up." Here two teams, the "Giants" and the "Yankees" played ball. Heinrich acted as pitcher for the "Yankees." After this there was storytelling, speeches, singing, a dinner and music by the Riley's Orchestra with dancing. Heinrich had been on the committee responsible for arranging the party.[40] Another happened 28 May 1929 with a variety of activities, including singing, telling jokes, acting, an absence of speeches and hinting at a dinner mystery, "...what became of the olives at Mr. Heinrich's table." Heinrich's daughter Virginia, at the time a tender five years young, was one of the "performers" at this event. At the time she was a pupil at Florence Greenland's School of Dancing and the photograph accompanying the article pictured her in ballet attire.[41]

2.12 Retirement

Heinrich retired from The Travelers on 27 April 1956. In contrast to the usual celebrity treatment, his retirement after 43 years working for The Travelers was a modest affair. A lunch in a local restaurant with the whole Engineering and Loss Control Division was the main event. During the 1940s, large dinner-dance celebrations had been the norm, but they had been "banned" and the sole "entertainment" were a few spontaneous speeches by the Engineering department management. This must have been a bit of a put-down for Heinrich, who was accustomed to being treated as a celebrity. Jesse Bird notes he "was treated like the proverbial prophet in his hometown." Vice-president George E. Peterson, the man who had persuaded Heinrich to join The Travelers many years ago, attended the celebration and thanked him for his 43 years of service.[42]

Upon retiring, Heinrich and his wife went on an extended trip to Spain, Italy, Turkey, Austria, and France (anonymous, 1956, p.7). After this vacation, Heinrich embarked on a new career. Retirement brought a return to his engineering roots when Heinrich became chairman and president of the Uniform Boiler and Pressure Vessel Laws Society, headquartered in New York City. This was a non-profit organisation founded in 1915 to promote the uniformity of laws governing the safe construction, installation, and operations of steam boilers and pressure vessels in the United States and Canada. Here he led the compilation of a book during 1957. Besides, he lectured on safety-related topics at New York University, which he did until his death. For his work and the preparation of safety courses, Heinrich frequently requested copies of talks and articles from the Travelers archives. Superintendent Burbank granted full access to the archives for these purposes.[43]

Together with Granniss, Heinrich worked on a revised version of *Industrial Accident Prevention*. Besides he would still regularly appear on conferences for safety speeches. In 1959, the fourth edition of *Industrial Accident Prevention* was published with Granniss credited as co-author. While the basis was still the same, for this edition the structure of the book was altered into several parts, and again the scope was expanded, this time also covering nuclear radiation. Apparently, the book remained in print for over a decade.[44] In 1980 a fifth edition appeared. This was an entirely revised version, authored by Dan Petersen and Nestor Roos, but still mentioning Heinrich as first author, even though much of the text was not by his hand anymore.

During the 1950s and 1960s, Heinrich collaborated with safety consultant Alfred R. Lateiner. Lateiner made some trips to Europe because Heinrich's age did not allow these travels anymore. There, he taught the "Lateiner Method," heavily influenced by Heinrich's work. Heinrich and Lateiner also launched a correspondence course under the umbrella of The Heinrich-Lateiner Management Training Center during the early 1960s and started working on a book together. Heinrich managed to finish his part – mostly a combination and slight update of the 1944 and 1949 texts on supervision – for about 80% before he passed away. Lateiner finished and edited the manuscript for the 1969 publication of *Management and Controlling Employee Performance*. This was a combination of two books, *Management* by Lateiner and *Controlling Employee Performance* by Heinrich.

In 1961, Heinrich received the ASSE Fellow designation from the American Society of Safety Engineers.[45] He had been a member of the society since 1924.[46] Occasionally, he played golf. Henry Elliott, one of Heinrich's golf companions, told Jesse Bird, "Heinrich did very well at 70," playing at the 100+ level. Heinrich resigned his position with the Uniform Boiler and Pressure Vessel Laws Society on 1 June 1962.

On 22 June 1962, Heinrich died aged 75. He was buried on Fairview Cemetery in lot 443 where he rests with his in-laws, James E. King and Minnie P. Keller who both passed away in 1961. Viola does not rest with her husband.

Heinrich was awarded several posthumous honours. In 1979, he was inducted into the Insurance Hall of Fame,[47] reserved for major contributions to insurance thought and practice. He was the first safety engineer to be elected for this, being honoured at the Insurance Hall of Fame convocation in Dallas, TX. In 1993, Heinrich was included in the Safety and Health Hall of Fame International[48] for his contributions to safety engineering and accident cause theory.

Notes

1 Checking Wikipedia on 3 June 2020 found English, German, Dutch, Italian, Portuguese, Arabic, Korean, and Japanese wiki-pages. The Japanese redirected to "Heinrich's Law" rather than being biographical. For a long time, also the English page was more of a summary of Manuele's (2002) book than a proper page about Heinrich.

2 And likely lack of effort to research properly, maybe even rhetoric.
3 This biography is an extended version from my thesis (Busch, 2019b). Heinrich only rarely told anecdotes of his personal life in his writings. One exception was his 1923 *Bedford Stone Club* talk, another are the post-war recollections in *The Beacon*. Unless otherwise indicated, the biography builds on information from the unpublished Heinrich biographies by former Travelers Insurance employees Jesse Bird and William Lischeid (Bird, 1976; Lischeid & Bird, 2008) and the recordings of the interview that Bill Lischeid did with Jesse Bird in the spring of 2009.

I cannot stress enough the help I have had from Bill Lischeid in compiling this biography. Apart from sending over rare material that he collected over the years, he went through the trouble of editing the almost two-hour interview with Jesse Bird in the spring of 2009 into manageable portions. Jesse was aged 97 or 98 at the time of the interview in a nursing home in West Hartford, CT. He was born on Christmas Eve 1911 and passed away 5 February 2010.
4 There was some confusion on the date of birth when I started my research. Some sources claimed 1886, while others (including Manuele) listed 1881. Heinrich's former colleague and unofficial biographer Jesse Bird insisted on 1881. He based this on a remark by Heinrich's close associate Edward Granniss that the year of birth in the obituary was wrong and Heinrich was much older. Some research on https://www.ancestry.com, however, produced a number of official documents (including the 1900 census) that all list 1886 as the year of birth. My warmest thanks go also to Bill Lischeid who made the effort to take a ride to Fairview Cemetery in Hartford to check Heinrich's headstone, and confirm 1886 as the "official" date.
5 The 1900 census (ancestry.com) is the main source for this paragraph, along with Heinrich's birth certificate.
6 The 1900 census mentions that Minna was mother to six, of which five children were still living. The census mentions nothing on the two other living siblings.
7 No tales of Heinrich's time as an apprentice or machinist seem to have survived. Kanigel (1997) discusses the subject extensively in his biography of Fredrick Taylor. One can imagine that much of this description still applied in Heinrich's days, about 30 years after Taylor. It shows that Heinrich and Taylor partly shared a common background, but one must note also important differences. Heinrich was working class, while Taylor was clearly upper class, with some academic education under his belt before entering apprenticeship and the machinist trade.
8 https://www.dol.gov/general/aboutdol/history
9 Beyer has some interesting reflection on the different approaches of early safety work in various countries.
10 https://www.ancestry.com/
11 So far, none of these has been located, and maybe they never will. Most articles in *Protection* were not credited to their authors.
12 Little information is available about Heinrich's wife (not even their date of marriage!) without engaging in some extensive genealogical research. On sources available through https://www.ancestry.com, her name appears as both Viola V., Emma V., and Virginia, which gives rise to the assumption that Virginia may have been her middle name which was passed on to the couple's daughter. Her year of birth appears as both 1900 and 1901. Most sources from https://www.ancestry.com agree on 1901. The reprint of the 1959 edition of *Industrial Accident Prevention* adds to the confusion by mentioning her as Viola E. Heinrich.
13 With access to only three of these, this statement by Aldrich (1997, p.357) confirms my suspicion that there must be several yet undiscovered Heinrich papers hidden in archives and libraries.

14 Sources do not mention why this project was started. One might speculate that the research into the indirect costs of accidents was so successful that one wanted to continue the approach with something similar. Neither do sources tell whether there is an overlap between the data sets of, respectively, 100 (HWH1925), 5,000 (HWH1926, 1927a, b), 10,000 (HWH1927c), 50,000 (HWH1929a), and 75,000 (HWH1928a, b) cases studied by Travelers Engineers.

15 Bird does not elaborate why. Most likely because they tend to be taken at face value without looking further what Heinrich said on the matter. Chapter 6 discusses the ratio and its consequences in more depth.

16 Popular Science Monthly (1928) Most Mishaps Avoidable. *Popular Science Monthly*, September 1928, 113 (3), p.56.

17 A quote found in several news reports, including The Evening News from Harrisburg, PA (10 September 1928), Reading Times (15 November 1928) and Muncie Evening Press (20 November 1928). We do not find the quote in one of Heinrich's paper. Most likely the first source was *Popular Science Monthly*.

18 It is interesting to see that this text already mentions "human failure," while the *Origin of Accidents* chart at this point still spoke of "causes of supervisory nature."

19 *Hartford Courant*, 31 December 1933.

20 *Hartford Courant*, 31 May 1934.

21 Aldrich does not explain why the cause code was "cumbersome," however. Most likely because it brought a greater "administrative" burden, necessary to get better knowledge about causation, circumstances, and other information of statistical and preventive value than previous methods.

22 *Hartford Courant*, 31 October 1937.

23 Previously, Stack had held a fellowship at Columbia University which had resulted in the publication *Safety Education in the Secondary Schools* in 1929. This was one of the basis documents of school safety programs.

24 Stack (1953) tells that Albert Wurts Whitney was one of the people behind founding the centre, together with Ned Dearborn, Director of the Division of General Education.

25 Apparently, the War was beneficial for the advancement of safety and consequently the spread of Heinrich's work: "safety engineering has now become a recognized profession," and "accident prevention is at last on a recognized educational plane" (NSC, 1942b, p.108) he said.

26 This quote is Granniss reading from the circular (NSC, 1942b, p.103).

27 Including 300 women, a little under 2%. At the time, safety (and supervision) was clearly mostly a male thing. But the war contributed to a change here too. Armsby mentioned an increasing participation of women.

28 Although we can tell now that Blake did not share that lucky fate.

29 *Hartford Courant*, 2 May 1941.

30 *The Palm Beach Post*, 16 July 1943, 18 July 1943, and 25 July 1943.

31 *Protection*, 29 March 1944.

32 *Protection*, 2 August 1944.

33 *Hartford Courant*, 26 June 1946.

34 After a successful exhibition in the American Museum of Natural History, the Safety Museum started in April 1907. It was part of a movement to "reform society through a greater regard for responsibility and awareness" (Wilson, 2017, p.937). The museum lost much of its significance with the rise of safety education at universities and societal changes but would continue into the 1970s.

35 *Hartford Courant*, 18 September 1952.

36 When comparing newspaper pieces about Heinrich to those about another safety pioneer from the same era, Lewis DeBlois, it is striking that Heinrich is

mostly presented in relation to his safety work, while DeBlois was much more of a public figure with a larger number of social activities written about.

37 The First World War draft registration card mentions under the item "Tall, medium or short?" simply, "Short." Hair and eye colour are also taken from this card, which is found on https://www.ancestry.com

38 From today's – and my personal – point of view. I admit slipping into a judgemental tone here.

39 This anecdote arguably supports the triangle...

40 *The Travelers Beacon*, 12 (6), June 1925, p.15.

41 *The Travelers Beacon*, July 1929.

42 *Hartford Times*, 26 April 1956.

43 Jesse Bird assisted finding and sending these copies of talks and articles. Heinrich appreciated the help greatly as evidenced by a letter from 14 May 1957 in which he wrote to Jesse that "In addition to being a generous personal gesture you may be sure too that this helps along the industrial safety crusade..."

44 Judging from a note in the colophon of the author's copy that Viola renewed the copyright in 1969.

45 Recently renamed to ASSP – American Society of Safety Professionals.

46 According to the foreword of *National Safety News*, May 1956, which features the Heinrich/Blake debate.

47 http://www.insurancehalloffame.org/laureateprofile.php?laureate=68

48 https://www.vetsofsafety.org/shhofi/

Chapter 3

Heinrich's work

This chapter gives an overview of Heinrich's work, including its development and main themes. It also explores what influenced Heinrich's work, traces how Heinrich's influenced later safety work, and reflects on the longevity of his ideas.

3.1 General overview

Heinrich was productive during his five decades of active work within safety. Best known is his classic book, *Industrial Accident Prevention*, which saw four different editions during his lifetime (1931, 1941, 1950, and 1959) and a posthumous update (1980). The book was translated and printed in Japanese, French, and Spanish and is among the most influential books in safety (Li & Hale, 2014).

It is less known that Heinrich also wrote and published two management books, aimed at supervisors, *Basics of Supervision* and *Formula for Supervision*. Posthumously, Alfred Lateiner finished and published a management book that Heinrich had been working on before he died. This appeared in 1969 as *Management and Controlling Employee Performance*.

He was a member or chairman of various committees, delivering, among others, the so-called "Heinrich Cause Code," the *ASME Standard Form for Use in Self-Appraisal of Industrial Plants*, and the book of the Uniform Boiler and Pressure Vessels Laws Society.

Heinrich wrote over 50 papers. Most of these are very hard to get hold of, so most likely there is more undiscovered Heinrich material somewhere in the archives. Some of his papers were very similar, being republished in slightly different form in another publication. Furthermore, he did hundreds, possibly thousands of safety speeches and lectures, including university courses. Only a few of those are still available today.

Heinrich's work concentrated mostly on industrial (occupational) safety. However, he did not restrict himself to this safety domain.[1] He made side steps into what we today would call process safety and traffic and transport safety. His books contain cases from (occupational) health; he touched on patient safety and wrote a paper on home safety (HWH1951). Obviously, one cannot forget the fact that he was very much a management consultant and many of his writings and speeches deal with management.

As told in his biography, Heinrich was an avid speaker with a wide area of interests. He also proved to be socially engaged.

3.1.1 Phases

One can identify chronological phases in Heinrich's work. The 1920s were the formative years for many of the main themes. He conceived (or rather adopted) and developed several of the subjects he would be known for, doing some of the studies to support them.

During the 1930s, he would consolidate most of the main themes, elaborate and refine them and gain wider recognition. He would also increase the structure of his work, e.g. by introducing the dominos as an accident model and the axioms as a framework for his approach to accident prevention.

The 1940s were an interesting period, characterised by war-time duties and Heinrich's concentrating more on management with a series of books and papers, besides his original concepts.

He returned to his original work in the 1950s and beyond, adding safety management as an explicit element in the third edition of *Industrial Accident Prevention*. He also paid more explicit attention to the professionalisation of safety.

3.1.2 Developments and changes

Studying Heinrich's work, reveals how ideas developed through his speeches and papers before they ended up in his books where they were presented to a wider audience. These developments – most of which will be discussed in more detail in the proceeding chapters – concern things as seemingly simple as wording. Heinrich varied several times between "incidental costs," "hidden costs," and "indirect costs," to mention just one example (e.g. 1926, 1927a, 1930c, 1931a). Also, subjects and themes got more or less prominent places in the various editions of the book. For example, the sections about hidden cost or accident proneness were moved around between editions.

Besides these cosmetic or editorial changes, Heinrich also altered the contents, description, presentation, and explanation of some of his themes – sometimes significantly. As an example serves the *Origin of Accidents* chart (see Chapter 6). This changed significantly from its initial conception (HWH1928a, b) to the 1941 edition of *Industrial Accident Prevention*, and after. One of the changes that stands out is Heinrich initially attributing 88% of accidents to causes of a "supervisory nature" (HWH1928b, p.124), while these 88% changed to "unsafe acts of persons" a decade later (HWH1941, p.19).

Another significant example is how the triangle evolved from the early "blocky" figure (HWH1929a, b, 1931a) to the much slicker drawing in later versions of the book (see Chapter 9). More importantly, however, the descriptions around the triangle changed between versions of *Industrial Accident Prevention* (Manuele, 2002). Also, principles and axioms changed throughout Heinrich's work.

It should not be surprising that subjects and themes change over four versions of a book with about a decade between versions. After all, the world changes, knowledge evolves, insights are rejected, and new ones gained. We can see this for example in the evolution of Heinrich's management books. Initially, these drew on earlier work presented in *Industrial Accident Prevention* and some papers. Then they grew into books of their own which then in turn influenced the 1950 and 1959 editions of *Industrial Accident Prevention*. As Manuele (2002) pointed out, however, often Heinrich did not offer explanations for some of these changes.

One of the most striking changes in Heinrich's work is probably from the 1931 to the 1941 version of *Industrial Accident Prevention*. The second version of the book appears to have a greater focus on the human, as stressed by the domino sequence and the changed wording in the *Origin of Accidents* figure. As mentioned above, in the 1931 version of *Industrial Accident Prevention*, Heinrich attributed 88% of accidents to causes of a supervisory nature and, in 1941, to "unsafe acts of persons."

According to the introduction no fundamental changes were made,

> Changes have been made in the revised text for the purpose of achieving greater consistency, clarity, and improved arrangement... In no way, however, has there been any alteration of the fundamentals on which the original publication was conceived.
>
> (HWH1941, p.v, vi)

One can of course debate what is "fundamental." Moving from "supervisory" to "man-failure" may feel like a rather fundamental change. Also the 1941 edition of *Industrial Accident Prevention* emphasised the role and responsibility of management, and the importance of other actions than supervision and enforcement. However, the message remembered by most is the focus on the human, because of the renewed 88:10:2 ratio, which now distinguished between unsafe acts and unsafe conditions, and the dominos, which in Heinrich's version were very much about the human.

However, in the text of the 1931 version, Heinrich already talked much about "man-failure." Also, many of the "causes" mentioned in the *Origin of Accidents* chart were rather attributable to "man" than directly to supervisory. So maybe Heinrich thought it was best to streamline this and present only direct or proximate causes. Supervisory causes would be of a more underlying nature, after all. Maybe that is what he meant by "greater consistency."

Additionally, the books are not all that is left from Heinrich's work. One can draw on the papers written in-between editions, and other sources. This is not the place to explore this change between versions for full, but the contents of several papers can help to partly explain the differences. Notably *Mastery of Machine* (HWH1933a), stressing that machines are not at fault, it was the people who made and used them, appeared to be one of the key-papers in his work suggesting reasons for changes in direction.[2]

The minutes of the September 1933 Meeting of the International Association of Industrial Accident Boards and Commissions in Chicago suggest political reasons for the change. In the discussion, Eugene B. Patton director of the Statistics and Information division of the New York Department of Labor, referred to Heinrich's attribution of most accidents to supervisory failure and recommendations to change cause codes accordingly. While Patton was sympathetic to the thought, he asked the critical question, "But do you think that the foreman who makes out that report, knowing that the accident may become the subject of a compensation hearing, is going to say that the accident was due to supervisor failure?" (BLS, 1934, p.190). Sharing his experiences from a meeting with representatives from insurance carriers,

> They almost hooted me out of the room. "We are not going to have our policyholders report to the Department of Labor of New York that this accident was caused by the fact that their foreman gave improper instructions, or failed to issue instructions, or that the machine was improperly guarded."
>
> (BLS, 1934, p.190)

While such an admission would not carry any financial consequences, the insurance carriers would not have it, flatly refusing, "...we are not going to require policyholders to report facts indicating that they are at fault" (BLS, 1934, p.190). Possibly considerations like these[3] moved Heinrich to change the text and attributions.

Many other changes between editions are harder to explain. Still, it may be understandable that Heinrich, or his editors, chose not to do extensive explanations in the new versions of the book. With a decade between the versions, chances are that this would only be confusing for new readers and rather distract than contribute.

3.1.3 Industrial Accident Prevention – an overview

The foundation for much of Heinrich's core message was laid in the first edition of *Industrial Accident Prevention*, based on several mid/late-1920s papers supplemented with other subjects. This basis was expanded and reworked from edition to edition with some substantial developments. The contents showed an expanding scope, shifting focus and priority for some subjects, as well as technical progress and other developments. The book grew steadily, going from 360 to over 470 pages. Each edition contained several appendices, as for example historical backgrounds, standards for guarding and illumination and statistics.

The second edition introduced ten axioms and the accident sequence (dominos) which Heinrich used as a framework for his basics. This edition furthermore added a chapter about *Creating interest*, incentives, and ten psychological characteristics to interest employees (and managers) in accident prevention. This item would stay in all future editions. Additionally, there were chapters about fatigue, occupational disease, and safety organisation, although the latter was limited to a description of some basic elements, safety committee, and first aid and hospital.

The 1950 edition of *Industrial Accident Prevention* added some new subjects to Heinrich's repertoire. As he mentioned in the foreword, the third edition was "revised to suit modern conditions. Basic principles remain unaltered and are reiterated" (HWH1950a, p.vi). The book opened with a picture of a safety management ladder and an entirely new opening, providing a new framework for the basics. The structure was akin to the second edition. The philosophy and principles of accident prevention were presented in the first chapters with the axioms and accident sequence as the substructure. Safety organisation moved to beginning of book, consistent with prominence of the safety management ladder.

Other new additions were the chapter *Formula for Supervision*, which was a clear influence from Heinrich's 1940s management books, and chapters dealing with new subjects: personal protective devices, motor vehicle fleets, and a short-form safety course.

The fourth, 1959 edition of the book opened with the prime objective "to treat the subject as both an art and a science" and point out "what accident prevention is, what it accomplishes, and how it is accomplished" (HWH1959, p.xi). For this, it brought a different three-part structure:

1   Basis and philosophy of accident prevention
2   Accident-prevention method
3   Special subjects

The first two parts corresponded with the previous editions, including some updated insights. The third part contained subjects as personal protective devices, illumination, and safety organisation, along with three chapters with new subjects: nuclear radiation (rather technical), small plant safety – a subject that had gotten some attention before in a paper (HWH1932a) – and fire protection.

The fifth, posthumous, edition in 1980 is discussed in the section on Dan Petersen.

3.2 Main themes

Heinrich's work deals with a wide range of topics related to safety management and accident prevention. In his work, one can identify some recurring subjects and ideas, which one may consider main themes. This book uses the following criteria for Heinrich's main themes. They:

- stand out in his work,
- return frequently,
- are subjects he considered important (as indicated by the way he stressed or repeated them).

This book is not the first to look this way at Heinrich's work. For example, Swuste et al. (2013) suggest six main themes, which they label "Models and Metaphors":

1 Hidden costs of accidents,
2 Accident cause analysis, including the 88:10:2 ratio,
3 Foundation of major injuries, including the triangle,
4 Accident process in dominos,
5 The foreman,
6 Scientific underpinning.

This list feels incomplete. Some themes are missing; others are presented narrower than they should. For example, Heinrich discussed the role of management broader than merely the foreman – notwithstanding the key role that Heinrich assigned them. Also, "hidden cost" was merely one, yet the most memorable and frequently communicated, subject related to the economic side of safety.

Based on thorough study of the available material, the following extended list is proposed:

Theme	Chapter
Scientific approach	4
Cost and efficiency	5
Causation	6
The human element	7
Role of management	8
Triangle and reacting on weak signals	9
Axioms of safety	8
Professionalisation of safety	10
Practical accident prevention	10
Safety management	8
Social engagement	10

This categorisation is not a strict taxonomy, but rather a practical division into themes. These sometimes overlap and are related to each other.

For example, the *Origin of Accidents* chart deals with both causes and the human element. Below is a brief introduction, to be discussed further in later chapters.

3.2.1 Scientific approach

Not a single paper has "science" as the main topic, but the sub-title of *Industrial Accident Prevention*, and the frequent mentioning of "scientific"/ fact-based principles warrant an inclusion in the list of main themes. Often the scientific approach is connected to either the role of management or the professionalisation of safety (main themes six and eight, respectively), or both. For example, when stressing the need for a fact-based approach, improved knowledge or arguing in favour of improved use of statistics.

3.2.2 The economics of safety

Cost of accidents features prominently in Heinrich's early work. As other main themes developed and gained prominence, costs lost prominence, but never fully disappeared. Best known in this main theme is probably the 1:4 ratio of direct to hidden costs. The theme also includes the argument that accident prevention is "good business" (HWH1931a, p.16), and that safety and efficiency go together. The monetary approach to safety was a "powerful stimulus to preventive action" (HWH1931a, p.17), in addition to humanitarian aspects. It had an important impact on safety work and the acceptance of safety by managers as something to prioritise.

3.2.3 Causation

Heinrich was convinced that knowledge of causes is central in order to find effective actions: "it is the cause of the accident that we must know in order to effect a remedy" (HWH1928b, p.122). Many early papers critique the assignment of "causes" at the time. He stressed the importance of seeing the different stages in an accident. The early sequence of cause(s), accident, and injury (e.g. 1928b) evolved into the dominos. Heinrich paid much attention to explaining better ways to do "accident cause analysis" (HWH1931a) and "fact finding" (HWH1941) and suggested improved cause categories.

3.2.4 The human element

Heinrich is widely known for his attribution of most accidents to man-failure. Besides, there are subjects like accident-proneness, safety psychology, fatigue, and the influence of age on accidents. Safety psychology had dedicated chapters in *Industrial Accident Prevention*, Heinrich lending great importance to the subject, "Psychology lies at the root of the sequence of accident causes" (HWH1931a, p.127).

Contrary to common beliefs connecting Heinrich to behaviourist approaches (e.g. Dekker, 2019), Heinrich did not advocate purely behaviourist actions, aimed at correcting the "unsafe acts." "Guarding and other action of an 'engineering-revision' nature often provide a remedy even for accidents caused chiefly by man-failure..." (HWH1940, p.112, 1941, p.35) and therefore, "...it is unwise to depend automatically and invariably on educational or supervisory methods..." (HWH1940, p.112).

Heinrich also acknowledged humans as an essential positive factor, advising to draw on their strengths, "...in many cases safety may be promoted by employing the ingenuity and mechanical genius of workmen in the average plant, in the revision of process and procedure" (HWH1931a, p.246).

3.2.5 The role of management

Although Heinrich is known for his focus on workers' behaviour, (top) managers were his primary audience, *the responsibility lies, first of all with the employer*" (HWH1931b, p.11). Heinrich paid much attention to foreman/supervisors through several books and papers (HWH1929d, e, 1930a, 1938a, 1944, 1949). In many of his papers, Heinrich emphasised that management, production, efficiency, quality assurance, and safety build on the same principles and go perfectly together, "it is therefore not only possible to prevent the great majority of accidents, but also to do it by *the exercise of the very methods* that make for *economy, greater production and greater profits*" (HWH1928b, p.130). Also, responsibility was a recurring subject in many publications. "Acceptance of responsibility is a vital factor in industrial leadership" (HWH1929e, p.23). According to his biographer, Jesse Bird, management was Heinrich's favourite subject (Bird, 1976, p.16).

3.2.6 The triangle and reacting on weak signals

The safety triangle, alternatively labelled "pyramid" or "iceberg" – this book uses mostly "triangle" – is one of Heinrich's most iconic metaphors. It poses that for any accident with major consequences there is a larger number of accidents with minor consequences and an even larger number of accidents without any consequence at all. He argued not to wait for the accidents with serious consequences, but instead react on "weak signals" (Weick & Sutcliffe, 2001). He said that "we are often misdirecting our efforts and ignoring valuable data" (HWH1929b, p.9) when we wait for the "big thing" to happen and only then act. Heinrich's main message with the triangle is one of *opportunity*: handle things while they are still small, do not wait for the rare, big ones before acting.

3.2.7 Axioms of safety

Heinrich's Axioms provided the framework for the 1941 book and later editions. While the Axioms only appeared relatively late, Heinrich provided throughout his work bullet point lists and principles as a guideline for safety and management work. These can be regarded as proto-Axioms.

3.2.8 Professionalisation of safety

Despite the safety movement from the first decades of the century (Aldrich, 1997), and the fact that companies like The Travelers had impressive staffs with safety engineers, in Heinrich's time, safety was in general not seen as a science or as "a separate domain" (Swuste & Sillem, 2018, p.27). Through his entire career, Heinrich tried to improve this situation, by suggesting improved use of statistics, knowledge of basic principles, definitions, proper causal analysis, and the like. Swuste et al. (2010) note that the publication of handbooks like DeBlois's and Heinrich's were a clear sign of professionalisation of safety.

3.2.9 Practical accident prevention

This is a bit of a "miscellany" category, but an important one, not in the least when one looks at the number of pages spent. For example, roughly one quarter of the pages of most editions of *Industrial Accident Prevention* are dedicated to machine guarding, illustrated with many pictures. Additionally, there were chapters about illumination, fatigue, occupational disease, nuclear radiation, and first aid. Besides concrete safety actions at the sharp end, Heinrich's work contained practical suggestions related to the organisation of safety, education, process, and procedure revision.

3.2.10 Safety management

Safety Management had been a part of Heinrich's work from the start. After all, he sketches the framework of "fundamental principles" for successful safety organisation already in the *Bedford Stone Club Speech* (HWH1923b). And while the theme had been implicitly present throughout his work (including the 1940s management books), it would become very explicit with the appearance of the safety ladder metaphor in 1950. From that moment on, safety organisation gained a more prominent place in the books as well. Other examples of Safety Management were the *Hazard Thru Track* (HWH1950a, 1959) and the 1947 ASME standard.

3.2.11 Social engagement

Scattered throughout his work, Heinrich demonstrated social engagement. Frequently he mentioned the humanitarian importance of accident prevention, and in his war-time writings there was patriotism. Outside his formal writings, Heinrich's social engagement was for example evidenced by Heinrich's ideas on creating work after the Depression.[4]

3.2.12 What about risk?

From a contemporary view, risk is one of the central elements in safety management. Heinrich hardly ever spoke about risk, and certainly not in the context how most safety professionals use it today. This may seem odd because some of the safety handbooks listed in Heinrich's bibliography do mention risk in our contemporary understanding, and this thinking about risk had been around earlier. For example, a 1916 article in *Safety Engineering* spoke repeatedly about risk, e.g. from raw material. The author even suggested risk perception: "many of those ordinarily interested are so close to and so accustomed to the risk involved that they are either ignorant or indifferent to it" (Schreiber, 1916, p.231).

Still, risk was not as much in focus in Heinrich's time to the same degree as it is today. Even though Cowee (1916), Hubbard (1921), Fisher (1922), and especially Williams (1927) used the term risk in quite contemporary ways, and DeBlois (1926a) spent an entire chapter on probability, it only entered occupational safety and health discourse after the Second World War (Swuste et al., 2019). This is one reason for its absence as a main theme.

More importantly, however, on the rare occasions Heinrich used the term "risk," he did so in an insurance context. In almost all cases he spoke of the insured object/company, for example saying that an accident-prevention engineer "must determine probable causes of accidents before he can hope to service the risk intelligently" (HWH1929c, p.173).

That he did not use the term as we would today, does not mean that Heinrich was a stranger to the concept. However, often he spoke about hazards, or used different terms. The triangle, for example, clearly expressed risk-based thinking for accident prevention when discussing "potential." He suggested the same in his *Warehouse* paper "Accident prevention is the work of *finding*, and then *eliminating* or *correcting*, the *personal* and *mechanical hazards* that will in all probability, cause future accidents" (HWH1943c, p.138).

In his discussion of *What Makes a Safety Engineer?* (NSC, 1942; HWH1943b) was a discussion of risk, without actually using the word: "The degree of a hazard is measured not only by its potential power to cause a serious injury, but also by the probability that it will cause several minor or major injuries" (HWH1943b, p.47), stressing the importance of "exposure" which he defined as "the probability of causing the most frequent or most serious accidents, and injuries, or in other words the greatest dollar losses" (HWH1948, p.2).[5]

3.3 Influences

Heinrich was to a large degree a self-taught man who did not follow a university or college curriculum but picked things up while we went along in various positions and through additional study. That presents us with some challenges when thinking about what influenced Heinrich.

Heinrich's biographer Jesse Bird mentions that Heinrich never read much safety history. He was strictly a man concerned with the present and his own accepted theories in the accident prevention field. We can take this assessment with a pinch of salt. While far from following APA standards, or what modern scholars would do, Heinrich did mention some work he drew on in his bibliography or footnotes.

Reflecting systematically about Heinrich's influences, there are some suggestions about his influences:

1   The Travelers, due to the intense, daily contact and the fact that they had a tradition with a professional safety environment and literature.
2   The works referenced in his bibliography and quoted in the text, appendix, and footnotes.
3   People he closely worked with outside of the Travelers.
4   Contemporary thinkers from the period, people he loosely interacted with, the network he moved within, and other literature.

3.3.1 "Best of" or original work?

Heinrich is often credited for the ideas that he presented in his books. Studying early safety literature, however, reveals that most of the ideas that Heinrich presented were in fact already there and discussed by other authors. Stone's review of the first edition of the book noted this:

> Students familiar with the best current literature and practice in the field will find little that is new in this treatise. It does, however, bring together in convenient, manual form the best current thought on the subject.
>
> (Stone, 1931, p.324)

This book contains several examples, including management responsibility, accident sequence, hidden cost, or man-failure as the main cause, discussed by other people at the same time or before him. Some, one can assume, influenced Heinrich's work because he mentioned them in his bibliography.

Others, however, are more uncertain, and may be coincidence. For example, there are parallels between Richardson's article on minor accidents and near-misses in the September 1916 issue of *Safety Engineering* and Heinrich's triangle. While it may be safe to assume that Heinrich knew the *Safety Engineering* journal, and read the article, it would be stretching the assumption too far that this small article consciously influenced his work. The main purpose of these examples is showing that many of the ideas were around at the time, and that most of what Heinrich wrote was not all his original work. Rather his work represents an overview of what was to be considered contemporary thinking in safety at the time.

Heinrich's speeches and papers of the 1920s and 1930s clearly indicate that the work was conducted by engineers of The Travelers Engineering and Inspection Division, under his supervision and leadership. Also, many of the ideas already appeared in Travelers publications from the 1920s. Heinrich's work thus collected work done by Travelers engineers and presented it to a much larger audience.

The conclusion that it was not all Heinrich's original thought is not a surprising one. We should not try to find ideas in the minds of people, but study discursive regularities (Howarth, 2000). Taking the perspective that ideas emerge rather than pop up like the well-known cartoon lightbulb, it makes sense to regard *Industrial Accident Prevention* mostly as a "best of" compilation of safety theory and practice of that moment.[6]

While the ideas often may have been around for a while, before Heinrich gathered them and wrote them down, he did make some very important contributions. First, as one can see through the various papers, he managed to present them in language adjusted to his audience. Second, he coupled the ideas to some highly memorable metaphors, visual representations, models, and ratios (Aldrich, 1997; Van Gulijk et al., 2009); "...because of their simplicity, metaphors have strong persuasive power..." (Swuste et al., 2014, p.16). Third, he illustrated and explained the ideas with many analogies and examples, enhancing understanding and anchoring of the message. Fourth, as Stone noted above, he combined the various elements in a framework that was manageable. Petersen agreed with this view many years later:

> Perhaps it was because Heinrich proposed a philosophy for safety that his work was so important. Before the publication of his book, safety had no organized framework of thinking. It had been a hodgepodge of ideas. Heinrich brought them all together and defined some excellent principles out of previous uncertain practices.
>
> (Petersen, 1971, p.11)

3.3.2 The Travelers

The Travelers must have been an important influence, through the day-to-day interaction with Heinrich for decades. Second, as described in the previous chapter, they had already a history and tradition with safety, a major professional environment with hundreds of safety engineers and many safety publications.

Some of the influence can be glimpsed from a May 1913 article in the *Travelers Standard*. This was based on an address by George Gilmour, Chief Engineer and Heinrich's department manager, about the elementary principles of accident prevention in factories, shops, and yards. One can assume that this had major influence on the young beginning safety engineer Heinrich, fresh with The Travelers. Many topics addressed here later returned in Heinrich's work.

Gilmour started emphasising the responsibility of employers and the important role of insurance companies, thanks to the recent adoption of workers compensation regulations.

> Manufacturer and other extensive employers of labor in our country have been slow to recognize their duties to their employees in this respect, but the subject of accident prevention has been taken up recently, with a good deal of zest, by insurance companies, engineering societies, federal and state governments, and business corporations, and quite a good deal of progress has been made.
>
> (Gilmour, 1913, p.141)

However, not all had realised yet "that the lives and welfare of the so-called working classes are of greater importance than increased factory production or extra dividends" (Gilmour, 1913, p.141). This touched on the age-old tension between safety and profit. Gilmour highlighted the "waste" through accidents and classified these into unavoidable, unforeseen accidents and those due to "curable causes." Based on German research he estimated that more than half[7] of the accidents fell in the latter group.

The necessity was both humanitarian and economic. Reliance on (new) laws was not enough. Practicable methods were necessary, which was a job for trained safety engineers. Here was an early call for professionalisation – one of Heinrich's main themes many years later.

> Safety engineering as a distinct profession is new to us. But it is broad and interesting, and as the demand for those skilled in it is certain to increase very rapidly, young men who are entering the engineering field may well consider specializing in this direction.
>
> (Gilmour, 1913, p.160)

Gilmour showed surprising progressive thinking about "causes," differing from, for example, Beyer (1916). One wonders whether Heinrich took inspiration from this view, capturing human elements, machines, and environment while hinting at organisational factors.

> It has been assumed, quite generally, that industrial accidents are chiefly attributable to the absence of safety-devices, around dangerous machines or in other places where danger is known to exist; but closer

examination has shown that a wider view of the subject must be taken than this, and that we must recognize many other causes also...

(Gilmour, 1913, p.146)

His discussion of causes emphasised environmental factors, showing some advanced thinking. One finds some established "safety wisdom" in his argument, including a basic hierarchy of controls, that safety by design is more effective and cheaper than later adjustments and that investments in working conditions served efficiency and productivity. However, "Safety devices are not magical things that can transform danger into absolute safety as the philosopher's stone of olden days..." (Gilmour, 1913, p.154).

As others did around the time, Gilmour mentioned "ignorance" and "carelessness" as "human causes". However, he saw "ignorance" not as a mental defect, but as a responsibility of the employer to give good and complete instruction about hazards and minimising or avoiding them. Regarding warning signs, he commented, "...they are really nothing but makeshifts. It is always far better to provide an effective protection than to merely announce the presence of danger" (Gilmour, 1913, p.159).

One of the elements that may have impressed Heinrich was the fact-based approach demonstrated by Gilmour, for example when discussing the influence of age on accidents, based on UK research by Sir John Brunner, and yet another subject to return in Heinrich's work many years later.

Obviously, many Travelers people that Heinrich worked with during the years must have left their influence, including his early mentor Dr Allan Risteen, Ralph Crosby with whom Heinrich would work even after he left the Travelers, and most of all his long-time collaborator Edward Granniss.

3.3.3 Bibliography

We can learn about Heinrich's influences from the limited bibliography for his 1931 book.[8] This is an imperfect and only approximate way on assessing influences, but it is what we can do. It also gives some insight in the general safety thinking at the time. Heinrich did not explicitly indicate his intentions with this bibliography. One assumes that these are works that have influenced and informed the contents of the book, but unlike other sources that he quotes directly, he does not refer these sources in the text.[9] Another possibility is that these are suggestions for further reading, but nowhere he says so.

Some titles are hard to follow-up, as unspecified National Safety Council "Safety Pamphlets" and the *Travelers Standard*. We can interpret this as that he was influenced by the current state of the art within safety. In addition to the titles discussed below, we find the *Safety Fundamentals* lectures given by the Safety Institute of America,[10] the *Safety and Production* study of the American Safety Council, and Michelbacher's book on *Casualty Insurance Principles*. The latter appears to have little effect on the text.

3.3.3.1 Safety handbooks

The first thing standing out from the bibliography, is a collection of safety handbooks, three of which can be counted among the early wave[11] and three from the late/mid-1920s. There is a clear difference. The early handbooks[12] by Ashe (1917), Beyer (1916), Cowee (1916), Tolman and Kendall (1913), and Van Schaack (1910) – the latter two not appearing in Heinrich's bibliography – concentrated mostly on what to do about specific situations (e.g. guarding) and contained little about basic principles, safety organisation and education. The post-1920 books also included practical solutions but were often on another level. Especially Heinrich's offered more of a framework and philosophy of how to approach problems.

BEYER

In 1916 two solid safety handbooks were published, presenting the state of the art then. Of these, Beyer's book provides the more structured approach with eight parts and 54 short chapters, paying attention to references and suggestions for further reading along with many pictures and drawings.

Beyer's book concentrated on the practical side of safety work, discussing hazards and how to deal with them. "...it has been the author's aim to make it an authoritative statement of accident prevention methods, proven by the 'road test' of actual experience to be effective and practical" (1916, p.v). There was little discussion of general principles or theories on safety. However, the first couple of chapters dealt with more general subjects.

The third chapter was about "Causes of Accidents," but Beyer seemed not to have had a specific causation theory yet. Its two-page list was rather uneven, blending injuries, activities, agents, and "causes." Heinrich would critique similar attributions as "so-called causes."

Other explicit connections to Heinrich's work were practical remedies (guarding), a section about costs and efficiency, and some discussion of foremen, but less than with other authors. Most explicit were the specific rules for foremen in chemical industry. The "human element" was not very pronounced, apart from reflections on carelessness. The clearest connection to Heinrich was obviously the book's title.

COWEE

Like Beyer's, this was a heavy tome of 27 chapters, with many illustrations and example forms. As the title indicated, it was a practical, clearly structured book with a detailed index, helpful in finding an answer to problems.

The book contained little method and theory or principles and relatively little on organisation. The first introductory chapter touched on several subjects that would return in Heinrich's work (and that of others) and possibly shaped his thought: "Safety engineering has become a well-established profession." and because "Education is the keynote to universal safety" it should become a part of the curriculum at technical schools. Accidents came with huge costs. They produced a "yearly waste" that was "incurred by preventable accidents". He claimed that "at least ninety percent of all work accidents are actually preventable," and therefore "safety is essential to efficiency," and "safety and efficiency go hand in hand" (Cowee, 1916, p.1–2).

The human element was scattered throughout the book. Workmen were expected to follow rules for safe practice, and if they did not, discipline should be applied. Cowee saw foremen as key persons. Among their tasks was "eliminating unsafe conditions and dangerous practices" (Cowee, 1916, p.10). This duality of causes resonated throughout Heinrich's work.

ASHE

Ashe's book was a rather slim volume, especially when compared to Cowee's and Beyer's. The book was a somewhat curious affair. The title suggested organisation as the main theme, but the introduction emphasised education. However, Ashe did not discuss that much either. Lacking an overall structure and guiding framework, the book appeared more a collection of loose ideas, discussing safe habits, education, the important role of foremen, medical and physical examination, and first aid. In the chapter about records, analysis and ratio curves, Ashe also had a small section on the importance of minor injuries, but his argument was mainly of having them treated quickly and professionally to avoid infection or worse. He did not think about using them in a proactive way.

3.3.3.2 Second wave

In the late 1920s/early 1930s, four American safety handbooks appeared a few years apart, by DeBlois (1926), Williams (1927), Lange (1927, but prepared in 1922), and Heinrich (1931). One might assume that the texts would not differ dramatically, but the books had different focus. Each stressed other elements of safety work. Heinrich had obviously the benefit of being last in the series and the opportunity to draw on the greatest assembly of safety knowledge.

DEBLOIS

DeBlois's book stands out from the others from the era. This may be partly due to his academic background. Also, his experience in DuPont and various safety organisations left their marks. DeBlois had a major influence on some of Heinrich's work, especially regarding causation and the accident sequence, but disappeared from his bibliography after the second edition.

The book's title suggested its focus and audience. Practical safety work as protection and guarding was rather absent and the focus was on safety organisation and some underlying principles of safety and accident prevention. The final chapter discussed the various motivations of safety – economical, moral, humanitarian, and spiritual. While Heinrich's approach would be more management and science oriented, DeBlois embraced safety as a way of thinking with leanings towards seeing of safety in terms of a religion.

The most outstanding chapters were those dealing with *Accident Investigation and Analysis* and *Chance and Probability*. The former certainly had a major influence on Heinrich's work, presenting an origin of thinking of accidents as a process. DeBlois clarified that, although connected, injury and accident were two distinctly different things with usually distinct causes.

Various subjects overlapped with Heinrich's, including statistics; "indirect costs"; maintaining interest; and the human element, which, for DeBlois, included discipline, sections on new employees, proneness, mental and physical examinations, and selection (or selective letting go) as tools to prevent accidents. Interestingly he even discussed "creating culture," here called "Safety Atmosphere," "...that indefinable influence which pervades the industrial plant that is doing conscientious safety work and reaches, instinctively, even the new employee himself" (DeBlois, 1926a, p.156). DeBlois drew strongly on Fisher's main categories of "mental causes of accidents," spiced up with many quotes from other relevant authors.

Another important likeness with Heinrich's work was the discussion of safety, production, and efficiency. DeBlois did not just accept the widespread belief that a safe plant was an efficient plant, but reviewed literature, building a qualitative argument based on various elements (accidents, routine, emergency, equipment, humans, and management) and concluding that safety and efficiency should correlate – while arguing for the need for fundamental research.

LANGE

Lange's book was a bit of a curiosity, standing between the first and second wave of safety authors. Written several years before its publication, in much of the writing Lange was more akin to the earlier authors, offering much detailed information for safety practitioners. The lengthy sections about locating hazards, guarding, illumination, and first aid may have inspired the practical parts of Heinrich's book.

Besides practical remedies, like guarding, illumination, and industrial health, Lange touched upon some themes in common with Heinrich, but often more superficially and briefly, including competence of safety engineers, efficiency, a brief mention of indirect costs, the responsibility and key role of foremen, and various aspects of the human element. Lange agreed with others that it was best to eliminate hazards and guard machines and discussed "engineering revision."

Lange paid relatively little attention to causes, and had a different understanding from Heinrich, often attributing the agent of the injury. While spending little attention on accident analysis, his thinking about multi-causal representations was quite progressive.

WILLIAMS

Williams's book was specifically aimed at "a man who devotes all or part of his time to industrial safety" (Williams, 1927, p.iii). For this, he drew on several of his contemporaries. All chapters contained a list of references – often NSC publications but also the work of DeBlois, Fisher, Beyer, and Heinrich, whose work on indirect costs of accidents had just been published.

Williams's focus was the organisation of safety, rather than safeguarding although there are some detailed discussions of hazards and remedies. Contrary to its claim of discussing basic principles, the book was mostly a how-to book of Safety Organisation strongly drawing on DeBlois, discussing committee work, inspections, safety meetings, and safety campaigns.

Other similarities between Williams and Heinrich were the importance of management and foremen, some attention for the human element, an emphasis on safeguarding machinery, and safety and production going together. Williams focussed on hazards rather than causes.

3.3.3.3 Men and machines

One book that most certainly must have influenced Heinrich's thinking, is Stuart Chase's *Men and Machines* (1929), being one of the few titles appearing in Heinrich's bibliography from 1931 to 1959. The book aimed to gain a better understanding of machines, and their relation to men, by exploring what machines are, what they do, their history, and what major effects their activities have on human beings, society, work, and the world in general. The book offered a reflected and multi-faceted discussion that included positive and negatives sides of the rise of machines.

One of Chase's returning main conclusions was that the problem was not so much that of the machine itself, but of man's problem to control them or use them wisely, "…there is nothing innately evil in the kind of work which most machines are designed to do. The evil, if any, must come in the abuse of function" (Chase, 1929, p.40). This resonated well with Heinrich's proposal in his *Mastery of Machine* (HWH1933a) paper. While Heinrich's duality of causes was reflected in the title, it is unlikely that Chase's book influenced this duality, since it was a common thought in early safety, and Heinrich presented it (HWH1928a, b) before the publishing of Chase's book.

3.3.3.4 The human element

FISHER

There are three sources from the 1931 book dealing explicitly with the human element. The first is *Mental Causes of Accidents* by Boyd Fisher, a unique contribution to early safety literature. Where the other safety handbooks dealt with organisation, technical remedies and approaches, discussing the human to a greater or lesser degree as just one factor, Fisher dedicated his entire book to the importance of the human mind for accidents. "Boyd Fisher has clearly shown that ill health, either physical or mental, lies at the root of many an accident" (Williams, 1927, p.154). Fisher set out for a "new view" of safety,

> In the matter of industrial accidents, we seem to take for granted that the 70 per cent of men who get hurt "of their own fault" do so wilfully. We set the cause down always as "carelessness" – which means that we assume that they knew better, but deliberately went wrong.
>
> (Fisher, 1922, p.4)

Instead of seeing an accident as a deviation, or violation, Fisher proposed that we "treated accidents not as delinquencies, but as forms of mental error" (Fisher, 1922, p.7), because, "Merely to say that a workman 'failed to use the proper safety device' does not sufficiently analyse the situation. If possible, we must get at and correct the state of mind which produced this error" (Fisher, 1922, p.7).

The word "mental" should not be interpreted as a disease. Although the author to some implied this, he discussed mostly causes that spring from mental processes, from the mind. Most chapters, therefore, discussed various "minds," e.g. "The Unguarded Mind," "The Puzzled Mind," or "The Diverted Mind." Fisher proposed 15 major (mental) causes of accidents, divided into five main groups: Ignorance, Predisposition, Inattention, Preoccupation, and Depression.

One wonders what Fisher's impact was at the time. Both DeBlois and Williams referenced Fisher's work in their safety handbooks. Heinrich, however, merely mentioned this book in his 1931 bibliography. Explicit links to Fisher's ideas are hard to find. This book may, however, be one reason for Heinrich's opinion of psychology's importance for safety.

SLOCOMBE AND BINGHAM

Men Who Have Accidents (Slocombe & Bingham, 1927), written by Dr C.S. Slocombe, safety advisor of the Boston Elevated Railway Company, and Walter Bingham of the New York-based Personnel Research Federation, is the only other title to make it in the bibliography of all editions of *Industrial*

Accident Prevention. Therefore, one is inclined to assume that it had a major influence on Heinrich. Heinrich quotes the findings from this paper in the 1931, 1941, and 1950 editions in the chapters on *Safety Psychology*. In the 1959 edition the reference and quote in the text disappeared, but an edited version was still there.

The paper should be seen together with the *Safety on the "El"* book. Heinrich drew on this book, which contained most of the paper, in the same chapter. The book described the recent work done by the Boston Elevated Railway on improving worker, passenger, and public safety through improved design and operations.

The book showed many similarities to Heinrich's work. For example, in the way it presented safety and production as two sides of the same coin. It also discussed the important roles of management and foremen, and promoted a fact-based approach and practical solutions. The book also regarded the human element as the next step in accident prevention, "After the Railway had spent considerable sums of money to better the operation of the system and make it safer, the next step taken was a study of the human factor in accident" (Boston Elevated Railway, 1930, p.53).

However, where Heinrich decided to look broader for "remedies," the Railway went down a more specific road. They hired the Personnel Research Federation to further their safety efforts by looking into individual factors of accidents with the aim of finding the "origin, causes and influences accident proneness" (Boston Elevated Railway Company, 1930, p.3). The preliminary study was presented in the Slocombe and Bingham paper.

DOW

Stay Alive! by Marcus Allen Dow was a bit of an outlier. It was an entertaining book, partly thanks to the drawings by H.S. Zoll. The character of Jim the Truckman tells 30 little stories about behaviour in traffic and Jim kicking the shins of the perpetrators. The telling was in a narrative style filled with slang and "rough" language. The spelling represented rather spoken language than correct syntax and grammar, as if Jim were telling the reader in a bar.

It is somewhat puzzling that this book appeared in the bibliography of the first three editions of *Industrial Accident Prevention* since it offered little in the way of theory or insights, but it may have confirmed Heinrich's views regarding the "human element" in accidents. Additionally, it may have inspired Heinrich in telling stories throughout his books.

3.3.3.5 Bibliography development

As the books evolved, the bibliography did. Since the 1931 bibliography is the most informative regarding possible influences, this section only gives a brief review of its development.

For the second edition, Heinrich dropped the pre-1920 safety handbooks from the bibliography. In the third edition, oddly only Lange remained (for no obvious reason). This could suggest that it had a more lasting influence that DeBlois and Williams – or may have been a slip. In the 1959 edition all the mid-1920s books had disappeared.

As the book expanded and added new subjects, the bibliography reflected this. The bibliography of the 1941 edition had two sections, one on "Industrial Safety," largely overlapping with the 1931 bibliography, and one on "Industrial Hygiene and Occupational Disease." Interestingly, the second section's size roughly equalled that of the first, while the discussion of these subjects was much more limited.

In the 1950 edition, the bibliography grew from not even one page in 1931 to almost three pages and three sections: "Industrial Safety," "Industrial Hygiene and Occupational Disease" (significantly expanded), and "Supervision and Psychology." Unsurprisingly, the latter category referenced Heinrich's 1940s management books along with Viteles's 1930 classic *Industrial Psychology*. The safety section featured many new safety books, including Blake's *Industrial Safety* and Stack's *Education for Safe Living*.

The 1959 bibliography expanded further, with a new section dedicated to "Nucleonics." The "Supervision and Psychology" and "Industrial Hygiene" sections were almost identical, but there were some minor changes in the safety section, among others *Safety Management* by Simmonds and Grimaldi, reflecting the evolution in Heinrich's work towards this subject.

3.3.4 Associates and acquaintances

It is likely, and almost inevitable, that Heinrich was influenced by the people he interacted with. This may have included other safety authors, but Heinrich does not discuss this. For example, it is very well possible that Heinrich and DeBlois have exchanged thoughts at some point. It is certain that they met, e.g. at the National Safety Congress in New York, October 1928.

We know for sure that Heinrich did not write his books alone. All editions contain contributions of others which Heinrich acknowledged. The most important of these contributors was obviously Granniss, contributing the appendices on *Chronology and Background of Industrial Safety* in the second to fourth edition, the chapter on *Personal Protective Devices* in 1950 and was credited as co-author for the 1959 edition.

Several others deserve special mention. Heinrich credited R.J. Crosby for his constructive critique to the basic principles, Elliot P. Knight who "fathered" the chapter on *Occupational Disease* (HWH1950a) and Walter A. Cutter from the Center for Safety contributed to the arrangement of the material from a pedagogical point of view (HWH1959).

Another educator Heinrich worked with was Herbert J. Stack of the New York University Center for Safety. Heinrich most certainly met and worked with safety education pioneer Albert Wurts Whitney. Apparently, they were friends (Stack, 1953). However, Heinrich never referenced to him, nor do we see clearly Whitney's work reflected in Heinrich's writings.[13]

During the Second World War, Heinrich and Robert Blake were responsible for the material for safety training. It is sure that Heinrich's work influenced Blake's, and at a later point also vice versa.

3.3.5 Others

Heinrich had an active network throughout his working life, interacting with the National Safety Council and other organisations, serving at various committees, and speaking at countless safety conferences, all of which must have left their marks. Interestingly, unlike most other early authors (including Beyer, Dow, and DeBlois), Heinrich never had a high-ranking function within the National Safety Council.

Heinrich was a gifted storyteller, and his books and articles were laced with stories and examples. Only rarely, however, did he refer to literary references. One of the rare exceptions is found in his books on supervision where he referred several times to Sherlock Holmes when discussing deduction as a method to check facts.

Even though he may have met with them, Heinrich appears not to have been part of the "Hartford Intelligentsia," like the poet Wallace Stevens or linguist and fire prevention engineer Benjamin Lee Whorf. Neither did Heinrich refer to known Hartford authors – save one single mention of Samuel L. Clemens (HWH1945a).

Given Heinrich's emphasis and interest on management, one may wonder what influenced him in that respect. The first references in that direction we find in *Formula for Supervision* (HWH1949) and in the 1950 and 1959 editions of *Industrial Accident Prevention* there was a bibliography section for *Supervision and Psychology*. These mainly refer to literature about supervisors, as publications from the National Foremen Institute although one also finds Viteles's influential *Industrial Psychology*.

As far as early management influences, Heinrich must have been familiar with the work of Frederick Taylor. However, he does not once refer to Taylor nor to Scientific Management, nor to any other author related to that school. The next chapter will discuss this more extensively.

Heinrich often spoke about the analogy between safety and quality and volume of product. However, there are no obvious links between his work and quality management thinkers of the era as Walter Shewhart. Despite his interest in statistics, nowhere did Heinrich refer to Shewhart or his methods (process charts or similar terms). If Heinrich was aware of this work, he did not mention it.[14]

3.4 Success and longevity

Heinrich's work was extremely influential on safety theory and practice, its influences lasting until today. It is puzzling that the work of other safety authors did not get the same traction. DeBlois had major influence on Heinrich and in some ways his work was more advanced. However, his name is largely unknown to safety professionals today. Many factors have contributed for sure, such as the timing and personal factors – DeBlois switching jobs, holding executive functions that may have restricted him and going through some private turmoil right after the publishing of his book. Most importantly, however, DeBlois's work lacked the stories, ratios, metaphors, and visual appeal of Heinrich's.[15]

Several factors were possibly in Heinrich's favour. For one, lucky timing. He published his findings about the hidden cost along with practical ways for accident prevention shortly before the Great Depression. He barely referred to the crisis in his writing, but while he did not offer magical solutions, his practical approaches may have had appeal.

Heinrich's work fit well into the emerging "science" or "profession" of Management, offering tools and methods to deal with one of the many problems that managers were confronted with. The good thing was that Heinrich's methods and tools supposedly also served other objectives, like quality, production, and cost reduction.

He was an optimist, convinced that with knowledge of causes and will to act, it would be possible to prevent accidents. His optimism shines even more through the continuous tone of encouraging "this is simple" and "you can do this" statements that one can find through his work, mostly directed towards managers: "...the fundamentals of successful accident prevention are simple. They are readily and inexpensively applied" (HWH1932a, p.28). Heinrich put great trust in technology. Much progress had been made, he argued, but "Still greater accomplishments will be forthcoming, moreover, because the surface of practical possibilities has merely been scratched" (HWH1931a, p.249, 1941, p.267). In 1934, he even wrote a paper titled *Safety Wins a Place in the Sun* in which he discussed "progress of startling significance" (HWH1934, p.112), and he was optimistic for the future of safety: "...there is progress and a constantly broadening vision" (HWH1940, p.118).

Ironically, the fact that he was "stuck" in his function as Assistant Superintendent may have contributed to him spending more effort on projects besides his main job. It is merely speculation, but had he been promoted to an executive position within Travelers Insurance, he may not have had the time to do so.

Aldrich attributes Heinrich's success to "easily remembered ratios" and the fact that there was "something for nearly everyone" (Aldrich, 1997, p.152) in his work, from managers to safety engineers and employees. "Heinrich's ideas were a departure from the safety thinking of the time. What he said, however, made sense to people in the field of safety, and his ideas were accepted" (Petersen, 1971, p.11).

Heinrich's legacy contains two of safety's most iconic metaphors – the dominos and the triangle. In terms of recognisability, only the Swiss Cheese Model (Reason, 1990, 1997) comes close. The metaphors appealed to common sense, were easy to understand and highly memorable. They were bound to be adopted by others. This contributed to spreading Heinrich's message. However, along with success thanks to visual strength came the risk of misinterpretation, possibly deviating from the intentions of the people they originated with (Le Coze, 2013; Busch, 2019a). Later chapters discuss several different interpretations.

> ...visual properties are specifically at the heart of their heuristic value and power to explain, to make sense and to perform (note also that their downside is that they are ways of not seeing, they lock users in certain interpretations).
>
> (Le Coze, 2018, p.90)

There are several possible explanations for Heinrich's lasting influence – over eight decades by now. Contrary to other early safety thinkers/authors, he remained active also after the Second World War (Van Gulijk et al., 2009), thereby bringing his thoughts and writings into a new era. None of the other safety authors from the 1920s and 1930s published after the War. Also, Heinrich updated some of his work, e.g. by adding a framework for Safety Management (HWH1950a).

The Heath brothers list several characteristics of ideas that "stick" (Heath & Heath, 2007). These characteristics can help explaining Heinrich's endurance. His work lives up to most of the characteristics identified by the Heaths. Metaphors like the dominos fill all or most of these simultaneously, being simple, concrete, credible, appealing to emotion, and illustrated through examples.

Mathews and Wacker open their book about business improvement storytelling[16] telling us, "the six most powerful words in any language were *Let me tell you a story*" (Mathews & Wacker, 2007, p.1). Contrary to the other safety authors, Heinrich tells lots of them, possibly contributing to his success.

The opening chapter of the first two editions of *Industrial Accident Prevention* is a good example.[17] Wrapped into a recognisable story about a management meeting, Heinrich explained the basic principles, summarising the book's main message. After this, subsequent chapters discussed the various approaches in more detail. Throughout the book Heinrich handed the reader memorable, practical tools as short lists of things to do, metaphors, ratios, and images, and many examples that readers could relate to.

Le Coze (2013) suggests another set of criteria for successful and persuasive models. They should be simple to understand, normative, generic, and therefore useable in different contexts, and have visual appeal. Seemingly, Heinrich had a knack for ticking these boxes, as Le Coze confirms explaining the success of the triangle,

> First is their capacity to be generic, whether in the chemical industry, in aviation or in the railways, both models adapt fairly well (1) and normative because they provide principle for assessing specific situations (2). They also have the ability to mobilise appealing metaphors (3), to be inscriptions (4) and to become boundary objects (5) with a performative dimension (6).
>
> (Le Coze, 2018, p.88)

Another useful way of thinking about Heinrich's longevity, is found in De Winter and Dodou's (2014) discussion of the persistence of the Fitts list. Drawing on Jacobs and Grainger (1994), they discuss six criteria: plausibility, explanatory adequacy, interpretability, simplicity, descriptive adequacy, and generalisability. Heinrich's metaphors do well when assessed on these criteria because they are rather simple, make sense, and do fit largely everyday observations.

An important note is that these criteria are meant for the evaluation of scientific models, which Heinrich's ideas are not (Swuste et al., 2014). Many of his ideas are rather in the realm of folk models (Dekker & Hollnagel, 2004). Still, these appear to be useful criteria to think about its longevity. Also, we need to keep in mind that the evaluation of Heinrich's metaphors and ideas has not so much fallen to scientists, but rather to practitioners and managers who were able to use them attacking their practical problems.

A final important factor for Heinrich's longevity is that others retold his stories, or expanded on his stories – which is a good sign for any story...

3.5 Heinrich's influence and developments

> ...his ideas were accepted. They were accepted so completely that even today we work largely within his framework. His work set the stage, in effect, for all safety work since 1931.
>
> (Petersen, 1971, p.11)

Heinrich's ratios, models, (visual) metaphors, and methods have become part of the "mainstream" safety curriculum and established safety wisdom/beliefs (Petersen, 1971; Manuele, 2013). Not necessarily in the form that Heinrich conceived them because others have adapted and changed them. These developments and critique offered over the years created continued interest in his work. Below, the most significant ones are discussed.

3.5.1 Lateiner

One of the first to take Heinrich's ideas further, was Alfred Lateiner (26 January 1911–8 May 1988).[18] Lateiner was a self-made man, who like Heinrich never had formal college education. During the War, he first worked at an aircraft manufacturing plant in New York, teaching supervisors. His work in safety kept him from the armed forces. At the Naval Shipyard in Brooklyn, he developed a five-day safety training program which he later would use commercially.

After the war, Lateiner contributed to popularising Heinrich's models in Europe during the 1950s (Swuste et al., 2010, 2019). Heinrich found himself too old to travel and instead Lateiner visited 11 countries, from Turkey to Norway, as part of the Marshall plan. He wrote several publications, explaining Heinrich's work in actionable terms for supervisors and managers. He was careful crediting Heinrich, but he did change some descriptions. He was the first to call the triangle an iceberg (Lateiner, 1958).

Heinrich and Lateiner worked closely together during the early 1960s. They offered a correspondence course for safety students and worked on a book together which Lateiner finished and published in 1969. Lateiner had his own safety and management consultancy during his life, moving through the USA and writing a book on supervision. First published in 1965 it went through at least 16 (revised) prints and was translated into French, Italian, and Japanese.

3.5.2 Bird and loss control

Frank E. Bird Jr. (19 January 1921–28 June 2007) served during the Second World War in the United States Navy Medical Department, initially wanting to become a medical doctor. After gaining a BSc from Albright College in Reading, Pennsylvania, in 1949, he soon after went to work for Lukens Steel Co. After some time at the shop floor, he joined the safety department and quickly rose to become their director of safety.

At Lukens, Bird led a seven-year study looking into 90,000 incidents, producing a new accident triangle with a ratio of one disabling injury to 100 minor injuries to 500 property damage accidents. This led to Bird's "all-accident control" concept which gained much attention, nationally and internationally. Together with Lukens's head of training, George Germain, Bird wrote his first book, *Damage Control* (1966).

Speaking frequently at conferences and publishing several articles made Bird an influential voice in safety. In July 1968, he accepted a new position as director of engineering services for Insurance Co. of North America (INA). Here, he created the well-known Bird pyramid, based on a study of over 1.5 million accidents reported by 297 companies. The pyramid brought a new ratio of 1:10:30:600. Property damage was the third level. This was the central driver for Bird's Loss Control concept.

Loss Control aimed at not only preventing injuries, but also at improvements to raise productivity, and businesses efficiency. The concept also included a cost iceberg, illustrating the hidden costs of accidents, and a loss causation model, which was a development of Heinrich's dominos, but emphasising management control and underlying causes rather than direct causes which Bird regarded as symptoms. Given the fact that Bird basically continued Heinrich's ideas, he did not credit Heinrich as much as he probably should have done.

In 1974, Bird founded the International Loss Control Institute (ILCI) at Loganville, Georgia, becoming its executive director. This helped spreading his work further. One key element was the acronym ISMEC – a shorthand for his management control function: Identify, Standardize, Measure (the central element), Evaluate and Correct. A collaboration with the South African Chamber of Mines led to the International Safety Rating System (ISRS), which would be the leading early safety management standard.[19]

Bird wrote several other books, including *Loss Control Management* (with Robert G. Loftus) and his classic *Practical Loss Control Leadership* (again with Germain) which first appeared in 1985 going through several revised editions. He would remain an influential speaker and author, nicknamed "the Billy Graham of Safety."

In 1991, Det Norske Veritas (DNV) purchased ILCI. DNV included virtually all of Bird's concepts in their programs, materials, and consulting services, adding to the further spreading of Bird's (and thus Heinrich's) ideas. Frank's son David Bird later founded International Risk Control America (IRCA).

3.5.3 Petersen and the fifth edition

Bird was also an influence on one of the most renowned and influential American safety scholars from the 1970s to the 1990s: Daniel "Dan" Petersen (4 March 1931–10 January 2007).[20] He earned his PhD in management at the University of Northern Colorado, with his dissertation *Human Error Reduction and Safety Management*. He worked with Dennis Weaver during the early 1960s. He would later become a safety consultant, develop an acclaimed Safety Perception Survey, and found the University of Arizona's Graduate Program in Safety Management.

Petersen wrote numerous safety books – including *Techniques of Safety Management* (1971), *Safety Supervision* (1976), *Safety Management: A Human Approach* (a continuation on Technique, seeing three editions: 1975, 1988, 2001), *Human Error Reduction and Safety Management* (three editions, 1982, 1984, 1996), *Safety by Objectives: What Gets Rewarded, Gets Done* (1996), and *Authentic Involvement* (2001) – and many articles, and produced a series of educational videos.

His debut, *Techniques of Safety Management*, took some of Heinrich's ideas as a starting point, offered critique, and expanded them. Petersen wanted to "fill some voids" and "challenge some outdated theories" and put the "spotlight on some new techniques." Instead of presenting a comprehensive textbook, he wanted to challenge thinking. The first chapter contained the present framework where he singled out Heinrich's contribution, "This text in industrial safety was revolutionary, for in it Heinrich suggested that unsafe acts of people are the cause of a high percentage of accidents – that people cause far more accidents than conditions do" (Petersen, 1971, p.11).

Petersen attributed much of the safety improvement until the 1960s to Heinrich's influence, "These achievements were made basically by doing what Heinrich has said to do. He had produced a formula that worked" (Petersen, 1971, p.11). However, achievements plateaued afterwards, which was Petersen's reason to re-examine the techniques. This included a less narrow look at accident causation. "To effect permanent improvement, we must deal with root causes of accidents" (Petersen, 1971, p.71), which often related to the management system.

He proposed five new basic principles centred around safety management, which was key for Petersen. Of these, "The key to effective line safety performance is management procedures to fix accountability" (Petersen, 1971, p.21) was "the single most important one (and most overlooked)" (Petersen, 1971, p.43) because "if no one is held accountable, in most cases, responsibility is not taken" (Petersen, 2001a, p.16).

Together with Nestor Roos, Petersen was asked making an updated edition of Heinrich's *Industrial Accident Prevention*. This was published in 1980, subtitled, *A Safety Management* approach, fitting Petersen's focus. In some ways the new edition was a fusion of the 1959 edition with Petersen's 1971 *Techniques* with some added elements.[21] The book retained the three-part-and-appendices-structure of the 1959 edition. Most chapters of Parts I and II returned, but some sections were seriously reworked, for example on collecting and analysing data. In many cases, the presentation of original subject was followed by more recent developments and often the presentation of an updated version of the subject. E.g. the original dominos were supplemented by several accident models, all showing a multi-causal approach.

Striking was the omission of chapters on *Safety Psychology* and *Process Revision*, while in the chapter on *Creating and Maintaining Interest* Heinrich's layman psychology of ten incentives was supplemented with new material on behavioural theory, including Skinner, McGregor, Herzberg, Maslow, and Likert. There also was a new chapter on *Motivation*, presenting a motivation model. The new edition also introduced new elements as information systems and computers and management decision techniques as Kepner-Tregoe.

Because they felt that the subjects were sufficiently covered elsewhere, Petersen and Roos replaced the entire 1959 *Part III, Special Subjects* with new chapters about safety law, insurance, liability, and risk. The appendices changed drastically to fit the changes in the text. The historical background was the only appendix that was continued, however entirely rewritten by Weaver. The bibliography was also expanded, and the separate chapters had references.

Petersen and Roos confirmed Heinrich's statement about the importance of mechanical guarding and engineering revision in accident prevention, but at the same time seemed to lean towards a behaviour-oriented direction lamenting the OSHA regulations:

> With almost universal belief in the principle that safety is primarily determined by people, the principle was almost totally rejected by the Congress, who chose to legislate a law based upon a totally opposite principle: that accidents are caused by conditions – by things.
>
> (HWH1980, p.60)

In the Preface of *Human Error Reduction*, Petersen said,

> This book is based on the belief that most accidents are caused by people doing things unsafely even though they know better, a belief that says people choose logically to act unsafely and that the choice is one that most people would make in the same situation.
>
> (Petersen, 1996, p.xi)

This book explored two of what Petersen called Heinrich's key insights, first that people are the prime cause of accidents and second that control of accidents is a management problem. While it spoke much of behavioural control, the book discussed extensively "system caused error" and management, culture, and design causes. Among other subjects, as culture and measurement, focus on the management system and the conviction that "safety is a people problem" (Petersen, 2001a, p.19) would be two core elements in Petersen's philosophy. As such, his call to behaviourists in his 2000 paper and the *Authentic Involvement* book were not surprising.

3.5.4 Behaviour Based Safety

Some authors state that behavioural approaches to safety are an offspring from Heinrich's work and have contributed to the continued popularity of Heinrich's teachings (Swuste et al., 2016; Dekker, 2019). The Behaviour Based Safety (BBS) movement indeed draws heavily on some of Heinrich's ideas, however, that is only part of it. Influenced by various developments, today, BBS covers a wide range of different approaches, including some that concentrate on safety culture, quality management, or awareness and attempts to renew, branch out or reinvent: for example by latching on to SIF (McSween & Moran, 2017) or "new view" approaches as HOP (Williams & Roberts, 2018). This section concentrates on the original form.

Behaviourism emerged in the early twentieth century as an alternative to the inward-looking, Freudian approaches of psychology. Instead, behaviourism focussed on the observable "outside" results of psychology, human and animal behaviour. The early behaviourists as Nobel Prize laureate Ivan Pavlov and John B. Watson (1913) made little impression on safety work, however.[22] This changed in the 1970s when the work of the new "star" of behaviourism, B.F. Skinner, was regarded as a new way of improving safety. Skinner (1953) regarded only observable behaviour and its environmental, social, and physiological conditions as valid for research. His behavioural approaches were successfully applied during the 1950s and 1960s in other areas, as quality assurance, production, education, psychological health, and crime prevention.

Frank E. Bird Jr. and industrial psychologist Lawrence Schlesinger may have been the first to write about the subject and propose "operant conditioning" techniques and "the fundamental insight that future behaviour is influenced by its immediate effects (reward or punishment)" (Bird & Schlesinger, 1970, p.16) to increase the frequency of safe behaviour and decrease the frequency of unsafe behaviour. Only later in the 1970s, others started making serious work of applying behavioural approaches in a systematic way, notably a team around Judi Komaki,

> By pinpointing desired behavior, in this case safe performance, one can not only have a more sensitive measure of the safety level within an organization, but one can also clarify and positively reinforce those behaviors, thus increasing their likelihood of reoccurrence.
>
> (Komaki et al., 1978, p.436)

Soon after, in 1979, Scott E. Geller coined the term Behaviour Based Safety.[23] It proved to be a huge success. If not for safety, it was for many consultants and companies who followed the new approach. Several proprietary programs (e.g. DuPont's STOP[24]) appeared since the eighties.

BBS approaches typically include the following elements (Brown, 1978; Komaki et al., 1982; Chhokar & Wallin, 1984; Geller et al., 1990; Krause et al., 1990; Geller, 1996; Petersen, 2000; Dula & Geller, 2007; Marsh, 2017):

- Identifying specific desired behaviours, typically of workers, that represent a safe way of performing a task,
- Identifying stimuli/reinforcers, often based on the Antecedent-Behaviour-Consequence (ABC) model of applied behaviour analysis. Of these consequences show to be more important than antecedents,
- Monitoring and measuring[25] behaviour through observations,[26]
  - Observing visible behaviour,
  - Establishing a baseline,
  - Setting goals,

- Maintaining the behaviour through preferably positive reinforcement, e.g. by providing direct feedback, frequently and following the behaviour as immediately as possible, or collective feedback through the presentation of statistics and trends,
- Expanding and revising the "behaviour inventory."

One advantage of these approaches was to move away from injury rates as a measure of safety, and instead becoming more proactive, providing measurements that told managers and workers the current state and what to do. However, widespread adoption of BBS-approaches led to controversies, e.g. with unions arguing that BBS was a way to deflect responsibility from management and turn it over to workers. The approaches were associated with blaming workers and victims, while the actions initiated by BBS programs often had no effect on risks outside the control of front-liners.

Some of the critique of BBS says (e.g. Hopkins, 2006) that it only addresses one of many possible causes in a network of multiple possibilities, and usually not the most effective because actions are aimed long way down the hierarchy of controls. Furthermore, BBS is in almost all cases directed towards frequent and visible behaviour, and ignores invisible, undetected and infrequent behaviour, which tends to contain more risk. Additionally, it may be difficult for an observer to distinguish between safe and unsafe behaviour. The judgement is only possible in connection of the context or as part of a longer sequence of acts instead of some discrete behaviour.

One might add to this critique that the connection between BBS and Heinrich is not as strong as generally assumed. An essential difference is found in the backgrounds of the approaches, Heinrich's safety engineering versus BBS's Skinnerian behaviourist psychology. Granted, there are certain similarities, like the assumption that it is perfectly possible to distinguish between safe and unsafe behaviour, and the attribution of most accidents to human behaviour (Komaki et al., 1978 Chhokar & Wallis, 1984; Krause et al., 1990; Geller, 1996; Petersen, 2000). Also, BBS-approaches adopted some Heinrichian thoughts and images, notably the triangle (e.g. Krause et al., 1990; Geller, 1996).

Still, one wonders whether these were fully understood and applied according to Heinrich's thoughts. Take for example Geller's statement, "Heinrich's well-known law of safety implicates at-risk behaviour as the root cause of most near-hits and injuries" (1995, p.93). This confounded the 88:10:2 ratio and the triangle. It also missed the point that Heinrich explicitly spoke of *direct and proximate* cases.[27]

As we will see in a later chapter, another important difference is that while Heinrich attributed most accidents to man-failure as the (direct) cause, he did not specifically aim at changing the behaviour. He advocated finding the reasons for behaviour and then first considering changing the environment or guarding machinery (e.g. Heinrich & Blake, 1956). Although

BBS-approaches advocate redesign of jobs or environment to maximise safe behaviour[28].

In conclusion, one may say that BBS is not the Heinrich spin-off as suggested, but rather a movement developing from a completely different background, showing some similarities.

3.5.5 Manuele

For a long time, Heinrich's work had been critiqued. An early example is the 1956 debate about the 88:10:2 ratio (Heinrich & Blake, 1956). This continued with people like Petersen in the 1970s and 1980s. At the beginning of this millennium Hale (2002) and Manuele (2002, 2011b, 2014a) were two notable safety scholars who offered their critique of Heinrich's legacy. Hale's thorough discussion of the triangle will be discussed in Chapter 9.

Of the authors who critiqued Heinrich's work, Fred Manuele was the only to dedicate a book to the subject, in addition to several papers.

> Heinrich was a pioneer in the field of accident prevention and must be given his due. Publication of the four editions of his book spanned nearly 30 years. From the 1930s to today, Heinrich likely has had more influence than any other individual on the work of occupational safety practitioners. In retrospect, the good done by him in promoting greater attention to occupational safety and health should be balanced with an awareness of the misdirection that has resulted from applying some of his premises.
>
> (Manuele, 2013, p.235)

Manuele delivered two main works with extensive critique. The first, *Heinrich Revisited: Truisms or Myths*, published through the National Safety Council in 2002. An updated version appeared in 2014. The second was his paper, *Reviewing Heinrich: Dislodging Two Myths from the Practice of Safety*, for the October 2011 issue of *Professional Safety*, the magazine of the American Society of Safety Engineers (ASSE). This paper revisited some of Manuele's earlier critique. It was later incorporated in the 2013 edition of *On the Practice of Safety*. In 2014, Manuele expanded his Heinrich critique in a paper countering Heinrich's premise that supervisors are best qualified to do incident investigations (Manuele, 2014b).

In *Heinrich Revisited*, Manuele summed up his critique in nine points:

1   Heinrich's sources are not available for review.
2   Are the data and ratios from the 1920s still valid in our time?
3   Heinrich placed disproportionate importance on "psychology."
4   The 88:10:2 ratio is invalid and has done great harm.

5   The 1:29:300 ratio is invalid and the changes between versions are not explained.
6   The belief that the predominant causes for accidents with major and minor consequences are the same is not supported.
7   The 1:4 ratio of indirect costs is unsupported, and it is not plausible that there is a universal ratio.
8   There is too much emphasis on individual unsafe acts as causal factors and not at other parts of the system.
9   The attribution of "root causes" in Heinrich's domino sequence (ancestry and social environment) is socially inappropriate in our time (Manuele, 2002, p.77–81).

The updated version left out the chapters on Heinrich's principles of accident prevention and his axioms – both rarely a subject of discussion in safety literature – but added a chapter about Heinrich's view on accident investigation along with six chapters dedicated to what Manuele considered as better approaches to safety management and accident prevention, unrelated to Heinrich's work.

Manuele's critique was thorough in some ways, notably his comparison of the various versions of *Industrial Accident Prevention*, and his discussion of unexplained changes between the first four editions.[29] With his critique, he delivered an important and influential contribution to safety science. Much of the post-2000 critique regarding Heinrich's work from others is based on Manuele's critique, although several authors have made original contributions.

Manuele's critique is not without flaws, however. For example, he did pay much attention to the various ratios but did not discuss the function they may have (see Chapter 4). Also, it is remarkable that Manuele only concentrated on the first four versions of *Industrial Accident Prevention* but discussed neither Heinrich's papers, of which at least the National Safety Council-papers should have been available to him, nor the reworked 1980 edition. These additional sources might have answered some of his issues. Besides, the 1980-update even featured some of the critique that Manuele came with – two decades earlier.

3.5.6 Serious injuries and fatalities

Following accidents like Texas City and Deepwater Horizon, some proponents of BBS started to "untangle" the safety triangle. Leading up to these accidents, the application of the safety triangle led to a focus on occupational safety, which distracted attention from process safety. This was seriously critiqued (e.g. Hopkins, 2008). Several authors (Manuele, 2008; Krause & Murray, 2012) noted that while there had been great reductions in workplace injuries, the number of serious injuries and fatal incidents

(SIF) showed a much slower decline. The conclusion was therefore that the premise of looking at the entire triangle (which according to them was what Heinrich had said) was wrong.

The "untangling" of the triangle led to the SIF-movement, which focussed on the "high potential" part of the triangle instead of minor, low potential events like slips, trips, and falls (Martin & Black, 2015). This development, to be discussed in Chapter 9, renewed interest in Heinrich's work.

3.5.7 "New view" safety

In the early 2000s a new safety movement emerged.[30] It argued that because "…workplaces and systems become increasingly interactive, more tightly coupled, subject to a fast pace of technological change and tremendous economic pressures to be innovative and cutting-edge…" (Safety Differently, n.d.), there was a need for new ways to work with safety. Traditional, linear approaches attempting to control risks through compliance and technical measures were insufficient.

The term "new view" as we understand it at this point in time was coined around 2002 (e.g. Dekker, 2002), but notions of it appeared much earlier, for example in the writings of Rasmussen who discussed complexity (Rasmussen & Lind, 1981) and problems with "human error" as a cause (Rasmussen, 1990). From 2012 onwards this "new view" gained wider traction with the emergence of related terms like Resilience Engineering (Hollnagel et al., 2006), Safety-II (Hollnagel, 2014a), Safety Differently (Hummerdal, 2014), or Human & Organizational Performance/HOP (Conklin, 2012).

Contemporary "new view" authors critique Heinrich and Heinrichian themes, sometimes harshly. In part this was due to progress made over the last decades. There have been many new insights in (safety) science regarding human factors, management, social psychology, and much more. However, it is not merely a matter of "better" knowledge.

During the early 2000s the developing "new view" drew on a variety of traditions (e.g. cognitive engineering and cybernetics) rather than having traditional safety approaches as a point of departure. Only after having established itself ideologically and theoretically, the "new view" turned against traditional safety elements. Heinrich became an interesting topic for "new view" authors when Safety-II and Safety Differently became a thing – although it not really added to their argument (Busch, 2019b).

Opposites and contrasts are powerful tools to convince and persuade (Cialdini, 1984; Pink, 2012). An example where we can see this at work is in the *Safety I and Safety II* book (Hollnagel, 2014a). Hollnagel uses half the book to decompose and point out downsides of traditional methods, before offering a "new view" (Hollnagel, 2014a). Apparently, there was a need to contrast and position "new" approaches against something "old." Most "new view" authors appear to be rather seeking and stressing opposition and posing views as contrasting, instead of presenting them as different and complementary ways to work on safety. Howarth, in his discussion of Foucault's archaeological approach, mentions the problem of "the relationship between discourses and the realities they claim to represent" (Howarth, 2000, p.70).

For authors seeking to establish another, newer view, it makes sense to emphasise the contrast between their own message/product and what came before, especially with the emergence of Safety-II and Safety Differently almost as brands of safety. Like Coca Cola versus Pepsi, it makes sense to contrast the positive sides of one's own "brand" and the negative sides of the "competition" – of which Heinrich happens to be one of the most recognisable flagships. Quoting Johan Bergström,[31] this may lead, in some cases, to situations where Heinrich is positioned as "the old view Satan." The "new view" did *not* need Heinrich for its emergence and development, but as a means of spreading its message and creating acceptance.

While highlighted contrasts may be a useful and powerful pedagogical device and an important tool in influencing and persuading others, over-communicating distinctions like "old" and "new" may contribute to separation and polarisation between various schools of thought, thereby limiting learning opportunities (Busch, 2019a). It should not be a choice between one approach and the other. "Old" and "new" views are complementary and both necessary (Hollnagel, 2014a).

3.5.8 Heinrich or not Heinrich?

Long before Heinrich was associated with Loss Control or BBS, Heinrich's work was incorporated in other safety handbooks. This was the case as early as Williams's book (1927), others include handbooks by Vernon (1935), Blake (1945), Stack et al. (1949), and Simonds & Grimaldi (1956). Heinrich's ideas and metaphors soon became part of the OHS curriculum, often in simplified, mechanistic, and adapted forms. McKinnon's book on near-miss management (2012) is an example of this, drawing on the work of Bird and Heinrich, combined with later insights.

The triangle and dominos became staples in safety education and inductions. Not necessarily following Heinrich's interpretation, but through the mere presence of the metaphor associated with him. Heinrich and Heinrichian ideas are rarely separated. Critique regarding the triangle is often directed at Heinrich, even when the critique addresses adaptations or derivates of the idea.

The success of Heinrich's metaphors probably contributed to the confusion. As these models and metaphors spread and were adopted by practitioners, consultants, organisations, and even regulators, they mutated and got new meanings. As Le Coze concludes is his discussion of popular safety models as the Swiss Cheese Model, "...by making the models their own, these users have become its promoters but also illustrates how models can potentially escape their designers' initial intention" (Le Coze, 2013, p.203). The result was like a game of Chinese whispers where the original idea or message comes out significantly altered.

Many practitioners, educators, and researchers do not go to the original sources, or at best superficially. For many people it may be hard to understand what mutation of a model or metaphor they actually talk about, and then it is natural for them to just refer to the "big well-known" idea, or to the "originator" – if it looks like the triangle it is probably Heinrich's triangle. Those who know of mutations but are uncertain or unwilling to invest time in finding out, may feel as the safest option. Although it is Heinrich's name that comes up, in many cases it is not necessarily *Heinrich's* interpretation when a Heinrichian subject like the triangle is discussed.

Robert Long[32] suggested an additional perspective. Maybe, when people mention Heinrich, they do not really intend to talk about Heinrich himself. They are not seeing Heinrich as the author, but rather as what he has come to symbolise and represent. For them, Heinrich stands as a symbol and shorthand for mechanistic control, reductionism, and positivism with metaphors that suit to support these philosophies – no matter what Heinrich himself ever thought, said, or wrote.

There is value in this perspective, but it comes with side effects. One problem with unfounded attributions and "symbolic use" of names or things (metaphors, subjects) may be that many readers will not get the symbolism, or have no way verifying the attributions. They do not study and have not read the primary sources, taking attributions and symbols at face value, especially when coming from a big, established name. This can lead to the creation of new myths (Busch, 2019a), some appearing in later chapters. Therefore, it is important to remain critical as readers – especially regarding non-peer-reviewed "guru books" – while authors could make a greater effort to explain nuances instead of creating polarisation.

3.6 Conclusion

This chapter argued that Heinrich's work has had major influence on the safety profession – for good and bad. One of his most significant contributions is a more proactive approach.

These days, Heinrich is seen as the prime example of traditional safety. As such, his work has attracted much critique and is often used as an illustration of all that is wrong with traditional approaches. However, "new" is not an absolute term, but relative. Remember, Heinrich was part of another "new view" back in his days. "Heinrich's ideas were a departure from the safety thinking of the time" (Petersen, 1971, p.11). Aldrich (1997)[33] agrees when he discusses how safety pioneers of the 1920s and 1930s introduced a view that saw accidents as events that were caused and could (and should) be prevented. Heinrich spoke even of "a change in trend of thought" (HWH1931a, p.260), discussing the growing recognition that previous cause attributions were not *really* about causes.

The following chapters contain detailed discussions of Heinrich's main themes. The final chapter will reflect on the value of Heinrich's work for contemporary safety.

Notes

1 Neither has the application of some of Heinrich's concepts been restricted to occupational safety.
2 This paper stands out in a number of ways. First, in the style of writing, using a lot of rhetoric and lively descriptions. Second, the presentation with no less than four captioned photographs. Few of Heinrich's publications in *The Standard* are illustrated, and never this lavishly. One photograph even spreads over two pages! Third, the repetition of the main message of the speech: We are surrounded by hazards of the modern age, yet machines are not at fault, people create them and use them, and that is the problem. "The machine is dangerous as man makes it so. It is the use of the machine – more correctly the abuse of it – that creates danger" (HWH1933a, p.14). This message is repeated no less than 12 times in the paper, sometimes wrapped in a rhetorical question as in the example of reckless driving: "Was machinery at fault?" (HWH1933a, p.12).
3 Aldrich mentions "outrage" about Heinrich's assertion that most accidents resulted from supervisory failure coming from the editor of the *American Mining Congress Journal* (Aldrich, 1997, p.243), opening for the assumption that there may have been more protests from employers.
4 Hartford Courant, 23 November 1932.
5 This is very similar to how we would define "risk" (and unlike we would define "exposure").
6 Interestingly, we find a similar reflection in one of the books in Heinrich's bibliography, "…invention is normally a social process, rather than the work of a few great men. We writers try to make our pages dramatic by concentrating on the geniuses, but inevitably we distort the story" (Chase, 1929, p.73).
7 A similar estimate re-appears in Heinrich's work without explanation how he arrived at this. We can only speculate whether Heinrich followed Gilmour's thinking.
8 The 1931 bibliography is assumed to be the most relevant because most main themes were formed at that time, so later additions to the bibliography have most likely affected changes and developments and added new subjects, but not contributed to the formation of the original themes.
9 Exceptions are the *Safety and Productivity* study and the Slocombe and Bingham paper. However, he does refer to and quotes from several other texts, including the work of Greenwood and Woods and of Farmer and Chambers on accident proneness.
10 *Safety Fundamentals* is available via archive.org. It is a collection of lectures on practical safety subjects as bodily injuries, protective clothing, eye and head protection, machine guarding, workplace arrangements, ventilation, illumination, and safety education. The latter chapter includes sections about "selling safety," an argument that "safety is good business" and indirect costs, interestingly framed as "hidden gains" rather than costs.

11 Several handbooks were published earlier, e.g. A. Wilson's *Common Accidents and How to Treat Them* (1889, London: Chatto and Windus).

12 Several safety handbooks were published just before the USA entered the First World War. This may partly have been coincidence. The safety movement was a decade and a half old, slowly maturing and developing. It gained wider acceptance which is reflected in these books. But the War may have played a role too. Preparations were well underway at the time, producing ammo and supplies for the Allied forces. According to Chase, the War was a positive force in the boost of the efficiency movement and reduction of "waste." It is not hard to imagine that this also may have a positive effect for the safety movement. After all, the safety movement helped to remove the waste of accidents and disturbances (and also, to control the workforce). A similar effect could be observed 30 years down the road with the Second World War.

13 Although Stack's 1953 biography of Whitney suggests that there were parallels in their thinking, e.g. "The same causes that produce accidents are also responsible for inefficient production" (p.48); "Foremen in factories... should be given special training in safety" (p.66) and "The unsolved problems today are to be found not in the physical world but in the human world" (p.87). None of these, however, are exclusive or in particular connected to Whitney's thinking, but rather reflect widely shared ideas at the time.

14 Their backgrounds make it unlikely that their paths would have crossed early on. Heinrich was a blue-collar guy who worked his way up to insurance engineer and assistant superintendent. Shewhart was an academic who worked for Bell's engineering department. There was no natural place for them to meet. After Heinrich's promotion and research in the late 1920, he became a wanted speaker, but mainly on safety/labour conferences. If Heinrich and Shewhart ever met, the most likely place would be during their contributions in war efforts (Heinrich for safety, Shewhart for quality). Even so, one would assume that a meeting with Shewhart would have left traces in Heinrich's texts for first-line managers/supervisors.

15 Somehow, the book failed to have the same impact as *Industrial Accident Prevention* (published by the same publisher) had only five years later, even though there seem to have been several reprints in 1929, 1934, 1935, and 1936. One reason for this book having been more or less forgotten in the course of time, and Heinrich's book still being cited, may be a matter of timing, another the way the subjects are presented. Heinrich's book has definitely a more practical approach that should appeal to managers and safety practitioners in need of solutions. In places DeBlois's book is probably more advanced, more reflective, nuanced, and philosophical. That makes it – despite its enthusiasm and passion for safety – less accessible and more difficult to use in the work of every day.

The end section may serve as another illustration why Heinrich's book may have had a greater impact than DeBlois's partly intellectually superior work. Where Heinrich gives a wrap up and an encouraging sense of "You can do this," and ends with something in the line of "now get busy," DeBlois ends philosophical and with a question. Heinrich delivered the better management self-help handbook, despite some deficiencies...

16 Although this book is mostly about corporate and brand stories, we can apply the principles in other areas.

17 Other good examples are the *Message to the Foremen* papers (HWH1929d, e, 1930a), starting with a story that supports Heinrich's arguments, then another illustrative story in the middle and laced with various examples his audience could relate to.

18 Most of the biographical information was provided by Lateiner's son Donald in personal e-mails to the author.
19 One cannot ignore similarities between the ISRS and the 1947 ASME standard.
20 The biographical information was taken from the EHS Today obituary, https://www.ehstoday.com/news/ehs_imp_44538 (Retrieved 4 October 2019).
21 Because of the editing and drastic changes, the fifth edition is a weak source to learn about Heinrich's work, but an interesting source to get an overview of the state of safety knowledge during the late 1970s.
22 Even Fisher's thorough discussion of psychology only briefly touches on behaviourism.
23 I have been unable to trace this claim to a concrete source. Marsh (2017) attributes it to Geller. Geller himself makes the claim in an interview with Sonni Gopal (Retrieved 29 January 2020 from https://www.redrisks.com/5-short-interviews-with-prof-scott-geller-behaviour-based-safety-pioneer/)
24 STOP = Safety Training and Observation Program.
25 Measurement is regarded as an essential element, "The only way to be sure an intervention has a desired effect is to measure the target behavior before, during, and after an intervention" (Dula & Geller, 2007, p.184).
26 Safety sampling (Pollina, 1962) was a technique introduced in the early 1960s. This can be regarded as a predecessor of BBS's observation schemes but inspired by the methods used by quality assurance and efficiency engineers. No reference is made to Skinnerian behavioural influence techniques. While safe and unsafe acts were the measure of this technique, its aim was not in the first place to change behaviour. The aim of Pollina's safety sampling was to offer a proactive way of "measuring" safety as an alternative to the reactive, traditional measuring by accidents and injuries, and as predictive aid.
27 A misreading that also found in other BBS-literature (e.g. Komaki et al., 1978; Chhokar & Wallis, 1984).
28 As did Taylor when he proposed his "one best way" approaches. Marsh claims, "The most effective behavioural approaches focus on the environment before the individual" (2017, p.xiii). Similar suggestions are found in the 1970 paper by Bird and Schlesinger, in Geller's work, and, more recently, in McSween & Moran, 2017. Many popular BBS-implementations ignore this, focussing mostly on behaviour.
29 Given the fact that he studied Heinrich's books closely, it is quite remarkable that Manuele does *not* comment the change of causes of a "supervisory nature" in the 1931 edition to "man-failure" in the 1941 edition.
30 This section draws heavily on my thesis. Only a summary is presented here, for elaboration on the subject (see Busch, 2019b).
31 Personal communication, circa May 2018.
32 Personal communication, 10 December 2017.
33 "That the vast majority of injuries were either due to 'trade risks' or the result of 'careless' worker behaviour was an article of faith among most nineteenth-century businesspeople. In either case, injuries were neither the fault of nor preventable by employers, or so most of them thought" (Aldrich, 1997, p.114). This thinking was opposed by the "new view" of safety thinkers of the 1920s and 1930s, including DeBlois and Heinrich.

Chapter 4

A scientific approach

Not a single paper has "science" as the main topic, but the sub-title of *Industrial Accident Prevention*, and the frequent mentioning of "scientific" principles warrant an inclusion as one of Heinrich's main themes. This chapter takes frequent critique as its point of departure, then exploring why it made sense to Heinrich to speak of "a scientific approach." This will include a discussion of possible links between Heinrich and the work of Frederick Taylor.[1] Finally, the chapter reflects upon the missing data.

4.1 Introduction

Heinrich discussed science relatively seldom explicitly in his work. The first mention was in the *Origin of Accidents* papers, "The prevention of accidents is a science, but it is not so recognized, nor is it treated scientifically today" (HWH1928a, p.9). The only mention of the term "Scientific accident prevention" came the year after: "Scientific accident prevention will solve the industrial accident problem, and in this study of existing preventive methods, as made by Travelers engineers, we are confident that there is a clue to more satisfactory progress" (HWH1929a, p.10). After a few early mentions, it took until 1950 for the subject to be explicitly mentioned again, "Accident prevention is both science and art. It represents, above all other things, control – control of man performance, machine performance, and physical environment" (HWH1950a, p.1, 1959, p.4).

When Heinrich published *Industrial Accident Prevention*, the preface said, "The use of the word 'scientific' in the subtitle of this book indicates the author's belief that science may be applied practically and successfully to the prevention of accidents" (HWH1931a, p.v).

Mentioning (scientific) principles returned in much of Heinrich's work. He saw science as an essential tool that had to be used in practice: "Accident prevention, when scientifically conducted, requires identical procedure…" (HWH1931a, p.8), "Science is applied to accident prevention in the same manner that it is applied to any other problem" (HWH1931a, p.10) and "The intention throughout is to present accident prevention as a problem subject to scientific solution…" (HWH1931a, p.15).

Later, in the preface of the fourth edition, he elaborated, "As in all art and science, one should expect to find governing laws, rules, theorems, or basic principles" and "After understanding comes application. What industry and its accident preventionists want most to know is how best to apply their knowledge" (HWH1959, p.xi).

Taylor's "scientific management" may have inspired Heinrich's subtitle. However, Taylor's ambitions appeared to be "higher" than Heinrich's. Taylor tried to find laws that were applicable to anything. Heinrich wanted to use scientific principles for solving practical problems and preventing accidents. "Accident prevention can be portrayed as a science and as a work that deals with facts and natural phenomena" (HWH1941, p.16). However, he did not quite go as far as Taylor, who stated that his scientific management was "applicable to all kinds of human activities" (Taylor, 1911, p.2). Heinrich wanted foremost practical solutions but not necessarily as strict and formalised as Taylor.

4.1.1 Critique

Labelling his books and work as "scientific" has drawn much critique, especially in recent years, from both safety scholars and practitioners. Many follow Manuele's (2002, 2011b) critique about the missing and thus unverifiable data. More about this later in the chapter. Besides, there has been some critique on the methods. For example, Dekker is very explicit, "…like Taylor, Heinrich announced his approach to be 'scientific', though a description of his method wouldn't pass scientific peer review today," and "…the subtitle of his book was *A Scientific Approach*. That would supposedly require him to at least divulge the basis for his selections or the statistical power behind his sample size" (Dekker, 2018, p.86). These statements are based on assumptions about what "scientific" means from a contemporary point of view. Dekker was not the first. Earlier, Metzgar wrote, "Heinrich subtitled his book 'A Scientific Approach'. Thus, it falls to him to demonstrate that he followed a scientific method in his development and writing" (Metzgar, 2002, p.27), and

> A scientific method would demand that the research for such a significant claim – including the data and the method for handling it – be published. Such publication presents the opportunity for others to try to replicate the experiment, confirm it or show how it could be improved.
> (Metzgar, 2002, p.28)

There are indeed problems with Heinrich's data, which was biased, and their analysis, which was arbitrary (see Chapter 6). This chapter, however, mainly aims to explore why it might have made sense to Heinrich to call his work "scientific," when many authors and scholars raise serious questions.

4.2 Why did "scientific" make sense?

If we employ the concept of local rationality (Woods et al., 1994) it is not important what *we* in retrospect think of something. It does not matter what *we* call scientific, eight decades down the road. What matters is how it would make sense to subtitle the book "scientific" at that time, given what came before, what the book contained, how it compared to the state of the art then and what direction it pointed. This chapter suggests five hypotheses why it might make sense to Heinrich to subtitle his book *A Scientific Approach* even though we today may find the work anything but scientific.

4.2.1 Understanding the term

A common approach to understand a subject is by starting to investigate what a term means, or how it is commonly defined. The *Oxford Dictionary* (n.d.) defines "scientific" as:

1   Based on or characterized by the methods and principles of science.
2   *informal*: Systematic; methodical.

The definitions found in the *Merriam Webster* (n.d.) agree:

1   of, relating to, or exhibiting the methods or principles of science.
2   conducted in the manner of science or according to results of investigation by science: practicing or using thorough or systematic methods.

The first definition of both dictionaries points us in the same direction of much of the critique. Scientific means that research should be done according to the scientific method. That much seems obvious to *us*.

But what did *Heinrich* mean? Did he mean something different than that definition? He referred only once to a definition of science, but following clues throughout his work enables one to form an idea of a possible definition for Heinrich. First, let us explore whether Heinrich should have understood how to do things "scientifically" such that his approach lives up to today's standards of scientific research.

The answer to that is no.

His biography tells us that Heinrich was not an academic. Even though he would teach at universities later in his life, Heinrich never had an academic education himself. Before he started working, he had six years of grammar school. His biographer, Jesse Bird, remarked: "Education at the university and doctoral levels was usually for professors, clergymen, barristers and people of leisure who inherited considerable wealth" (Bird, 1976, p.28). Heinrich did not fit these categories and rather was a self-made man who added knowledge and education as he went along. In this, he differed from other Hartford intellectuals working in the insurance industry and making important contributions to science or arts, as poet Wallace Stevens or linguist and fire prevention engineer Benjamin Lee Whorf.

So, should he have known? He may have had some knowledge about scientific research and experiments, but not the formal training to adhere strict scientific guidelines. It was therefore not natural for him to write in an academic way or use academic approaches. He was a practitioner and wrote as one – even though he worked with, and was assisted in his work by, academics like Dr Risteen and Edward Granniss.

Heinrich's writing style was not academic. His books were foremost aimed at managers and practitioners. We see this in his language, the examples he used, the stories he told, and by the relative absence of quotes and footnotes. Heinrich did not quote any philosophers or academics. When he quoted, he mostly drew on definitions from standards, for example the definition of accident causes (HWH1931a, p.41–42), or practical safety texts, as the lengthy passage on illumination, taken from the work of another Travelers safety engineer (HWH1931a, p.250–256).

One can just compare the limited number of references and use of footnotes by Heinrich (HWH1931a, 1941) and DeBlois (1926a) – both engineers and practitioners – with the academic Vernon (1935) who had one or more footnotes and references on almost each page. In America, safety was mainly a practical/applied field. Swuste et al. note that American safety authors all had an industry background and therefore had a "more applied and managerial interpretation of occupational safety" (Swuste et al., 2010, p.1009) in contrast to more academic European authors like Vernon. There is even reason to believe that Heinrich would have good cause to shy away from an overly academic approach. Among others, Donkin (2001) and Stewart (2009) suggest an "antipathy" against academia among US entrepreneurs[2] – Heinrich's prime audience.

Not having had a strict academic/scientific training, Heinrich may have had the impression that he actually did follow a scientific method. And in a way he did, implementing an approach, observing effects, and drawing conclusions, although arguably not in a rigorous way that lives up to the standards applied to scientific research today.

This opens for another hypothesis: although some practices in natural sciences had been around for a long time, the scientific standards from the 1920s and 1930s were most likely not the same as today. When Heinrich and the Travelers engineers did their research during the mid- and late 1920s, Fisher's influential work on the design and analysis of scientific experiments was only just published (Yates & Mather, 1963). Even more, "proper" scientific testing of the efficacy of treatments did not enter medical science until the second half of the twentieth century (Tetlock & Gardner, 2015; Blauw, 2018), which was several decades *after* Heinrich published his work for the first time. It is therefore reasonable to assume that applying these scientific methods was not best practice within safety during the 1920s and 1930s, and it is quite unreasonable to require this in hindsight. Safety was not even widely recognised as a profession let alone as a science (HWH1928b).

Finally, there is a nuance going beyond mere semantics. The word *approach* is open for a less strict interpretation than *method*. Stewart contrasts scientific/science (according to the scientific method and living up to criteria for scientific research) with a "scientific attitude" which he loosely defines as "...a disposition to test hypotheses against facts through controlled observation" (Stewart, 2009, p.48). However, as Stewart argues, a scientific attitude alone does not make something (e.g. a construct or an activity) scientific – using the example of grocery shopping. There is good reason to value Heinrich's statements in that light. He had a scientific attitude towards accident prevention, but that alone is not enough to make his work proper safety science.

4.2.2 Structured method, facts, and principles

The second definition from the consulted dictionaries fits very well with the interpretation of the subtitle as discussed in the previous section, meaning "with a scientific attitude" or following approaches inspired by science, but not necessarily through today's scientific peer-reviewed processes. The second, more informal, definition stresses the use of systematic and methodical approaches based on facts, whether they fully live up to scientific methods, or not.

Exploring why it made sense to Heinrich to speak of "a scientific approach," a good start is looking at what a dictionary entry would have told in *his* time. While the entry for "scientific" looks the same – "Agreeing with, or depending on, the rules or principles of science" (Webster, 1930b) – "science" was first defined as "knowledge of principles or facts,"[3] and then as "Accumulated and established knowledge, which has been systematized and formulated with reference to the discovery of general truths or the operation of general laws; knowledge classified and made available in work..." (Webster, 1930a).

This interpretation appeared in the 1932 review of *Industrial Accident Prevention*,

> It is the author's contention that science may be applied practically and successfully to the prevention of accidents, *using the term 'science' in the sense of knowledge of principles or facts*. The principles laid down have been applied so extensively that they indicate beyond a doubt their practicability and effectiveness.
>
> (Hayhurst, 1932, p.119, emphasis added)

This seems to correspond with Heinrich's intention. In the preface, he stated that "...if accident-prevention work is to be conducted scientifically, it must be founded upon well-established principles or facts that have been proved by application" (HWH1931a, p.v).[4] On various occasions, he stressed that the book intended to offer principles and fundamentals, not specifics. This is one way in which Heinrich's work distinguished itself from many of the earlier safety handbooks by, for example, Beyer (1916), Cowee (1916), Ashe (1917), and Lange (1926). Many of these texts discussed to a greater degree specific cases and applications, while Heinrich aimed mostly at fundamental processes and methods. Fundamental principles were also found in the work of Lange (1926) and especially DeBlois (1926a), while Heinrich discussed specifics (like the chapters on guarding and illumination). By and large, however, *Industrial Accident Prevention* was characterised by offering an "organized framework of thinking" (Petersen, 1971, p.11) within accident prevention.

Additionally, Heinrich emphasised the use of "logic" (HWH1928a, b), explained as first finding out what is "wrong" before applying a solution. According to him, this approach was applied to other business problems, but not for safety.

The importance of following a structured approach to safety work is found in Heinrich's work as early as his first writings, e.g. the *Bedford Stone Club Speech* (HWH1923b) where he offered the audience ten "fundamental principles" for successful safety organisation, some of which can be linked to his future axioms.

Like other things in Heinrich's work, the principles and their descriptions changed throughout the years. Not all variations and nuances are presented here but some are highlighted to illustrate Heinrich's systematic and methodical – and thus "scientific" – approach. In the first edition of *Industrial Accident Prevention*, we find "four principles of scientific accident prevention":

1   Executive interest and support
2   Cause-analysis
3   Selection and application of remedy
4   Executive enforcement of corrective practice (HWH1931a, p.6)

He called this "successfully applied modern accident-prevention methods in a business-like way" (HWH1931a, p.6), and therefore, "…inefficient plans and devices should be discarded because research and investigation have brought to light new modes of procedure, based on correct practice, which are producing highly satisfactory results" (HWH1931b, p.5).

The principles would change slightly and evolve over time and be adjusted in some papers. In the *Safety in the Small Plant* paper (HWH1932a), for example, he emphasised that the principles of accident prevention between small and large plants were the same, only the details differed. The principles were then presented in a somewhat simplified form, probably because of his audience, which he did not want to give unnecessary burden:

1   Executive interest, support, and action.
2   Knowledge of accident facts
3   Appropriate and effective action based on these facts (HWH1932a, p.30).

In a paper specifically aimed at managers, stressing the importance of planning in accident prevention, Heinrich even called the principles "planned executive work" (HWH1932d). Yet another version appeared in the *A Place in the Sun* paper where he chose to elaborate on the various steps:

1   Recognition of responsibility by the employer, and his supervisory staff, followed by his resolve to act.
2   Investigation and recording of the two major facts of accident occurrence; namely (a) the specific unsafe act of some person, and (b) the physical or mechanical hazard at fault.
3   Corrective action directed against the unsafe act or the physical hazard (HWH1934, p.118).

In the second edition of *Industrial Accident Prevention*, the principles were streamlined once more,

1   The creation and maintenance of active interest in safety.
2   Fact finding.
3   Corrective action based on the facts (HWH1941, p.6).

It is possible to see elements and similarities between applying Heinrich's principles and following the Shewhart/Deming-cycle (Deming, 1986). The plan-do-check/study-act cycle can be seen as a scientific approach as well, corresponding nicely to the scientific method. Elements of both PDCA and the scientific method can be found in Heinrich's principles, yet incomplete.

Central in Heinrich's principles, was the knowledge of facts. He mentioned for example, "knowledge as to specific unsafe practices and unsafe physical or mechanical conditions, and the reasons for the existence of them" (HWH1931a, p.98), and "investigation and recording of the two major facts of accident occurrence..." (HWH1934, p.118). This leads to yet another view of science, namely as something that is fact-based. "His books had as subtitle 'a scientific approach', referring to an approach based upon facts, and less on believes" (Swuste et al., 2013). Heinrich emphasised the importance of facts frequently,

In this age of exact knowledge, when facts of proved value are replacing theories and when business, under pressure of economic necessity, must concentrate upon the things that count, it is startling to find that the efforts of thousands of individuals are misdirected and that the attainment of a worthwhile objective is seriously delayed through failure to recognize an obvious truth.

(HWH1928a, b, p.9)

One of the most "flowery" occasions was in the *Reward of Merit* paper (HWH1930b), opening with a discussion of "outworn beliefs," myths, and superstitions, contrasting these to beliefs supported by facts, drawing on such examples as the fear of falling off the edge of a flat earth, the Salem witch trials, medical blood-letting, and other superstitious practices, "One of the most beneficent results of intellectual progress in the human race is that which evidenced by the smashing of popular idols – the overturning of pet beliefs unsupported by fact" (HWH1930b, p.141).

An earlier paper, discussing the safety problems of the construction business (HWH1925), contrasted Heinrich's approach to that of "quacks." Heinrich argued that one must know the problem in order to be able to address it effectively, "Accident prevention engineers must know *why* accidents occur; they must know the *causes* of accidents and direct their attack at the removal of these causes, instead of selecting a remedy blindly or arbitrarily" (HWH1931a, p.268).

For Heinrich this was also a reason emphasising the importance of better statistical data frequently. Not as a bureaucratic enterprise, but in order to get better knowledge and have a greater evidence-based foundation for preventive actions. "Conclusions derived from judgment or from one or two serious cases, are never as valuable as are those drawn from a review based on greater exposure" (HWH1935, p.32), he wrote many decades before the term "big data" caught on.

This focus on facts – which still permeates our contemporary thinking – was very much a sign of the time and may partly have been inspired by Taylor, "...scientific in Taylor's context meant a decision-making process based on rational arguments" (Swuste et al., 2010, p.1004). Muller notes in his discussion of Taylor's legacy, "Decisions based on numbers were viewed as scientific, since numbers were thought to imply objectivity and accuracy" (Muller, 2018, p.35). Heinrich's approach fit this scheme, by stressing the knowledge of facts, through use of better statistics, and by creating ratios.

4.2.3 Practicable and useful

As if to underline that he was no "ivory tower" academic, Heinrich stated in the second edition of *Industrial Accident Prevention* that the principles offered by him "...are not based on theory as much as on time-proved practice" (HWH1941, p.6). This speaks of perceiving science as something practical, not merely theoretical. Therefore, he stressed that

"Practicability and common sense must prevail in safety as in other things..." (HWH1941, p.139).

Heinrich was an engineer. Engineering is an applied science with a strong preference for practical applications rather than pure theory. As the 1929 quote mentioned, "scientific accident prevention" had a purpose: progress. On several occasions, he mentioned that mere knowledge or critique of inefficient practices is not useful. One must come with suggestions for improvement, and these must be practicable: "This information concerning accident costs may be interesting, but it is not necessarily of tangible value unless it is supplemented by concrete suggestions as to how best the costs may be reduced" (HWH1931d, p.97).

The *War-time Motion Study and Foremanship* paper (HWH1942) shines an interesting light on Heinrich's relationship with science. Again, it was confirmed that he was no scientist. First, he was a practitioner who wanted practical results for his audience, and he understood them well: "Foremen don't care much for the term 'motion study'. It sounds unnecessarily 'high-brow'" (HWH1942, p.177). Therefore, they should rather use their "...God-given common sense to motion study..." and "...find it no hodgepodge of theory but merely the practical application of knowledge and facts that you have long made use of in connection with the mechanical[5] phases of your work" (HWH1942, p.178).

Heinrich was not alone in this view. Common sensical everyday understanding, interpretation, and use of the term "scientific" does not necessarily relate to the first definition from the aforementioned dictionaries. Most people do not necessarily think of the scientific method and peer review. For many lay persons, the understanding of "scientific" corresponds more to that the second definition from the *Oxford Dictionary* and how the magazine *Popular Science* advertises itself:

> Popular Science gives our readers the information and tools to improve their technology and their world. The core belief that Popular Science and our readers share: The future is going to be better, and science and technology are the driving forces that will help make it better.[6]

Therefore, we can also regard (applied) science as "problem solving" and "what works" (Davies et al., 2003). Arguably Heinrich's way worked better and was an improvement over what was before.[7] The way he proposed causal analysis and implementation of measures for prevention went beyond many of his contemporaries. Even if we see problems with his approach today, many were convinced of the improvement, not the least himself.

Others share Heinrich's view on science as something that needs practical application. The Australian committee tasked to develop a body of knowledge for the OHS professional stated, "OHS is an applied science and the focus of OHS professionals should be on solving problems" (Pryor, 2019, p.20). Hudson, discussing implementation of a safety culture program, wrote,

> Fine distinctions of theory, the daily fare of the fundamental scientist at the cutting edge, are too fragile to base a system on if that system is to work. If an approach does not work in the long run there will be a lot of problems created, including the believability of the academic world. So this means that we need to reframe the role of the scientist as implementer as more like that of the engineer, who knows how to take a wide range of established scientific knowledge, plus hard won experience, and design and construct something people can bet their lives on.
>   (Hudson, 2007, p.719)

Dekker acknowledged that Heinrich "...was probably a practical man. He needed to find things that could work, that the insurers' clients could use in their daily practice, and that could ultimately save his company money"[8] (Dekker, 2018, p.86). Focus on practical application is something that one also might expect from "new view" authors, given their interest in "work-as-done" (Hollnagel, 2014a). This leads several "new view" authors rather to write books for practitioners than for academics. Hollnagel says of *Industrial Accident Prevention* that the "...background for the book was practical rather than academic..." (Hollnagel, 2014a, p.35), and then follows Heinrich's example:

> This book is intended for practitioners rather than researchers (...) The reason for this partiality is simple – *it is the practitioner who can make changes to practice, not the academic.* The intention is that the practically minded reader should be able to read the book without constantly consulting the references and even without caring much about them.
>   (Hollnagel, 2004, p.xiii, emphasis added).

Many other publications as the *Safety Anarchist* (Dekker, 2018) and the *Pre-Accident Investigation* series (Conklin, 2012, 2016, 2017) would fit in the "practical" rather than the "academic" ("science") category. One might be tempted to say that many contemporary authors are striving after similar goals as Heinrich, just in different eras and with different approaches.

Few "professions" are more interested in practical solutions that management. "The aim of management thought, say the guru's supporters, is not to produce scientifically rigorous theories but to supply practical tools and concepts for the managers of the world" (Stewart, 2009, p.247). It must therefore be stressed that *Industrial Accident Prevention* is not a scientific text as such. It is mostly a management book. In the summary, Heinrich identifies (top) managers as his primary audience: "It is chiefly to the employer, therefore, that this work is directed" (HWH1931a, p.271, 1941, p.365, 1950, p.401, 1959, p.402). As such, the subtitle of the 1980 version of *Industrial Accident Prevention* was rather suitable: *A Safety Management Approach* had replaced the "scientific" tag.

Not only should one regard *Industrial Accident Prevention* as a management book first, and as a safety book second. There are even elements that might justify regarding it as a self-help book. Some of the "stories" in the 1931 edition, and the regularly appearing encouragement of "this is simple" and "you can do this"-like comments surely remind of self-help literature. Interestingly, the big sellers from the era that lasted until today, Napoleon Hill and Dale Carnegie, came only *after* Heinrich's first book (1937 and 1936 respectively).

Typical for the self-help management side of Heinrich's work, is the story that opens the 1931 edition, that in a nutshell presents some core concepts, among which hidden costs and the principled approach, in a management meeting setting. Here Heinrich was clearly "selling" his method, which indicated from the start that this was more a management book than science. Speaking of "selling," "scientific" is also a fashionable and marketable word, as we will explore in the next suggestion why things made sense to Heinrich.

4.2.4 Marketing

The first suggestion why Heinrich might have chosen the "scientific" claim in the subtitle of his book may be plain and made intuitively by many people. It was a tag to attract attention and market the book. In recent times, the adjective has an almost universal positive connotation to it. If something is deemed "scientific," it is regarded as progressive, objective, rational, and beneficial. Adding a scientific touch to something to sell it is a well-established marketing practice for toothpaste, washing powder, and so on. Starting with Taylor (1911), also management books have a long-standing tradition of presenting themselves as works based in science.

Another reason for Heinrich to add the subtitle may have been to set his book apart from the 1916 book by the same title by David Beyer. Beyer's book was much more a "how to" book with specific examples of tools, hardware, equipment, machines, and environment with attention to specific industries and hazards. In contrast, Heinrich's book was more concerned with a philosophy of accident prevention and principles and methods to guide this work, although there was some similarity to Beyer's book in for example the long chapter on guarding and the chapter on illumination. It has been a long-established practice – also in science – to stress that what the author presents here is "better" than what came before. In this case by adding the "scientific" label in the subtitle.

Many people associate Heinrich with Taylorism. As will be argued below, Heinrich should not be regarded as a "hardcore" Taylor adept. There are some clear similarities, however, for example analysis as a basis for action, choosing from a set of causes, pre-defined "best" actions, and so on. Therefore, it is not surprising that many people associate the subtitle, "A Scientific Approach" with "Scientific Management." An easy explanation for the subtitle is therefore that Heinrich was inspired by Taylor's successful management method. If readers made the same link back in Heinrich's time, connecting to a popular management "fad" must have been an attractive prospect. It is worth exploring the possible association between Heinrich and Taylor some more.

4.2.5 Scientific management and Heinrich

Scientific Management, often called Taylorism because Fredrick Taylor was its main promoter during the early years of the twentieth century, is a highly successful and widespread management theory. It "changed the workplace for good" Donkin (2001, p.144) remarks, while at the same time noting that the principles were not new at all. Military had drilled soldiers since the dawn of time and the breakdown into separate tasks had also been practiced in many industries and activities for centuries, but probably not at a scale and with a rigour as launched after Taylor and Ford.

Taylor (1911) stated "the best management is a true science, resting upon clearly defined laws, rules, and principles, as a foundation." According to Taylor, any human activity can be improved through "rigorous analysis" (Stewart, 2009, p.30) and finding the "best way" to perform the activity. For this he adopted a mechanistic approach, taking "...the logic of the machine shop into the human world. It rested on a freighted analogy between the technology of machines and the technology of human organisation" (Stewart, 2009, p.28). In fact, Scientific Management was not too scientific, but it employed some practices associated with science, as observations and measurements.

> ...Taylor's impact was like that of Darwin, Marx, and Freud. Each brought a deeply analytical, 'scientific' cast of mind to an unruly, seemingly intractable problem – Darwin to the chaos of life on the planet; Marx to the vagaries of social and economic systems; Freud to the swirling depths of the mind; Taylor to the physical, economic and psychological complexities of human work. That none of them was, in every way, truly 'scientific' made their impact less profound.
>
> (Kanigel, 1997, p.9)

Taylor (1911) specified four essential elements of scientific management:

1   The development of a "scientific approach" to work by reductionism and breaking work down into separate tasks, making the task measurable

through time-motion studies and production criteria,[9] simplification and standardisation of tasks, specifying steps, and finding "one best way" of doing the work.

2 Selection and training of workers according to these "scientific" principles.

3 Cooperation with workers,[10] to make sure tasks are being performed according to these "scientific" principles.

4 A division of work and responsibility (accountability) between management and the workforce; thinking/planning and doing are separated.

Many of Heinrich's critics place him squarely into Taylor's camp (e.g. Long, 2013, 2014; Dekker, 2018). However, they do not offer any concrete explanation or support for their claims by pointing out specific references or practices advocated by Heinrich.

The association of Heinrich with Taylor is probably based on two things. First, there is the subtitle of Heinrich's book, reminding of Taylor's "scientific management." Second, Heinrich's book was written in the late 1920s/early 1930s and therefore common wisdom suggests that Heinrich must have been influenced by the successful major management movement of the period. This is seemingly confirmed by the fact that Heinrich wrote about controlling "unsafe acts."

These days, Taylorism is often seen as a synonym of "bad," especially in Human Factors and "new view" circles. Authors writing critically about traditional management practices, such as Stewart (2009), Peters and Pauw (2004), Schwartz (2015), and Pink (2009), are very critical as well. Dekker's discussion in *The Safety Anarchist* is relatively neutral, yet critical, classifying Taylorism as "dehumanizing" (Dekker, 2018, p.43) and speaking of "Orwellian managerial control" (Dekker, 2018, p.103). Linking something, or someone, to Taylor is for many people not a positive association. Hollnagel notes, "Today Taylorism is unfortunately often used in a negative sense meaning an oversimplified and 'mechanical' approach to work management" (Hollnagel, 2014b, p.41 f1). Interestingly, he offers a surprisingly nuanced brief discussion of Taylorism, stating "Scientific Management rightly deserves to be seen as a kind of proto-ergonomics" because of techniques like bottom-up task analysis and time/motion studies (Hollnagel, 2014b, p.41).

Reading Heinrich's collected work the answer whether he was a Taylorist must remain somewhat inconclusive. Heinrich did not reference to Taylor or any of the other Scientific Management authors in his work. However, there are some elements that can be connected to Taylorism. He discussed job analysis, "break down the job into its several operations and show the hazards of each..." (HWH1931a, p.96), as a guide for worker selection (Glenn, 2011). On several occasions, he suggested that there are "best ways" of doing a job, and separated management and workers; between them who do as they are told[11] and "executives who are paid to do their own thinking"

(HWH1932d, p.72) with management in an enforcing role (Chapter 5 of the 1931 edition is titled "Executive Enforcement of Corrective Practice"), echoing Taylor:

> It is only through enforced standardization of methods, enforced adoption of the best implements and working conditions, and enforced cooperation that this faster work can be assured. And the duty of enforcing the adoption of standards and of enforcing-this cooperation rests with the management alone.
>
> (Taylor, 1911)

There is another similarity between Taylor and Heinrich, which might suggest that Heinrich was inspired by Taylor. When Taylor (1911) introduced his aims, he stated the following:

> First. To point out, through a series of simple illustrations, the great loss which the whole country is *suffering through inefficiency in* almost all of our daily acts.
>
> Second. To try to *convince the reader that the remedy for this inefficiency lies in systematic management*, rather than in searching for some unusual or extraordinary man.
>
> Third. To prove that the *best management is a true science, resting upon clearly defined laws, rules, and principles*, as a foundation. And further to show that the fundamental principles of scientific management are applicable to all kinds of human activities, from our simplest individual acts to the work of our great corporations, which call for the most elaborate cooperation. And, briefly, through a series of illustrations, to convince the reader that whenever these principles are correctly applied, results must follow which are truly astounding.
>
> (Taylor, 1911, p.2, emphasis added)

Heinrich followed the same structure. He opened *Industrial Accident Prevention* with a tale of waste and hidden costs, even connecting inefficiency and safety (HWH1931a, p.3). Then, he suggested a structured, systematic approach based on principles and said that the same principles used production and quality could be used for safety (HWH1931a, p.5). This resonated well with Taylor's build-up quoted above. Still, it did not make him a Taylorist per se.

The most obvious link to Taylorism, or rather to Scientific Management,[12] is found in the title of the paper *Keep 'em Moving (War-time Motion Study and Foremanship)* (HWH1942). In this text, he promoted a "one best way" approach: "...there is but one right way - one best set of production motions to do a given job - and who knows that way to be the safest and quickest" (HWH1942, p.175), "They are in a certain sequence and *must be done in that sequence*" (HWH1942, p.177) and "...every job or task is made up of a number of separate acts performed in a correct sequence..." (HWH1942, p.177). He even used a men-as-machines analogy in the preceding paragraphs, and summarising his point towards the end, he stated,

> ...there are right and wrong motions of persons. Know the right ones and the safe ones. Analyze each personal task, write down the motions, if this will be helpful, watch for violations, instruct your men, and see to it that they do the job the way you want them to do it; namely the best way, which is invariably the safe way.
>
> (HWH1942, p.178)

On the other hand, however, Heinrich had a somewhat loose interpretation of "motion study." In his view "motion" roughly equalled "act" and he did not stress *measuring* time and motion, which was essential in Taylor's approach.

Next to the above quotes that seem to make Heinrich a follower of Taylor, we also find passages in Heinrich's work where he positions himself apart from Taylorism. While making the distinction between management and workers, Heinrich did not promote a strict Taylorist distinction. Like many contemporary safety authors, he suggested that one should ask the people who actually do the job. They knew best, Heinrich agreed: "...in many cases safety may be promoted by employing the ingenuity and mechanical genius of workmen in the average plant, in the revision of process and procedure" (HWH1931a, p.246).

Jesse Bird tells in his biography about an experience Heinrich had during the First World War that taught him about the advantages of non-Taylorist approaches and a positive view of human expertise: "detailed instructions are apt to do more harm than good with experienced personnel" (Bird, 1976, p.4). This is one of several examples that Heinrich did not want "ox-men" (Taylor, 1911) who unthinkingly follow procedure. Instead he appreciated, encouraged, and even expected competence, initiative, and adaptation. This is a very non-Taylorist attitude. According to Stewart (2009), Taylor did not want or care for the initiative of the workers.

> In our scheme, we do not ask for the initiative of our men. We do not want any initiative. All we want of them is to obey the orders we give them, do what we say, and do it quick.
>
> (Kanigel, 1997, quoting Taylor on p.169)

Other examples are found in the anecdote that opens the *Message to the Foremen* papers (HWH1929d, e), praising the initiative of an operator and the positive effects that sprung from it. The opening paragraph of the 1940 *Unsafe Habits of Men* paper saw him making the non-Taylorist choice of preferring "...capable, and experienced men who work under unsafe conditions" (HWH1940, p.112) over untrained and unthinking working under safe conditions. This is almost prototypically resilience engineering thinking: people who have the capability to handle variations in uncertain environments. At some places Heinrich also had a rather nice way to dismiss a "one size fits all" approach, "Obviously a suit of clothes that can be hung on a hundred men of varying proportions, cannot be expected to prove a good fit for all of them" (HWH1937a, p.104).

Concluding, the attribution whether Heinrich was a Taylorist, or not, requires much nuance, study, and context. The most correct stance is probably that Heinrich moved along a Taylorist spectrum depending upon the requirements of the situation – as probably everyone does. In that respect we must acknowledge that anyone writing or working post-1911 has been influenced by Taylor's ideas or some branch of Scientific Management[13]:

> Taylor's thinking, then, so permeates the soil of modern life we no longer realize it's there. It has become, as Edward Eyre Hunt, an aide to future President Herbert Hoover, could grandly declaim in 1924, 'part of our moral inheritance'.
>
> (Kanigel, 1997, p.7)

4.3 What about the missing data?

This chapter discussed several suggestions why Heinrich was talking about science, but not as we tend to know it. That covers the book's subtitle but still leaves the issue of research. Heinrich frequently mentioned research done by Travelers engineers which was the basis for the ratios of hidden costs (the 1:4 ratio), the *Origin of Accidents* (the 88:10:2 ratio), and the triangle (the 1:29:300 ratio) (e.g. 1926, 1927a, 1928a, 1929a, 1931a, 1941).

Several authors (e.g. Manuele, 2002; Conklin, 2017a; Dekker, 2018) point out that it is not possible to verify and replicate the findings because Heinrich's data is not available.[14] This is usually the core of their critique on the scientific element of Heinrich's work, and mostly drawing on Manuele (2002, 2011b). Keeping the previous suggestions of alternative understandings of "science" in mind, one could wonder whether this is relevant at all. Sure enough, we cannot go back and check how Heinrich got his ratios from the data available to him. Over the years, several studies have indeed been done to replicate, verify, or disprove the ratios, some of which can be found in Chapter 9.

Doing just a little mind game... What if Heinrich's data were retrieved in a corner of the Travelers' archives and was made available for re-evaluation. What then? What if we found the same ratios? What if we did not? Would it prove anything? Would it merely start another round of bickering over the way the replication was done? Maybe we should reflect on whether the ratios should be regarded as a result of research, or rather as an idea and argument in themselves.

First, should the "exact" ratios (which are, after all, the result of the data) be replicable? Most of the critique seems to suggest this. Regarding the cost and triangle ratios, Heinrich himself points out repeatedly that the numbers presented by the ratios are approximations and averages. Therefore, variations may be found (e.g. 1926, 1927a, b). About "hidden" costs, he said,

> It is not contended that the four-to-one proportion holds true for every industrial accident, nor every individual plant, and it is granted that in nation-wide application the ratio may vary; yet it has already been tested sufficiently to provide approximate confirmation.
>
> (HWH1931a, p.17)

Some papers suggested for example that cost in the construction industry may be higher, "In construction work it was estimated by the Associated General Contractors of America that the ratio was probably closer to 6-to-1 than 4-to-1, due largely to the effect of accidents on delays..." (HWH1938, p.374).

Regarding the triangle, Chapter 9 discusses the attribution of "fixed" ratios, pointing out that the 1:29:300 ratio was merely an average based on numerous unique ratios. Heinrich provided many examples of cases from which his average ratio is determined. All these examples have wildly different ratios. This not only illustrates that ratios are different, but also that they are scenario or context specific, as also suggested by Manuele (2002), Hale (2002), and Hollnagel (2004).

In a reply to many misunderstandings, Heinrich even found it necessary to explicitly emphasise, "...the actual ratio may never be known exactly" (HWH1959, p.30). Of course, this was not new knowledge that suddenly surfaced, Heinrich had said this long before, "The number of such no-injury or potential-injury accidents in comparison to actual injuries has always been a nebulous quantity, and it probably will never be known exactly" (HWH1929a, p.4).

Heinrich did not make any of these caveats about the 88:10:2 ratio. However, also here it is of little use to discuss the numbers because there are more fundamental issues that render them irrelevant. After all, Heinrich was trying to measure something that cannot be measured objectively – causes are constructed and selected[15] and thereby inherently subjective. Besides, Heinrich talked about direct causes which from a contemporary point of view are the least interesting point of attack (e.g. Manuele, 2002). Not for Heinrich, however, as will be discussed in Chapter 6.

Second, instead of focussing on the concrete numbers derived from the data and their validity,[16] seductive as this might be, one should look at the principles. As Hollnagel says in *Barriers and Accident Prevention*, "the importance lies not in the actual numbers, but their meaning" (Hollnagel, 2004, p.24). So, what was Heinrich essentially saying?

- The 1:4 ratio says that there will be costs that are not accounted for directly, and often cannot be accounted for, but the total costs of accidents will be larger than the direct costs from medical treatment and insurance.
- The 88:10:2 ratio tells us effectively that almost every accident has a human component.
- The 1:29:300 ratio teaches us that there are more near accidents than accidents; most (major) injuries are preceded by precursors that can be acted upon before actual harm is done. It suggests an opportunity for more proactive accident prevention.

One does not need numbers to understand these basic concepts. Read the above bullet points once more, skipping the ratios. The message is still there.[17] However, numbers do have some major advantages over no numbers, and therefore we should see the numbers and ratios for what they are. They are not scientifically determined "laws," even though the 1:29:300 ratio is labelled by some as "Heinrich's Law."[18] Just as the stories Heinrich told throughout his work, the numbers and ratios are tools intended to *convince* and *persuade*. This applies also to data and the research done to reach these numbers. Most likely, the research was done to confirm the common sensical hunches about these issues, to find a ratio that was not totally off the mark (instead of merely guessing or estimating) and to make the principles more credible by backing them with research, no matter how questionable. Besides, the ratios provided easy to remember "anchors" for the concepts.

Third, remember who Heinrich was writing for. He was not a scientist writing for other academics. As we saw, despite the "scientific" sub-title, *Industrial Accident Prevention* was mostly intended as a practical safety management book. He explicitly mentioned (top) managers as his primary audience (HWH1931a, 1941[19]). It was them he tried to convince, and the ratios and numbers were powerful tools of persuasion. Numbers and mention of research lend credibility to the concepts – just as an actor in a white coat tends to give a veneer of credibility to a toothpaste advertisement on television. These practices are not necessarily fraud, but neither do they represent "science" in the "pure" sense of the word. The fact that Heinrich's ratios are still around today says something about their persuasive power. As does an anecdote documented by Heinrich: "As one employer recently stated when confronted with his direct monetary loss, 'It may not be four-to-one, but I can see that it is something-to-one – and whatever it is, it is too much'" (HWH1926, p.247, 1927b, p.47).

Petersen and Roos would remark something that echoes this manager's sentiments in their update of *Industrial Accident Prevention* many years later, "attempt to actually quantify hidden costs is an almost impossible task, and probably not worth the effort. If management believes in the concept, it is often unnecessary to have to quantify" (HWH1980, p.90).

So, should we be worried that we do not have access to Heinrich's original research data?[20] If we insist that what he did was proper exact science in the strictest sense of the word, maybe we should indeed. However, by interpreting the meaning of "a scientific approach" differently and seeing the numbers and ratios as something else than scientifically determined laws of nature, we can relax and rather focus on the greater picture and lessons to draw from this.

One could compare this to Sanne Blauw's discussion of the problems with Alfred Kinsey's late 1940s research into sexual behaviour. She concludes that his research was basically activism packed in a scientific guise of diagrams and tables (Blauw, 2018, p.93). A similar argument can be made about Heinrich; he was very much an activist trying to promote what he regarded as better approaches in safety.

Notes

1 This chapter draws heavily on the chapter on "science" from my thesis, combined with the discussion of Heinrich and Taylor (Busch, 2019b).
2 Taylor wrote an essay on the subject in 1908, titled *Why Manufacturers Dislike College Students* (Proceedings of the Society for the Promotion of Engineering Education, XVII, 1909).
3 Heinrich seems to refer to this in the preface of the first edition of *Industrial Accident Prevention*.
4 We find a later version in the fourth edition: "Accident prevention can be portrayed as a science and as a work that deals with facts and natural phenomena" (HWH1950a, p.14).
5 Not the subject of this chapter, but from a human factors perspective, some critique would be in place here. Humans are after all not quite the same as machines!
6 https://www.popsci.com/
   Even renowned philosophers/scientists subscribe to this notion,

   The real test of 'knowledge' is not whether it is true, but whether it empowers us. Scientists usually assume that no theory is 100 per cent correct. Consequently, truth is a poor test for knowledge. The real test is utility. A theory that enables us to do new things constitutes knowledge.
   (Harari, p.289)

7 In the fourth edition, he claimed,

   Largely because of the recognition of basic principles, such as originally given in the first edition of this book and as herein reiterated, accident prevention has progressed from uncoordinated and arbitrarily selected activities, often ineffective and wasteful, to interrelated steps based on knowledge of cause, effect and remedy, i.e. to an effective, practical, scientific approach.
   (HWH1959, p.xi)

8 Indeed, this may have been one of the main drivers. However, reducing it to a matter of money is oversimplifying the matter. As noted in Chapter 2, it is not

intuitive for readers today to understand the role of insurance companies in the early decades of the twentieth century. Stone's review of the first edition of *Industrial Accident Prevention* discussed both the financial aspect and the important role of insurance companies in the formation of safety theory and practice:

> It is fortunate the casualty insurance companies have found it so distinctly to their economic interest to reduce accident claims by taking an aggressive part in promoting safety engineering. The insurance companies provide in major part the support of the National Safety Council. They are active in promoting the work of standard safety code committees. They publish and distribute, free of charge, educational literature of high merit. They have promoted the statistical reporting of accidents. They have themselves maintained a large corps of safety engineers for consultation work. These companies, then, are largely responsible for the extension and improvement of accident prevention methods.
>
> (Stone, 1931, p.324)

9   Stewart (2009, p.54) notes that Taylor initiated a "dogma of a singular metric," e.g. efficiency, or sales, which is another form of reductionism since there are in general many different, competing objectives (Rasmussen, 1997).

10   This point may seem remarkable since Taylorism is usually understood to be connected to mechanistic organisation, and command and control towards the workforce. Instead, Taylor speaks several times of "friendly cooperation" between management and workers. Still, Taylor's cooperation has to be understood in a paternalistic way and with the underlying aim that each should do that he is best suited for: management planning and preparing, workers doing, well, the "dumb" work. Cooperation then could rather be seen as "implementation" than an open exchange between thinkers and doers.

11   Even more, on some occasions he said that "Employees are paid to work safely" (HWH1931a, p.77) and therefore employers have the right to expect that they do so. That expectation extended to foremen as well, "...employers clearly have a right to demand of them that proper measures be taken to control accident-producing conditions" (HWH1931a, p.78).

12   For Taylor, motion studies were only a by-product or a means to an end. The real "inventor" and promoter of motion studies was actually Taylor's former associate, Frank Bunker Gilbreth, who seems to have first recorded "motion study" in 1884 or 1885. Gilbreth had a less "scientific" (i.e. he did not time the motions) and more common-sense approach to efficiency that had as main focus studying the individual motions a job consisted of in order to eliminate "wasteful" and unnecessary motions and find one "best" way to do the job (Gilbreth, 1911).

   In several aspects, one could say that Heinrich was more like Gilbreth than Taylor with his blue-collar background and common-sense practitioner approach.

13   Besides, as Kanigel suggests several places throughout his biography of Taylor – Taylorism soon became a loose and ill-defined concept (1997, p.502), meaning that just about anything could be categorised as being influenced by Taylor to a greater or lesser degree.

14   Some have suggested that Heinrich did not do any research but that the ratios were just made up. This is unlikely. Even though he did not offer the data, he discussed his research rather consistently through various papers and books, and it is even mentioned in Annual Reports from the Travelers Insurance. If it were made up, it would be an extremely elaborate hoax, which is not plausible.

15   Even Heinrich said as much on several occasions. See Chapter 6.

16 Which especially in the case of the 88:10:2 ratio can be questioned on basis of the fact that the data (insurance files) most likely was biased in one direction or the other. However, what incident report is *not* biased? Even contemporary reports that we may consider far superior to what probably was delivered in the 1920s will contain several biases, like the selection of "facts" presented to us. Typically, it is just one story and a selective description of reality.

17 A similar comment is found about Taylor's work in Dean's paper about Gilbreth's *Primer of Scientific Management* (Dean, 1997, p.39).

18 It must be noted that Heinrich in fact does refer to natural laws in the summary of *Industrial Accident Prevention* when discussing that injuries are preceded by "hundreds of accidents" (HWH1931a, p.269, 1941, p.363). So, he may have contributed somewhat to this misunderstanding.

19 The second edition was also intended as more of a textbook (HWH1941, p.vi) where he includes some others to the primary audience, "employers, the industrial executives, and the directors of safety" (HWH1941, p.72).

20 Another question to ask, but rarely raised nor answered: do other branches of science have access to the data from the 1920s and 1930s? If not, is it reasonable to expect this from Heinrich's work?

Chapter 5

The economics of safety

This chapter discusses the economic side of safety in relation to Heinrich's work. The cost of accidents was the subject that initially made his name, and which may have been the steppingstone for everything after. First, some historic background about the early safety movement and financial considerations. From there we turn to how safety and efficiency were seen in the early twentieth century, then the concept of "hidden costs" is addressed and after a brief discussion of Heinrich and the Great Depression there is a brief look on the subject in the period after Heinrich.

5.1 Early safety and economics

Safety and profit have been considered natural enemies ever since safety (and working conditions) became an issue in modern society. The industrial revolution brought a great mechanisation of workplaces and introduced new and greater sources of energy. The direct consequence was an increase of health and accident hazards for workers (Muntz, 1932). Control of these hazards was often seen as driving costs or to be conflicting with production and profit. An example of the effects of this perspective was the delayed implementation of the 1833 Factory Act by a decade over a controversy of fencing machinery.[1]

In the nineteenth century and first years of the twentieth century, under the common law system, almost always injured workers bore the costs of an accident. They were compensated only in case of employer negligence, but if partial responsibility for the worker could be proven the employer was "off the hook." And even if workers were compensated, most of the money went to cover expensive lawsuits (Eastman, 1910; Beyer, 1917). Consequently, there was little incentive for employers to engage in accident prevention.

This changed thanks to growing public and social awareness – as evidenced by some progressive employers who actively improved workplaces – and especially the advance of worker compensation acts. The first compensation act was ratified in Germany in 1884. The adoption of similar regulation in other countries followed soon. This legislation turned the tables. Because injured workers were compensated, regardless who was to blame for the accident, there suddenly was a strong incentive for accident prevention. Employers had insurances for the compensation, and accidents added to the cost of his insurance. Safety measures thus became investments that often paid back. By the 1920s, nearly all states had workmen's compensation laws. A few southern states lagged behind.

Some of the early authors emphasised the humanitarian and moral angle, possibly because of some of the drivers of the early safety movement had been public awareness and outrage of the number of accidents and fatalities in certain industries,

> In the early stages of the factory system, the interest of the employer in the protection of his employees was primarily humanitarian. The number of workmen in a factory was small. The employer was in many instances a fellow workman and, in any event, was in close contact with those in his employees.
>
> (Alexander, 1926, p.6)

Larger organisations added layers which removed top management from the work floor. Still, many were driven by humanitarian reasons, the author argued, in addition to safety and compensation legislation. Others had more pragmatic thoughts, however. Melville W. Mix, president of the Dodge Manufacturing Co. addressed the Safety Congress of the National Council for Industrial Safety in October 1914, stating that "Safety First is Not a Philanthropic Movement." He suggested that safety was an investment that should be regarded as "a hard practicality of business extension" that should be considered from a financial viewpoint.

> It cannot always appear in the inventory; it may not be accounted for in the financial statement; it may not be a known basis of credit, yet in its very earning power it becomes one of the most productive elements of a business.
>
> (Mix, 1914, p.296)

As DeBlois said, "...if industrial safety is to survive and prosper, it must prove its economic worth" (DeBlois, 1926a, p.1), safety professionals needed to make a case to prove their worth.[2] This was best done by connecting their work to the primary objectives of the companies. There were two main (and sometimes overlapping) roads in which safety professionals tried to argue that investing in safety was worthwhile: increased efficiency and reducing "waste" created by accidents.

The Manager of the Safety and Claim Departments of the American Car and Foundry Company, asked "Does Accident Prevention Pay?"

> Hitherto, the two chief factors considered in a producing organization have been material (including machinery) and personnel (including

organization). To these I would now add a third factor, namely, accident prevention. Accident prevention directly affects both the material and the personnel; and it affects them in four directions: (1) Increased Production (2) Decreased Overhead[3] (3) Decreased Labor Turnover (4) Saving in Money Compensation.

(Orth, 1926, p.20)

Orth argued forcefully that safety was a business problem with major sums at large. His answer to the question posed, "Does accident prevention pay? Nothing else pays so well" (Orth, 1926, p.24). Agreeing on the fact that safety was good business but using a somewhat different argument, Albert Whitney wrote in the *Safety and Production* report,

When the safety movement was started, it was felt that the correlation in question might even be negative. It was alleged that the use of safety devices, and the precautionary acts required of employees, interfered with production. This state of mind has almost completely disappeared and has been replaced by a very general feeling that safety and efficiency of production go together.

(American Engineering Council, 1928, p.8)

The need for safety was recognised in management literature. Lansburgh opened his *Industrial Safety* chapter with a section titled "The need for safety," mentioning thousands of fatalities and millions of non-fatal accidents each year, "...which make them miss days of employment. What does this mean in terms of production delays, of injured organization morale, of increased costs and narrowed markets?" (Lansburgh, 1928, p.171). The costs were enormous, and, in the author's view, employer and employee must work together to reduce accidents, emphasising, "It is the one phase of industrial management of which no one can question the desirability" (Lansburgh, 1928, p.171).

5.2 Efficiency

"It is positively good business and profitable business to prevent accidents" (HWH1923b, p.3), Heinrich said in the *Bedford Stone Club* speech and would repeat similar statements often. He used both main arguments to illustrate the benefit of safety to businesses. When he spoke of efficiency, he drew mostly on the work of others; spoke of accidents (and other losses) in terms of inefficiency[4]; or leaned towards a Taylorist approach, equating the safe method of doing a job with the best and most efficient way.

Heinrich was most likely influenced by the focus on efficiency during the first decades of the twentieth century. Following the efficiency craze of 1911, the breakthrough of Scientific Management and the World War, which contributed significantly to the advancement and growth of the management movement,[5] numerous safety authors linked safety and efficiency. A quick browse of some volumes of *Safety Engineering* magazine illustrates this (Kennedy, 1914; Howard, 1917; Cousins, 1918; Heyne, 1918; Kennedy, 1920). Many of these articles link safety and efficiency in a positive way: "We must systematize the safety work and reduce it to a commercially efficient basis…" (Kennedy, 1914, p.480), one author urged; "An Accident Is an Inefficiency" (Howard, 1917, p.65), the president of the Commonwealth Steel Company stated. Another author claimed that "Experience and statistics have proven that safety is the foundation of efficiency" (Kennedy, 1920, p.216).

A safety historian observes that "job analysis was used to bring out risks just as it was being employed to enhance output" and "was also used to fit the worker to the task" (Aldrich, 1997, p.159). While Scientific Management often not is seen in the light of humanitarian approaches, authors at the time were more optimistic,

> It is certain that the general efficiency of any business is at a higher standard when the employees feel that their lives, their health, and their interests are matters of importance to the management than when this feeling is absent.
>
> (Tolman & Kendall, 1913, p.11)

However, Tolman and Kendall did not close their eyes for the tension there might be between safety and efficiency. On one hand, lack of safety hinders efficiency, because workers are uncertain about actions, or extra careful. On the other side, a mindless drive for greater efficiency and treating workers as machines, wears them out and leads to accidents. Their conclusion was therefore that efficiency and production needed safety management,

> Any efficiency movement which is not based on the sense of security due to a thorough system of accident prevention, the maintenance of health through sanitary conditions of work, and mutuality, or those reciprocal relations of goodwill which must obtain between capital and labor, or any industrial system which ignores these fundamentals is foredoomed to failure.
>
> (Tolman & Kendall, 1913, p.10)

Worth noting are nuanced views regarding Taylorist efficiency and safety. When DuPont started experimenting with efficiency, they did not incentivise personal production because they found that the production of explosive materials and rushed work were a bad combination. Also, their production relied rather on flexible teamwork than individual specialisation. In DuPont's view, a "best way" to do things would have to include a higher degree of safety (Stabile, 1987), something which also reflected in Heinrich's writings.

The 1928 American Engineering Council report *Safety and Production* was probably one of the most important publications on safety and efficiency.[6] Heinrich quoted from this work in all editions of *Industrial Accident Prevention*. It seemed to be his main source when discussing efficiency. The study aimed to test generally accepted, but unproved hypotheses,

- That a safe factory was an efficient factory (and vice-versa).
- That safety and efficiency of production were both aspects of good executive control.
- That a right organisation and proper managerial effort, bring about efficient and safe production.
- That neither could be satisfactorily achieved in isolation.

It was suggested that improved production resulted from good management of production processes and that accident prevention under production changes also could result from good management. The study concluded that in most industries accidents were controlled at the same time that production increased, and that safety did not interfere with production. A "safe" factory was found to be 11 times more likely to be productive than an "unsafe" factory. One finding was that increasing production required much management attention and leadership while recognising that reducing accidents required exactly the same. Whitney wrote, therefore,

> This, then, raises the question: *Is not the key to the safety movement of the future the chief executive himself?* Up to the present time the safety movement has in general been, essentially, a second-rank movement. It has had the knowledge and approval of executives, but it has not entered into the consciousness of most executives as a matter demanding their personal interest and attention.
>
> (AEC, 1928, p.7, emphasis added)

Aldrich critiques the study as not being a scientific study but rather a document to persuade top managers. Whitney said as much, "Frankly, the primary purpose of this study is to interest executives in the subject" (AEC, 1928, p.7). Interestingly, the report also mentioned the shortly before published 1:4 ratio without direct reference,

> There is undoubtedly a direct relationship between safety and production that is of considerable importance. The disturbing effect of an accident upon business is now known to be much greater than has been generally supposed. In fact, the effects of an accident that are commonly insured against, probably constitute not more than a fourth or fifth of the entire economic loss.
>
> (AEC, 1928, p.9)

The report emphasised that the future of safety management was in adopting this "new view" and having top managers take their responsibility. More about this in Chapter 8.

> Action in line with these recommendations will bring in a new safety movement which will far eclipse the old. There are hidden sources of strength in executive control which hitherto have not been applied to the accident situation. If American industrialists will adopt the same executive policy toward safety which they have fully developed toward production, we may confidently expect a decreasing number of deaths, permanent disabilities, and temporary disabilities, with their attendant costs.
>
> (AEC, 1928, p.36)

5.3 Hidden costs

Heinrich's main contribution to the theme of economics and safety was the ratio of direct and indirect costs. Heinrich experimented with the term. One finds several variations, including incidental, indirect, or hidden costs. By this, he meant secondary, non-insurable losses, in addition to direct liability claims and medical costs: "the costs of accidents, other than as represented by compensation, liability, and medical payments, is by far the greater expense" (HWH1926, p.245).

5.3.1 The idea

The idea of hidden costs was not new when Heinrich published his work. Beyer (1916) suggested it in his book. Lange's book also contained a discussion of "intangible" costs of accidents (Lange, 1926, p.8). Williams (1922) even did a study on the costs of accidents which he presented in September 1921 at the Annual Meeting of the International Association of Industrial Accident Boards and Commissions in Chicago. In this study he attempted a rough quantification, but also mentioned forms of economic loss that defied quantification. However, he assumed that these indirect costs as a result of "interference with production" and effect on morale of workers were even larger than those he had managed to calculate.

Travelers Insurance publications had also featured the subject before Heinrich wrote about it.[7] The November 9, 1921 edition of *Protection* even foretold Heinrich's ratio on page 156, "In the opinion of one of the best authorities in the country the actual cost of insurance represents not more than 25 percent of the total economic loss due to accidents."

This referred to a report of the Committee on Elimination of Waste in Industry. This committee was created by the Federated American Engineering Societies at the suggestion of Secretary of Commerce Herbert Hoover.

In 1925, future President Hoover was again presented in the *Protection* of July 8. The article told the reader,

> Mr. Hoover Has Paved the Way
>
> Industrial accidents have been indicated as one of the big causes of industrial waste. Secretary Hoover's committee which made an exhaustive study of the subject has recommended that industry must look to trained engineers to help cut down this particular loss.
>
> Manufacturers are beginning to realize the serious losses that accidents entail. Formerly most of them thought that their compensation insurance premiums fully covered this expense. Now they are coming to see that the compensation premium is only one item, and frequently not the most important, in the total accident cost of their plants. Some of these other costs are:
>
> - Lost time on the part of the worker while receiving first aid treatment for minor injuries.
> - Impaired production on the part of the worker, due to minor injuries, particularly eye and finger injuries.
> - High turn-over costs due to the necessity of replacing workers who have been seriously injured.
> - Loss of time on part of all workers in a department in which a serious accident occurs.
> - Disrupting effect of serious accidents on the morale of a department with a consequent loss in the production of that department.
> - Difficulty in obtaining good workers in a plant where the accident frequency is high.
> - High compensation insurance costs necessary to cover the costs of unnecessary accidents.

The article then connected this to the services of safety engineers from the Travelers inspection service, helping to reduce these costs of industrial accidents. Undoubtedly this influenced Heinrich.

5.3.2 Research and the 1:4 ratio

The first mention of Heinrich's study into the costs of accidents was during the New York State Safety Congress on 2 December 1925. The paper links safety and production positively linked together and contains a lengthy passage about cost. At the time, Heinrich based his preliminary results on a study of 100 random cases (HWH1925).

The first results were published in December 1926 (HWH1926, 1927a, b), and this time the study built on a review of 5,000 cases through which the Travelers engineers had tried to systematically measure all costs associated with accidents. The number of cases was soon after doubled to 10,000

(HWH1927c, 1930c). The study showed that, according to the factors used by the researchers,[8] the indirect costs of accidents were on average four times greater than the direct (insured) costs. Although he would present the ratio time after time, Heinrich was careful to present it as an average,

> The four-to-one ratio of hidden to direct accident costs does not apply, of course, to each individual accident. Many serious injuries and fatalities result in huge compensation payments, with relatively small hidden costs. Thousands of trivial injuries, on the other hand, create greatly disproportionate monetary loss.
>
> (HWH1931d, p.96)

In his work, he mentioned several other, independent studies into the costs of accidents (e.g. 1933b, 1938b) and provided additional estimates. Because the larger impact "…the small plant has a higher-than-average hidden cost" (HWH1932a, p.29), and for contractors he mentioned an estimate that was closer to six-to-one (HWH1938b) with cases mentioned with ratios as wide as 1–1,000, 1–375, and 1–4,400.

5.3.3 The study's importance

Even though it was not innovative, strictly speaking, Heinrich's study resonated with many people, most likely supported by the fact that he had managed, through what looked like scientific research, to attach a number to what previously had been a common-sense notion. Heinrich's paper was re-published several times and was featured in newspapers, making Heinrich's reputation.

To Heinrich the monetary approach to safety was an additional, powerful incentive for employers (and the State) to engage in safety work, in addition to humanitarian aspects: "a powerful stimulus to preventive action" (HWH1927c, p.222, 1931a, p.17). It seems that it has had an important impact on safety work and the acceptance of safety by many managers as something to prioritise.

This is an important point, because Heinrich was not interest in the ratio just for the sake of knowledge. Heinrich thought that knowledge or criticism in itself was not useful. One had to come with suggestions for improvement, and these must be practicable, "This information concerning accident costs may be interesting, but it is not necessarily of tangible value unless it is supplemented by concrete suggestions as to how best the costs may be reduced" (HWH1931d, p.97). Because Heinrich spoke first to management, it made sense emphasising the economic side of safety. Accident prevention was "good business" (HWH1930c, p.1122, 1931a, p.16), and safety and efficiency went together.

The reception of the work by Heinrich's peers was very positive. According to Aldrich, "Heinrich's 'rule of four' was the answer to a safety expert's prayer; by quantifying what had been vaguely asserted for years, the rule justified their efforts and helped solidify their corporate position" (1997, p.151). The discussion, following Heinrich's presentation (HWH1930c) at the annual meeting of the International Association of Industrial Accident Boards and Commissions in September 1930, was documented. Joseph Plumstead, vice president of the Delaware Safety Council stated,

> How can we help our young executives to convince their chiefs of the crying need for money and action? Accident-cost information and publicity for each industry would seem to be the answer. Are there other more direct and effective methods of meeting this task? [...] ...the sooner the State can bring the employer to a realization of the total cost of industrial accidents, the sooner another great step forward will be made toward safety in industry.
>
> (BLS, 1931, p.181)

Dr Leonard Hatch member of the Industrial Board of New York, concluded his comments,

> It seems to me Mr. Heinrich has presented us with a very practical document to be used in connection with State appropriations for safety work. I have always felt that appeals to State legislatures for money for accident prevention were based too much upon sentimental arguments or offhand suggestions about the sufferings of workmen; that is reasonable enough, but it has been said so many times that it is not making the impression it should. We know enough from such papers as Mr. Heinrich's and other studies to make it possible to go to the legislature and say, 'This is not an appropriation to take more money out of the community with no return; this is simply an appropriation to spend more money in order to save still more money'.
>
> (BLS, 1931, p.183)

The indirect cost study had also important side effects. In all likelihood it inspired the *Origin of Accidents* study, and it led to the "discovery" of a ratio between similar accidents with major and minor outcomes, which turned into the accident triangle.

5.3.4 After Heinrich

Cost of accidents featured prominently in Heinrich's early work, mentioned as early as 1923 and in all the early papers and speeches, and opens the 1931 book with an entire chapter dedicated to the subject. As other main themes developed and gained prominence, costs moved to the back,[9] although the subject never fully disappeared. Perhaps Heinrich thought he had gotten all there was from the subject and moved on to things he thought more important.

Others picked up where Heinrich left off, and the (hidden) cost of accidents became a standard feature of many safety handbooks, including Blake's (1945) and Simonds and Grimaldi's (1956). The latter even spent several chapters on the subject, building on and expanding Heinrich's foundation.

During the 1960s and 1970s, Frank Bird Jr. expanded on Heinrich's work, focussing on "loss control" and picturing it as the "iceberg principle of hidden costs" (Bird & Loftus, 1976, p.49), showing the direct costs of accidents (medical and compensation) as well as indirect costs like material damage, production delays and interruptions, and hiring and training of replacement workers. Bird would even do an entire book dedicated to the subject of safety as good business (Bird & Davies, 1996).

Throughout the years the subject of hidden costs kept turning up in safety literature. Niskanen and Saarsalmi (1983) discussed indirect costs in their research. In 2001, Liberty Mutual reported on indirect costs as being three to five times higher than direct costs which was seen as a justification of safety cost (Masimore, 2007). Rivers (2006) calculated the costs of accidents, including "indirect" costs, finding a 17:1 ratio which he attributed to greater accuracy than Heinrich's estimates. Recent Australian research suggests that "employers commonly underestimate the true financial impact" of accident, illustrating that little has changed in argumentation since the mid-1920s:

> While the financial impact of construction accidents can pose a very real threat to the success of construction companies, clear identification of their scale can be leveraged to encourage the internalisation of true accident costs and positively inform the improvement of safety programs.
>
> (Allison et al., 2019, p.886)

Not only were many attempts made to replicate, expand, or update Heinrich's findings, over the years also critique has been offered. In the fifth and final edition of *Industrial Accident Prevention*, the authors smartly remarked: "to actually quantify hidden cost is an almost impossible task and probably not worth the effort. If management believes in the concept it is often unnecessary to quantify" (HWH1980, p.90). In this edition one also finds a brief critical discussion whether safety and efficiency always go together. Sometimes they do, but the authors also note exceptions, stating that one cannot know for sure, unless one knows all the factors. They considered the question "somewhat academic" for practical purposes. Safety management, according to them meant achieving maximum safety within the given constraints, while "arbitrary preaching" that "safety is good business" might do more harm than good[10] (HWH1980, p.90–91). Interestingly, Petersen made no mention of costs or efficiency when he formulated his ten "New principles of safety management" (Petersen, 2001a, p.15). On another occasion, he remarked that "safety is small potatoes" (Petersen, 2001b, p.54) compared to other business costs and objectives.

Manuele discussed also Heinrich's cost ratio in his critique, pointing at differences between elements included in various studies, that ratios were invalid because situations had changed over time, and because of differing scopes. Regarding Heinrich's original research, he again critiqued the missing data.

> [It] can be attention-getting and convincing, provided the data are plausible and can be supported with suitable references. Unfortunately, little research and hard data exist to support the frequently used ratios of indirect to direct costs that appear in safety-related literature.
>
> (Manuele, 2011a, p.39)

Aldrich, an economist, comments, "Heinrich's accounting is flawed. It includes overhead and is thus an average, not a marginal concept, and it double-counts lost wages and lost output" (Aldrich, 1997, p.357).

5.4 Heinrich and the Great Depression

Heinrich laid the groundwork for his work in the mid/late-1920 and then made a breakthrough to a wider audience, publishing the first edition of *Industrial Accident Prevention* during the Great Depression. One wonders how this affected his work. It is not unreasonable to assume that he even may have benefitted from this.

Peter Petersen of John Hopkins University is one of a few who have written about the development of safety in this period[11]:

> Understandably a company's survival in the short term was a paramount concern along with worker anxiety about unemployment. Nevertheless, safety management professionals progressed in developing a theoretical foundation and effective practices. While the Great Depression brought out increased academic interest and first-rate work from many dedicated safety management professionals, their innovations and accomplishments often were overshadowed by poor safety practices caused by desperate short-term cost containment measures.
>
> (Petersen, 1990, p.382)

Aldrich (1997) notes that one might have expected organisations to cut back on safety in depression times. It is a fact that the National Safety Council lost many members in the period. Aldrich suggests, however, that the depression may have given safety "a boost" because it led to the closing of older and less safe places, and various companies "stepped up" their safety efforts to cut costs. Besides, "Throughout the Depression, many workers and employers continued to be optimistic in thinking that, prosperity was just around the corner" (Petersen, 1990, p.382). The timing of Heinrich's work may therefore have been excellent, offering solutions to save money[12] and deal with problems in practical ways.

Interestingly, in his work, Heinrich makes almost no reference to the financial crisis. The opening of *Safety in Gas Manufacturing* might hint at the Depression, "At no time in history of business development has there been a keener or more justifiable interest in the prevention of accidents to employees than now exists" (HWH1931d, p.93). His presentation at the Annual Meeting of the American Society of Mechanical Engineers in December 1932[13] and the *Conservation as Essential in Industrial Recovery* paper were the first time to openly reference to the economic crisis: "The depression places economy in the spotlight of managerial consideration, but tends to stifle creative and constructive action" (HWH1933b, p.52). However, both economic use of resources and innovation are necessary, according to Heinrich. This provided a new angle. Before, he spoke mostly of accident prevention and reducing costs. Now, he emphasised, "...business recovery depends as much on initiating constructive procedure as on eliminating wasteful methods" (HWH1933b, p.52).

Despite the fact that Heinrich wrote little explicitly about the Depression, he was definitely interested and engaged in this subject. A lengthy 1932 article in the Hartford Courant[14] describes a work creation plan devised by Heinrich. Suggesting that neither charity nor unemployment was necessary, he proposed that idle workers could work for contributors to relief funds. The plan aimed to bring together people without work, work that needed to be done, and funds for wages. The next year, Heinrich was loaned out by the Travelers and appointed as state safety director of CWA, the job creation program established by the New Deal, in Connecticut. The program distributed about $9.5 million to over 600,000 people.[15]

Several years after the Depression, Heinrich published the short, one-page article, *A Break in The Vicious Cycle* (HWH1937b), which seemed to reflect on the crisis. The first half concentrated on negative side-effect of "protective measures" in periods of financial depression. Often these actions, while on the short-term making sense, made things only worse, leading to a downward spiral. In contrast to these measures, Heinrich offered something that paid off always without contributing to a vicious cycle: accident prevention. His message was: safety is good business and good for the economy.

5.5 A continuous theme

Soon it is 100 years ago that Heinrich published his work about hidden costs. The message is still relevant and useful. However, from today's perspective, one can also argue that the argument may have been too simplistic.

Surely, accidents bring costs (direct and indirect), but safety costs as well. In the end safety is about striking a balance between various objectives. Heinrich spoke mostly of the gains of safety, but little of its cost, even though he acknowledged, "Safety at any cost, safety at all costs, is not to be desired" (HWH1930b, p.151).

The message around the four-to-one ratio omitted these nuances, possibly to gain persuasive strength. The simple version, "avoid accidents thereby avoid the costs of accidents" is a truism. But it is only part of the entire picture and can serve as a "benefit façade" (Marriott, 2018, p.38). Besides, a quote attributed to Russel Ackoff says, "Getting less of what you don't want doesn't get you what you want," meaning that avoiding loss may not give you profit.[16]

For safety measures, a challenge is that cost comes before benefits. One has to invest certain money into uncertain returns. Especially when budgets are tight and focus is on financial targets, managers may prefer certain short-term effects over protection against uncertain outcomes. Apart from this, there is the eternal challenge of how to measure safety – the absence of accidents is no guarantee of safety (Conklin, 2012; Townsend, 2014).

Risk homeostasis is another issue. Greater safety (as in better protection, or reduced risk) often allows for more, faster, and cheaper production. As an everyday example, take straight and well-maintained roads. When looked upon in isolation, as many often do, these are arguably safer than narrow, bended, bumpy roads and allow for faster and more efficient transport/travel. This, however, may mean that risks are merely transferred to another place or another actor (Wilde, 2014). For example, when higher speeds mean more risk for cyclists and pedestrians, or cause more noise replacing a safety problem with an environmental/welfare problem.

Another factor is that people have the tendency to "use up surplus safety." This may manifest itself through continuous optimising and "fine-tuning" (Starbuck & Milliken, 1988). People may also use the increased safety for other goals until they have reached a level of risk that they are comfortable with – some authors talk of a "risk thermostat" (Adams, 1995; Wilde, 2014). This brings us to the next challenge, safety as a control problem.

As Rasmussen (1997) taught, "surfing the boundary" is the best place to be when you want to achieve an optimal – but extremely unstable – balance of costs, workload, and safety. There are two main problems connected to the boundary of safe performance. First, people operating in the space do not know exactly where the boundary is – until they crossed it. Based on their experiences they may have some sense of its location, of course. That is why they often watch a safety margin. After all, keeping from the edge prevents you from falling off it.

Second, the boundary itself is dynamic and can vary over time, e.g. due to degradation of barriers, influences from other actors (Rasmussen, 1997), or random influences from external factors or unexpected combinations within the system – comparable to the holes in Reason's Swiss Cheese that continuously move over the place and vary in size (Reason, 1990).

Many contemporary authors agree with Heinrich that there is not a question of safety *or* production. The question should rather be about how to achieve *safe production*. Safety is good business, much of the time. When it is not, there are certain boundaries that must not be crossed. A model of hidden costs of accidents is therefore a useful tool, but better even may be more recent models that include workload, efficiency, and financial constraints as drivers for a migration towards the boundary of what can be considered acceptably safe (Rasmussen, 1997), or describe how too much safety can lead to bankruptcy, while too much focus on the financial part leads to accidents (Reason, 2008).

> Safety has financial implications that cannot be ignored, and it is understandable that costs do have an influence on the choice and feasibility of safety measures. It is all the more understandable because safety costs are immediate and tangible whereas the benefits of safety investments usually are potential and distant in time.
>
> (Besnard & Hollnagel, 2012, p.11)

Notes

1 There are certainly richer and better sources, as Hutchins and Harrison (1911) freely accessible online, but the reader with a casual interest can glimpse some of the troubles around these early laws from https://en.wikipedia.org/wiki/Factory_Acts

2 A strong statement to illustrate this was made by Sidney Williams. "If industrial safety meant simply the saving of some 23,000 lives annually and the elimination of a few million non-fatal injuries, the movement would neither have reached its present development…," he said.

> Safety in industry actually means far more than this, because it is directly correlated on the one hand with efficiency and economy of production and on the other hand with the establishment and maintenance of proper relations between employers and employees—and these two are probably the most important problems facing any American industry today.
>
> (Williams, 1926, p.8)

3 Several modern scholars would disagree and point toward the bureaucratic burden created by safety (e.g. Dekker, 2014, 2018). Orth's argument was:

> Overhead is decreased by accident prevention in substantially lessening both the cost of insurance and the cost of compensation. But there is a further decrease. The cost of accidents is far more than the cost of compensation to the injured worker. The loss in time due to the disorganization which follows an accident, the delay in "speeding up" the plant afresh, represent losses not easily calculable in figures, yet they usually end in loss of often many hundreds of hours for which there can be no return. These working hours are utterly wasted, so far as production is concerned. Accident prevention avoids such a loss, and this is a definite gain in overhead decrease. It, therefore, pays.
>
> (Orth, 1926, p.21)

4 For example, emphasising the need to understand "...that safety and production are positively linked together; that accidents result from inefficiency; that, in short, accident prevention pays" (HWH1925, p.257).

5 For a short while, management and efficiency were almost equated. Lansburgh offered some interesting critique of this phenomenon (Lansburgh, 1928, p.25).

6 The following paragraphs are partly based on Roger Brauer's digest of this document in *The Archives of Safety and Health*, January 2020.

7 The following paragraphs draw on Jesse Bird's unpublished biography due to lack of access of the original sources.

8 In the papers (HWH1926, 1927a, b) Heinrich presented a 9-point list which grew to 11 points in *Industrial Accident Prevention* (HWH1931a). He added a disclaimer that even these lists did not feature all items possible for consideration.

9 Compare for example the place in the 1931 and 1941 versions of the book.

10 Regrettably, the authors only spend half a page on this quite controversial and important subject, and not clearly build their argument. One of the examples they present (highways would be very safe with a 6-mph speed limit, but this is hardly acceptable to society) suggests that things are safe when risk approaches zero. Whether that is a useful definition of safety, is open for discussion. However, this short section in the 1980 book could make a nice introduction for Amalberti's (2001) discussion of ultra-safe systems and the limited or non-existent return of investment in safety in those systems.

11 Oddly, Petersen's paper makes no reference to Heinrich's work who is arguably the most visible author (to this day) from the period. Petersen refers much to DeBlois and somewhat to Williams. He seems to have his main focus on the industrial psychological side (Münsterberg, Viteles, Shellow, and Fisher) along with an uneven selection of subjects, possibly due to the author's background outside of safety.

12 See for example the newspaper article from Hartford Courant, 8 December 1932, headlining "Heinrich Offers Plan to Prevent Accident Waste."

13 This speech must have made quite some impression as it was featured in several newspapers the following days. Among these, The Philadelphia Inquirer, The News Palladium, and The San Francisco Examiner from 8 December 1932, and Pensacola News Journal and Reading Times from 9 December 1932.

14 *Hartford Courant*, 23 November 1932 – making the frontpage.

15 *Hartford Courant*, 31 May 1934.

16 Nor, as some "new view" safety scholars will argue, does preventing accidents necessarily give you safety (Hollnagel, 2014a).

Chapter 6

Accidents are caused

This chapter deals with one of Heinrich's most central themes. Arguably also one of his most important contributions to safety theory and practice. Subjects to be discussed include various causes, his duality of causes, the accident sequence. The chapter concludes with a more complete model of what Heinrich wrote about causation than the well-known dominos metaphor.

Causation was the first main subject that Heinrich turned to after his breakthrough. He mentioned it for the first time explicitly in the last of the hidden cost papers, noting a similarity of causes for fatalities and minor injuries (HWH1927c). Causes were at the core of the *Origin of Accidents* papers (HWH1928a, b), in which he introduced the accident sequence for the first time and critiqued contemporary causal attributions. Ever since, discussion of causes would be an essential part of Heinrich's work.

6.1 The importance of causes

Heinrich was convinced that knowledge of causes was necessary to find effective actions: "it is the cause of the accident that we must know in order to effect a remedy" (HWH1928b, p.122). Many of his earlier papers critique the assignment of "causes" at the time. He emphasised the importance of separating the different stages in an accident.

> Above all, however, if we assume (as it seems proper to do) that the chief purpose of analysis is to furnish a clue to accident prevention, it is vital to know the *cause* of the *accident* itself as distinguished from the cause of the *injury...*
>
> (HWH1928b, p.122, 123)

Heinrich placed causes at the core of fact-based "scientific accident prevention." He made the analogy to a physician treating a patient, who did not "...base his selection on guesswork or on mere judgement founded only upon limited information" (HWH1931a, p.11). The physician made a careful diagnosis to determine what illness he was dealing with, where the cause was and what the best treatment would be. "In accident prevention, likewise, there is need for diagnosis or analysis of the problem before deciding on a particular remedy" (HWH1931a, p.11).

He made an engaged appeal adopting better practices at the Seventeenth National Safety Congress in October 1928,

> I emphatically recommend the revision of present accident-prevention practices. I urge that the term 'accident cause' be clarified; that a more systematic and determined effort be made to obtain the complete story of accidents from which the real causes may be more readily determined; that we set up a list of probable accident causes such as I offer you today (or another list, if a better one can be found); that we establish in our plants the practice of assigning accidents to such causes...
>
> (HWH1928c, p.130)

He argued that better information about causes would improve accident statistics and increase their usefulness (HWH1929c). Heinrich spent much space in his books on explaining better ways to do "accident cause analysis" and "fact finding." Heinrich mentioned many different kinds of causes: "basic," "real," "true," "so-called," "direct," "proximate," "subcauses," "indirect," "underlying," and "reasons." Heinrich did not clearly define all of these,[1] but often, one understands the meaning from the context and examples given.

On a few occasions these terms changed meaning. "Basic cause" was the same as "real" or "true" cause in the *Origin of Accidents* papers (HWH1928a, b). In the *Safety Psychology* chapter of *Industrial Accident Prevention*, however, it was used as a synonym for "underlying" cause. This is comparable to the contemporary understanding of "root cause," as used in accident analysis tools, as Tripod (Groeneweg, 1992). The term "real" and "true" causes changed meaning as well between the first two editions of *Industrial Accident Prevention*.

6.1.1 Early 1900s causation

When safety became a field for serious attention, so did accident causes. However, in the early twentieth century there were only rough ideas about accident causation. Many regarded accidents as a part of the work. They were something inevitable that came with the trade. Alternatively, it was contributed to carelessness of workers (Swuste et al., 2019). Therefore, the realisation that accidents were caused and not just happening because of "bad luck" or divine intervention, was elementary for safety work. If accidents were caused, they were controllable and preventable.

Eastman wrote, "No thorough study of the causes of accidents had been made in the United States when this investigation was begun" (Eastman,

1910, p.15). The first part of her report for the *Pittsburgh Survey* dealt with this, trying to look at the "how" and "why." The study presented a great variation of causal explanations for (fatal) accidents emphasising conditions. Long working hours, guarding, planning, and maintenance, were important factors while she had a very critical stance towards the prevailing belief that "95 per cent of the accidents are due to carelessness" (Eastman, 1910, p.84).

"Causes" appeared regularly in early professional literature. Some of these illustrate that the thinking about causes had not matured yet. Downey, writing in the November 1916 issue of *Safety Engineering* about the relative importance of accident causes in industries, mostly discussed objects, machines, tools, and corrosive materials involved in accidents or events. None of these factors would be regarded as "causes" today.

Beyer's book serves as another example. The chapter, *Causes of Accidents*, offered a two-page list with an uneven mixture of injuries, activities, agents, and "causes" as "Caught between belt and pulley," "Blasting and drilling," and "Falls from fixed ladders" (Beyer, 1916, p.9–10). Lange (1926) presented a superficial section about accident analysis, paying little attention to causes, mostly mentioning the agent of the injury (e.g. a machine part) as the proximate cause. This way of assigning causes was even regulated, as illustrated by Bulletin 276 on *Standardization of Industrial Accident Statistics* which lists under the header of "General Cause Classification" categories as "Machinery," "Falls of persons," and "Handling of objects" (BLS, 1920, p.36).

There were also differing and more nuanced views as Eastman's, or Gilmour's lecture, mentioned in Chapter 3 (Gilmour, 1913). Writing in the same year as Downey, Schreiber (1916), started his paper in a way that fits well into contemporary thinking: "As much as we may wish to ascribe each accident to a specific cause, we all know that many causes have probably contributed to the result," suggesting, "Every single factor which may contribute to an accident thus becomes something of interest and importance" (Schreiber, 1916, p.230). This corresponds well with modern thinkers, as Pupulidy (2015), stressing the importance of context.[2]

Five year later, Hubbard agreed that it was necessary to study the causes of accidents in order to prevent them. He offered three groups of factors: mechanical/physical, humans, and the "conditions of the environment." While this looks more advanced than Heinrich's later dual approach (act and condition), Hubbard was unclear how conditions factored into the equation since he concluded "Suffice it to say that the human element in accident causation is about 90 per cent, and that the mechanical element is about 10 per cent" (Hubbard, 1921, p.25).

None of the early safety authors did have a framework for causes, and neither had they an accident model or a specific causation theory to help them. DeBlois did some important groundwork in his 1926 book, heavily influencing Heinrich. Quoting C.J. Rutland, he indicated that accidents were learning opportunities, "We cannot get the full value out of accident experience unless we are able to interpret the meaning of individual accidents, and in particular, uncover their true causes" (DeBlois, 1926a, p.46).

DeBlois clarified that, although connected, injury and accident were distinctly different, with usually distinct causes. While an injury was the consequence of the accident, the accident did not need to end up in an injury. This thought reappeared in Heinrich's writing, most importantly in the triangle. DeBlois also displayed a progressive view on multi-causality:

> Most accidents are not the result of a single, well-defined cause but of a train of events or combination of circumstances each of which contributes in some degree to cause the final accident and consequent injury. Many superficially simple industrial accidents arise out of a highly involved network of condition and circumstance. They appear simple solely because we do not make the effort to trace the causative relationship to its source and are content merely with what is termed *proximate cause.*
>
> (DeBlois, 1926a, p.47)[3]

6.1.2 Real causes?

In his early work, Heinrich frequently spoke of "true" or "real" causes. This may strike as odd, especially when coming from a constructivist view (Le Coze, 2012). Hollnagel calls "root" causes, "an oversimplification, since a cause is an attribution after the fact or a judgement in hindsight, rather than an unequivocal fact" (Hollnagel, 2000, p.40). Davies et al. argue that causality is a property of the human mind, not a property of the natural world,

> Causes, events and consequences are thus not properties of the universe but properties of people; a moveable feast whose starting point is governed not by any physical reality but by the pragmatics of problem solving. The *now* event is selected because it is the thing that strikes us as the most salient from our own subjective viewpoint; the *cause* is so called because it represents the point at which we feel action is necessary, possible and/or affordable; the *consequence* is an entirely evaluative dimension (did we like it or not?) on the basis of which we decide whether we want to change things or not in order to prevent recurrence. In other words, the tripartite division of the world in causes, events and consequences is a motivated act of construction that helps us deal with things, and which we impose on a continuous and undifferentiated process of 'what led to what' which otherwise goes back to the birth of the universe.
>
> (Davies et al., 2003, p.34)

According to Rasmussen, causes are constructed, not found. They depend on one's "categorization of human observations and choices" (Rasmussen,

1990, p.451). Hollnagel (2004) says that causation is inferred from observation but cannot be observed directly. The cause is constructed from understanding the situation, rather than found. Even more, a cause is selected from a set of possible causes, so it is rather the result of an act of inference rather than an act of deduction.

Through choosing causes, one creates a model of how one thinks the accident happened. In this process one most likely includes unspoken and implicit assumptions about what is true or taken for granted. For example, when dropping a ball, one may select "letting go of the ball" as cause, while not mentioning gravity as this is taken for granted. Because of the stance that causes are constructed, the notion of a "real" or "true" cause must be rejected. From a constructivist view, no such thing can exist, because a construct is subjective and therefore there cannot be just one truth.

This is akin to Heinrich's wording in the early papers and the 1931 edition of *Industrial Accident Prevention*, e.g. "we find almost invariably that accidents are assigned to...," "...which results in the assignment of accidents to...," "...when an accident is clearly chargeable to...," "...should therefore be charged..." (HWH1928b, p.122–126), and many more. Whether this choice of language was a conscious constructivist approach or something else will not be explored here. His language became more deterministic in the 1941 edition.

A good example of causes being constructs rather than "real" phenomena is found in Mowery's article in *Safety Engineering* (1915). Where most of his contemporaries attributed 80% of all injuries to human carelessness, the author reached a radically different conclusion and put the direct causes of these events in the context of the situation.

6.1.3 Why did "real cause" make sense?

There are two explanations for use of terms like "real" or "true" cause throughout Heinrich's writings. His early work used the terms "real cause" and "true cause" as opposed to what Heinrich labelled "so-called causes" (HWH1928a, p.10). Heinrich's paper on the *Relation of Accident Statistics to Industrial Accident Prevention* was in this respect very illustrative, "In short, the so-called cause-of-accident code is not a cause code at all. The title is misleading. Nor are there any other codes that deal with actual accident codes" (HWH1929c, p.172).

One important point from the *Origin of Accidents* articles (HWH1928a, b) is the difference between the event and its cause(s), and between the cause of an accident and the cause of the injury. In Heinrich's opinion, accident prevention of his time was "inaccurate and partially effective practice" (HWH1928b, p.131), because many safety practitioners (and others) muddled up these things.[4] Heinrich therefore distinguished between "basic causes" (in the book renamed to "real causes") and "so-called causes."

The latter, assigned by Heinrich's contemporaries, were often descriptions of the event or mechanism, like "hit by" or "fall from." One finds such classifications in many safety texts and statistics of the first decades of the twentieth century.

So, while it may look strange from a constructivist point of view to talk about "true" causes, Heinrich's intention was to distinguish his way of attributing cause from the commonly used categories. With that intention in mind, introducing a "new view," it made sense to talk about "real" or "true" causes.

Heinrich kept this use of the terms "real" and "true" cause up until the first edition of *Industrial Accident Prevention*. The examples discussed on pages 50 to 54 of this book showed "incorrect analysis" and "incorrect cause," naming the event ("sliver in finger") or consequence ("sprain"), according to the "existing inaccurate and partially effective practice" (HWH1931a, p.50) as opposed to the way recommended by Heinrich.[5]

Heinrich would maintain the argument against in his eyes "incorrect" causes, "...so-called 'accident-cause codes' have not really been *cause* codes at all, but have merely indicated accident *types* or the agencies involved in the occurrence of accidents" (HWH1941, p.330). However, in his later work, he would no longer refer to his way as "true" or "real."[6] At some point in time he started using these terms for something else, without explaining that or why he did so.

The terms "real" and "true" cause appeared much less frequently in later editions of *Industrial Accident Prevention*. Compared to the first edition, the term reappeared relatively late with only few mentions of "real" causes early in the second edition, indicating that they were "specific" (HWH1941, p.4). The term "real" or "true" cause was discussed strictly in relation to remedy and corrective action. Heinrich argued that a reversal of cause would provide a useful measure to prevent reoccurrence. For example, when someone slipped and fell and a slippery floor was chosen as causal factor, then reversing cause would provide guidance for corrective action: a non-slippery floor. Like many other things in Heinrich's work, this line of reasoning appealed to common sense and might work for many situations. However, it is not without problems, e.g. by opening for counterfactual reasoning.[7]

Apparently, the later use was merely a "label" that Heinrich had used before, changing the former meaning as a contrast to other practices to a new meaning where "real" or "true" causes were the ones providing a remedy when reversing them, while other causes did not provide that explicit guidance. Partly, this explanation was also applicable for the earlier use of "real" and "true" cause, because the "so-called" causes of his time (e.g. fall from height) rarely provided guidance. This new use of the term did not come entirely out of the blue. Already in his earlier use of the terms, Heinrich hinted at the fact that causes should give concrete clues of what to do, "Basic causes must be the guide for prevention" (HWH1928b, p.137).

"True" cause in the later meaning makes some sense. Take for example the often-heard conclusion that an accident happened because of "a poor safety culture." In general, these conclusions ("causes") do not give concrete clues to useful actions and are merely a substitution of one term or label for another (Dekker & Hollnagel, 2004). Pointing towards more specific factors (e.g. conflicting objectives, scarce resources, power issues) – "true causes" in Heinrich's terminology – might actually give useful suggestions.

6.2 Dualism of causation

6.2.1 Origin of accidents

While Heinrich's breakthrough came with the indirect cost of accidents, one could argue that 1928 was his defining year when he presented several of his most important ideas to the world in the *Origin of Accidents* papers. Prepared in March, and published in June and July 1928, these papers presented the results of new research by Travelers engineers led by Heinrich.

Heinrich did not tell what initiated this research. One possibility is the desire to find "real" causes of accidents and the greatest contributor. This is supported by how he started his discussion, suggesting that "physical hazards are becoming less and less of a real factor" (HWH1928b, p.126) but no research was available to support this assessment. Another possibility (not mutually exclusive and suggested by the text as well) is that the research was initiated to look into the preventability of accidents. After all, the first result that Heinrich presented in both the text and the chart, was the percentage of preventable accidents.

Heinrich's description was not extensive but gave clues about the research (HWH1928b, c, 1931a). His team drew 12,000 random closed-claim cases from the Travelers files. These covered a wide variety of industry and territory. This data was supplemented with 63,000 cases from the records of plant owners. Analysing these 75,000 cases, the actuarial records, engineering reports and with cooperation of employers, Heinrich's team determined that 98% of the accidents were of a preventable kind. This was the first conclusion from the study – most accidents could have been prevented. The remaining 2% would later be labelled as "Acts of God," although Heinrich himself rarely used this term.

This first conclusion was encouraging for prevention work, but the second would have a greater impact. From the case descriptions the Travelers engineers determined causes. Heinrich acknowledged that this was not easy, "the existing system of reporting accidents has not been so planned as to provide reliable data upon real causes" (HWH1928c, p.129). Causes were divided in two main categories: supervisory and physical. Of these two categories *only one was allowed* in this research. Heinrich described how they "corrected" the numbers and reduced an initial 25% of physical causes to 10%. If they came across a case where both causes were possible, they determined the *"chief"* cause of the two. Heinrich later commented,

> Admittedly, judgement must be used in selecting the major cause when a mechanical hazard and an unsafe act both contribute to an accident occurrence. Personal judgement may lead to error, but it is defensible and necessary and in the majority of cases results in fair conclusions.
>
> (HWH1941, p.20)

The research thus concluded with "a total of 88 per cent of all industrial accidents that can be prevented through the enforcement of proper supervision" (HWH1928b, p.126). The remaining 10% was attributed to "physical or mechanical causes." The results were graphically displayed in a flowchart (Figure 6.1), displaying the total amount of accidents, of which 98% was preventable, on the top. This total was split over the two main causal categories, supervisory and physical which then flowed to the control: primarily by the employer and top management, and secondary by employees. Heinrich emphasised at that both main causal categories were "within the power of the employer to remedy" (HWH1928b, p.137) especially because one could do so "by the exercise of the very methods that make for economy, efficiency, greater production, and greater profits" (HWH1928b, p.130).

The ideas presented in the *Origin of Accidents* papers were not new. Similar messages were found in earlier safety literature:

- Tolman and Kendall divided the causes of accidents in "two great classes"[8]: namely those related to machines and "those touching the human factor" (1913, p.194).
- Jesse Bird's biography mentions a 1919 *The Travelers Protection* article about an analysis of over 185,000 accidents by the U.S. Steel Corporation claiming "90% of accidents preventable" (Bird, 1976).
- Hubbard (1921) suggested that most accidents were preventable, and as we saw, he also offered a division similar to Heinrich with a clear emphasis on man-failure.
- DeBlois (1926a) did not have a dualism of causes in his accident sequence, but he seemed to acknowledge men and machine as two main categories, however without giving this any special attention.[9]
- Chaney described clearly a duality of causes, but was reluctant to discuss the human factor, because "subjective causes are obscure, difficult to observe and still more difficult to measure" (Chaney, 1926, p.41).

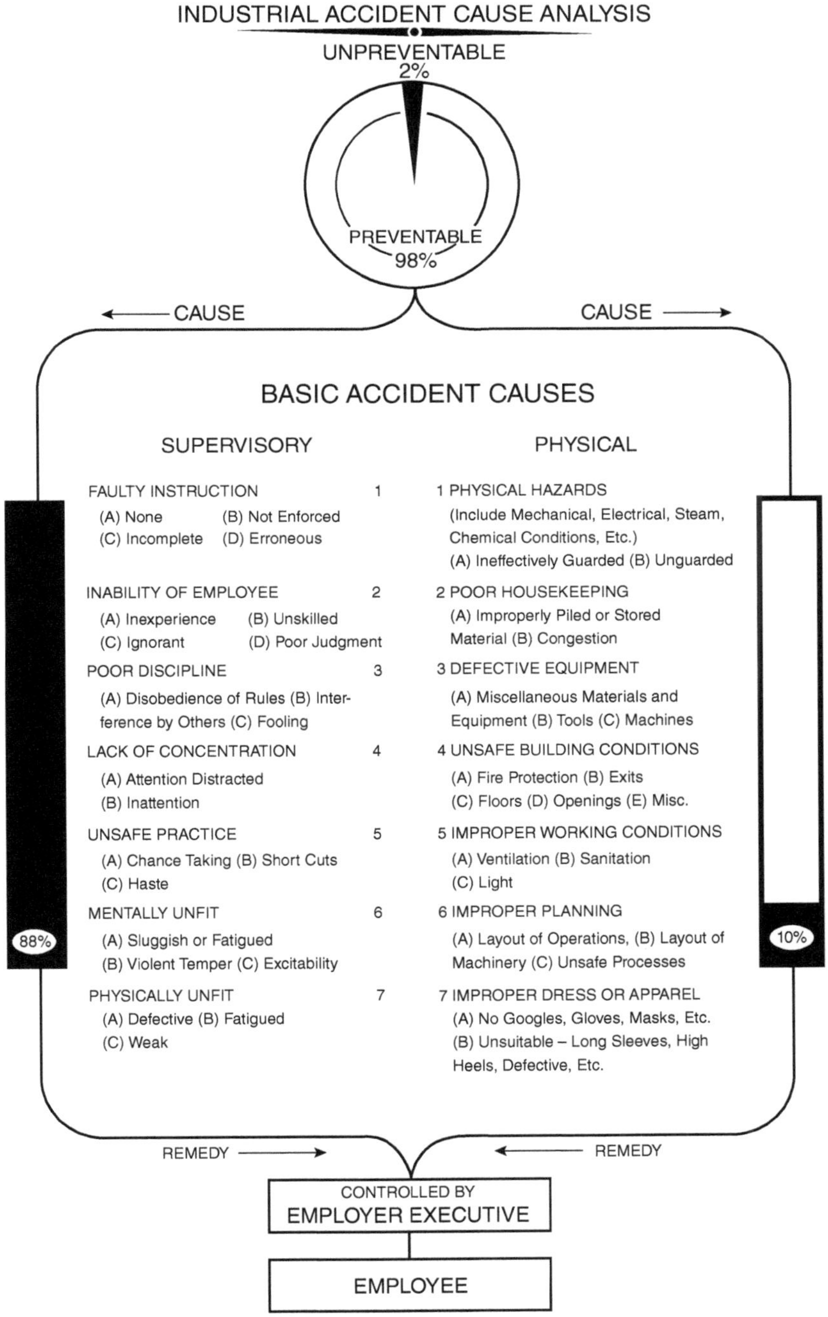

INDUSTRIAL ACCIDENT CAUSE ANALYSIS

UNPREVENTABLE
2%

PREVENTABLE
98%

◄──── CAUSE   CAUSE ────►

BASIC ACCIDENT CAUSES

SUPERVISORY	PHYSICAL
FAULTY INSTRUCTION 1 (A) None (B) Not Enforced (C) Incomplete (D) Erroneous	1 PHYSICAL HAZARDS (Include Mechanical, Electrical, Steam, Chemical Conditions, Etc.) (A) Ineffectively Guarded (B) Unguarded
INABILITY OF EMPLOYEE 2 (A) Inexperience (B) Unskilled (C) Ignorant (D) Poor Judgment	2 POOR HOUSEKEEPING (A) Improperly Piled or Stored Material (B) Congestion
POOR DISCIPLINE 3 (A) Disobedience of Rules (B) Inter- ference by Others (C) Fooling	3 DEFECTIVE EQUIPMENT (A) Miscellaneous Materials and Equipment (B) Tools (C) Machines
LACK OF CONCENTRATION 4 (A) Attention Distracted (B) Inattention	4 UNSAFE BUILDING CONDITIONS (A) Fire Protection (B) Exits (C) Floors (D) Openings (E) Misc.
UNSAFE PRACTICE 5 (A) Chance Taking (B) Short Cuts (C) Haste	5 IMPROPER WORKING CONDITIONS (A) Ventilation (B) Sanitation (C) Light
MENTALLY UNFIT 6 (A) Sluggish or Fatigued (B) Violent Temper (C) Excitability	6 IMPROPER PLANNING (A) Layout of Operations, (B) Layout of Machinery (C) Unsafe Processes
PHYSICALLY UNFIT 7 (A) Defective (B) Fatigued (C) Weak	7 IMPROPER DRESS OR APPAREL (A) No Googles, Gloves, Masks, Etc. (B) Unsuitable – Long Sleeves, High Heels, Defective, Etc.

88%   10%

REMEDY ────►   ◄──── REMEDY

CONTROLLED BY
EMPLOYER EXECUTIVE

EMPLOYEE

Figure 6.1 Origin of Accidents chart.
Source: Travelers Standard XVI, June 1928. Used with permission.

6.2.2 Developments

Heinrich's 1928 papers were the basis for a section in *Industrial Accident Prevention*, including the chart. The text shed some light on why Heinrich chose to emphasise the human element over the mechanical hazard,

> Mechanization of industry created new machines and new dangers. At first, these additional hazards received little attention, and the greater number of industrial accidents were caused by the use of unguarded machine equipment. Under these conditions it was perfectly proper to charge accidents to specifically named machines, parts of machines, or mechanical equipment, and to state that these things were the actual causes of accidents.
>
> (HWH1931a, p.43)

In the meantime, much had changed, however. Improvement in guarding and environment had increasingly been introduced and therefore "man-failure became the predominant cause of injury" (HWH1931a, p.43).

For the second edition of *Industrial Accident Prevention*, Heinrich made some significant changes. One was the renaming of the two main categories and the underlying categories in the chart. It is surprising that the percentages were unchanged given these drastic changes. Also, Heinrich did not comment the changes specifically and one can only guess the reasoning behind. As suggested in Chapter 3, this may have been a matter of coherence, along with possible political reasons.

Heinrich changed the flow in the chart's new version (Figure 6.2). It now started with management which controlled "man-failure" which caused or permitted "unsafe acts" or "unsafe conditions" which then caused accidents, 98% of which were "of a preventable type" and about half of these "practicably preventable."

Besides changing, the figure moved. In the 1931 book, it had a central place in Chapter III, *Accident Cause-Analysis* (HWH1931a, p.46). For the second edition, it was part of *Section 3 – Man vs. Machine* in Chapter II (HWH1941, p.18), where it stayed for future editions. The positioning under this header may give a clue of the greater focus on the human in the 1941 book.

The second half of Section 3 was largely made up of material from the 1933[10] paper *Mastery of Machine*, and it may have been elements from this paper that influenced strongly Heinrich's thinking. He wrote, "...we do not die nor are we hurt, because of machine fault" (HWH1933a, p.9). It seems that Heinrich realised that machines in themselves do not do a thing. To some degree we find similar reasoning in Chase's book (1929). It was man who made them, and it was man who used them.

In Section 3, Heinrich also discussed other studies, by the National Safety Council and the State of Pennsylvania. These had found roughly equal parts for acts and conditions. The differing ratios were mostly because these other

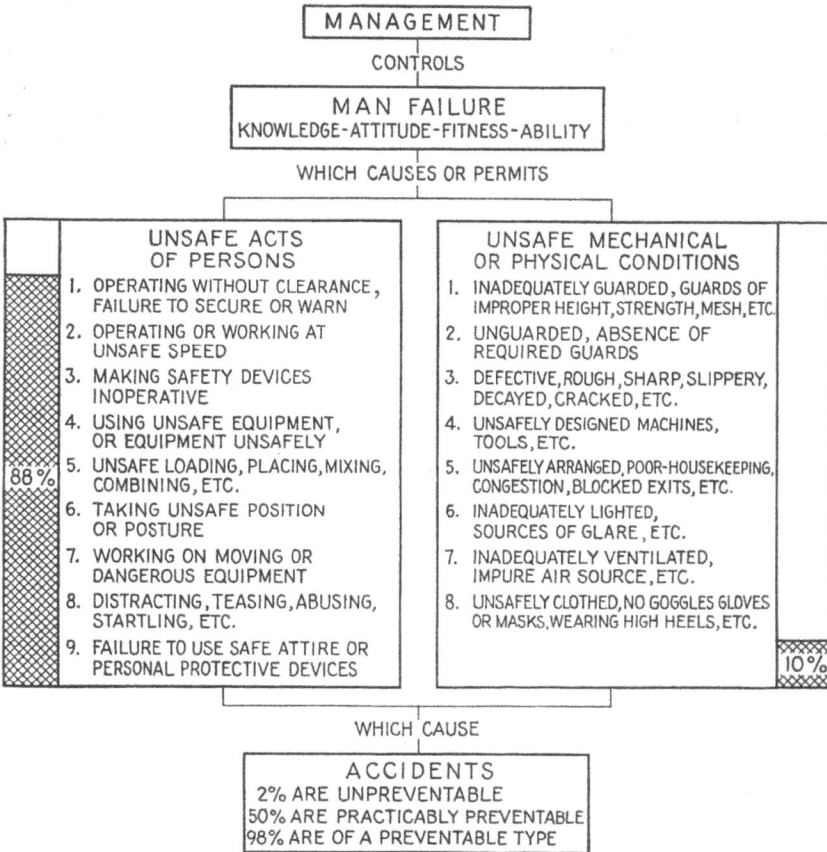

```
                    ┌─────────────────────┐
                    │     MANAGEMENT      │
                    └─────────────────────┘
                            CONTROLS
              ┌───────────────────────────────────┐
              │           MAN FAILURE             │
              │ KNOWLEDGE-ATTITUDE-FITNESS-ABILITY│
              └───────────────────────────────────┘
                     WHICH CAUSES OR PERMITS
```

	UNSAFE ACTS OF PERSONS	UNSAFE MECHANICAL OR PHYSICAL CONDITIONS	
	1. OPERATING WITHOUT CLEARANCE, FAILURE TO SECURE OR WARN	1. INADEQUATELY GUARDED, GUARDS OF IMPROPER HEIGHT, STRENGTH, MESH, ETC.	
	2. OPERATING OR WORKING AT UNSAFE SPEED	2. UNGUARDED, ABSENCE OF REQUIRED GUARDS	
	3. MAKING SAFETY DEVICES INOPERATIVE	3. DEFECTIVE, ROUGH, SHARP, SLIPPERY, DECAYED, CRACKED, ETC.	
	4. USING UNSAFE EQUIPMENT, OR EQUIPMENT UNSAFELY	4. UNSAFELY DESIGNED MACHINES, TOOLS, ETC.	
88%	5. UNSAFE LOADING, PLACING, MIXING, COMBINING, ETC.	5. UNSAFELY ARRANGED, POOR-HOUSEKEEPING, CONGESTION, BLOCKED EXITS, ETC.	
	6. TAKING UNSAFE POSITION OR POSTURE	6. INADEQUATELY LIGHTED, SOURCES OF GLARE, ETC.	
	7. WORKING ON MOVING OR DANGEROUS EQUIPMENT	7. INADEQUATELY VENTILATED, IMPURE AIR SOURCE, ETC.	
	8. DISTRACTING, TEASING, ABUSING, STARTLING, ETC.	8. UNSAFELY CLOTHED, NO GOGGLES GLOVES OR MASKS, WEARING HIGH HEELS, ETC.	
	9. FAILURE TO USE SAFE ATTIRE OR PERSONAL PROTECTIVE DEVICES		10%

```
                        WHICH CAUSE
              ┌───────────────────────────────────┐
              │            ACCIDENTS              │
              │ 2% ARE UNPREVENTABLE             │
              │ 50% ARE PRACTICABLY PREVENTABLE  │
              │ 98% ARE OF A PREVENTABLE TYPE    │
              └───────────────────────────────────┘
```

Figure 6.2 The 1941 *Origin of Accidents* chart (HWH1941, p.19).

studies allowed for multiple causes per case. Heinrich acknowledged that "in the majority of cases there is both a personal and a mechanical contributing cause" and that "both kinds of causes should be eliminated as far as is practicable" (HWH1941, p.21). Still, it was his opinion that one of both often was "more directly responsible than another" (HWH1941, p.21).

At this point, it seemed that the concept had reached its maturity and for the 1950 edition of *Industrial Accident Prevention* no further changes were made. Apart from a slight graphic overhaul, some rephrasing and one added category, regarding unsafe processes (chemical, nuclear, etc.), under conditions, also the 1959 version was mostly the same. The main addition in 1959 to Section 3 was a summary of the Heinrich/Blake debate.

A variation of the *Origin of Accidents* chart was published between the books as a 1954 Travelers brochure (Figure 6.3). This pictured "Management and Supervision" as controlling two different streams – one "Employee Performance" and the other "Physical Environment" – including some ways of exercising that control as selection, training, indoctrination, and safety organisation on the employee side, and methods, procedures, design, planning, and safeguards on the other. Interestingly, it then says that employee performance and physical environment "control occurrence of substandard situations" regarding unsafe acts and conditions.[11] As before, these unsafe acts and conditions caused or permitted accidents to happen. A note under the chart cautioned about the presence of additional causes.

6.2.3 Cause code

Heinrich's duality of causes would be highly influential. It would influence his domino model and the various investigation methods that came from this, as well as accident standards as the ANSI 1941 and providing a basic structure for many future accident report forms. The 1954 Travelers brochure tells about the *Origin of Accidents* study leading directly to the Heinrich Cause code (Travelers, 1954, p.3).

Heinrich had been vocal in the need for better distinguishing cause, accident and injury, and improving causal data for statistics (e.g. 1929c, 1930c, 1931a, 1932c, 1935). In the mid-1930s, he was invited to participate in and lead a sub-committee of the American Standards Association to provide a standard accident cause code. Heinrich firmly put his stamp on the work. He "furnished the basic rubrics" (Kossoris, 1939, p.526) and looking back, one finds the basic approach for the cause code already in the first edition of *Industrial Accident Prevention*.

During his address at the Twenty-fourth Annual Safety Congress in October 1935, Heinrich already presented a template for accident reports, developing the relatively simple "Cause of accident" (HWH1931a, p.236) into two separate points: "Fault of machine or machine part" and "Fault of persons – unsafe acts" (HWH1935, p.31). Both were elaborated upon with sub-categories. Heinrich noted,

> ...the two items of most significance are fault of machine and fault of persons. Safety engineers should bear in mind that no preventable accident has ever occurred or can ever occur unless one or both of these two conditions are present.
>
> (HWH1935, p.31)

In 1937, the sub-committee delivered a provisional draft of what was "popularly known as the 'Heinrich Cause Code'" (Kossoris, 1939, p.526). The underlying philosophy of the code was that industrial accidents were

```
                    ┌─────────────────────┐
                    │     MANAGEMENT      │
                    │         AND         │
                    │     SUPERVISION     │
                    └─────────────────────┘
                            CONTROL
```

EMPLOYEE PERFORMANCE	PHYSICAL ENVIRONMENT
Selection, indoctrination, training, instruction and placement, Safety organization and morale.	Structures, machines, tools and equipment. Methods and procedures, planning, arrangement, design, safeguards, and maintenance.

**WHICH CONTROL OCCURRENCE
OF SUBSTANDARD SITUATIONS**

88%

UNSAFE ACTS OF PERSONS	UNSAFE PHYSICAL CONDITIONS
1. Operating without clearance, failure to secure or warn.	1. Unguarded tools, machines, equipment and mechanical exposures in general.
2. Unsafe work method in handling materials, operating, loading, feeding, mixing, combining, etc.	2. Inadequalety guarded, guards of improper size, height, mesh, strength, meterial, not locked or interlocked, etc.
3. Using unsafe equipment or using equipment unsafely.	3. Defective tools, structures and equipment, broken, splintered, sharp, slippery, cracked, etc.
4. Tampering with safety devices.	4. Unsafely designed machines or tools.
5. Operating or working at unsafe speed	5. Unsafe or inadequate equipment, for materials handling, unsafe processes and procedures.
6. Exposing self unnecessarily, toxing unsafe grip, position or posture, failing to look or observe.	6. Unsafe arrangment, poor housekeeping, congestion, blocked exits, aisles, etc.
7. Oiling, adjusting machinery or equipment while in motion.	7. Inadequale lighting, glare, etc.
8. Distracting, teasing, startling, abusing, quarreling, practical joking, etc.	8. Inadequale ventilation orcontrols for dusts, fumes, vapors, radiations and noise
9. Failure to use personal protective devices.	9. Unsafe dress, long sleeves, cuffs, high heels, etc.
10. N.E.C.	10. N.E.C.

10%

**WHICH CAUSE OR PERMIT
THE OCCURRENCE OF**

ACCIDENTS
98% are of a preventable kind

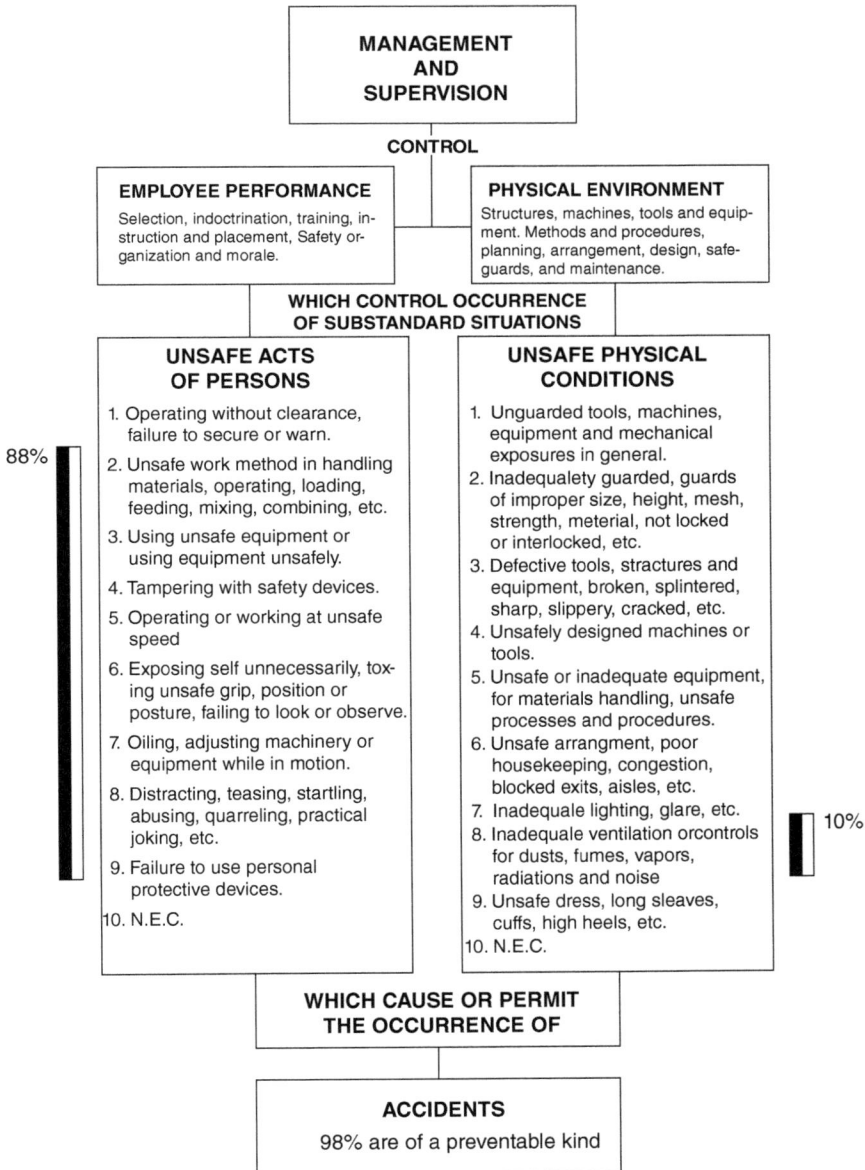

CHART OF DIRECT AND INDIRECT ACCIDENT CAUSES

NOTE – The items shown are broad general groupings within which, additional unsafe acts, unsafe physical conditions and indirect causes may be considered.

Figure 6.3 The 1954 *Origin of Accidents* chart.
Source: Travelers Insurance, 1954. Used with permission.

due to unsafe conditions and unsafe practices, "the classification of accident factors is not intended to deal with obscure causative factors, or factors too far removed in the accident sequence to be definitely ascertainable" (Kossoris, 1939, p.527). Preventing unsafe conditions and practices was assumed to prevent recurrence of similar accidents. Six factors were to be recorded: the agency, the agency part, the kind of accident, the unsafe act, the reason for that unsafe act, and the unsafe condition.

The information was expected to provide safety engineers with data to guide accident prevention efforts. "It should go a long way towards meeting the growing and insistent demand for more and better factual data of value in the prevention of industrial accidents" (American Statistical Association, 1937, p.716).

The draft was published for trial use, for example by The Department of Labor and Industry of Pennsylvania which started using the new standard from 1 January 1938. In 1939 a new committee was assembled to prepare a final draft based on feedback received. The new standard was approved on 1 August 1941 as ANSI Z16 and would be used in that form for at least two decades (American Standards Association, 1941a, b). Later, Heinrich used the same principles to propose a Boiler Cause Code, found in the 1950 and 1959 editions of *Industrial Accident Prevention*.

6.2.4 Critique

The belief that most accidents are caused by humans had been around since the earliest days of the safety profession and proves to be very persistent and prevalent among safety practitioners. It is often used as one of the rationales for BBS approaches and made its way into many safety documents ever since, as illustrated by publications as the *Dirty Dozen* flyer (FAA, 2012) or a recent report from a Health and Safety Inspectorate,

> Analysis of fatal occupational accidents shows that human error and behaviour play an important role; errors which often could have been prevented, if employees and management within the organisation had better safety awareness.
>
> (I-SZW, 2017, p.4)

As we saw earlier, this belief has also seen serious critique since the early days, for example Eastman provided a strong argument to disprove the stance that 95% of all accidents were due to "carelessness" of the workers. In her time (pre-worker's compensation regulations) blaming the workers was a powerful tool for employers to deflect liability. DeBlois was another who was critical of "the oft-repeated statement that 75 to 90% of industrial accidents are the fault of the men themselves." He underlined this critique by concluding, "...the statement as usually made is not only fallacious but, by causing resentment on the part of those whom it is intended to correct, has hindered rather than hastened the progress of the safety movement" (DeBlois, 1926a, p.60).

While Heinrich merely intended to point towards the "most important" cause without blame in thought, Eastman and DeBlois already pointed out some of the problems and possible misunderstandings, namely that this thinking might lead to blame and misdirection of safety efforts. The successful 88:10:2 ratio contributed to spreading this thinking. After all, it seemed to provide a scientific backing for an existing belief.

Over the years, the ratio drew more and serious critique. One remarkable occasion was the public debate on the 88:10:2 ratio between Blake and Heinrich, published in *National Safety News*, May 1956. The four-page pro and con article presented both sides. Blake opened his critique, "In my opinion the very concept of a ratio between physical hazard and unsafe action in injury causation is both fallacious and harmful" (Heinrich & Blake, 1956, p.19). He explained his position by proposing that an injury[12] was impossible without a hazard present. He therefore quipped, "Use a 100-to-100-ratio if you must have a ratio, for that and only that is valid" (Heinrich & Blake, 1956, p.19). Furthermore, he argued how the belief in man-failure as the main cause for most accidents served as an excuse to avoid expensive machine guarding by rather focussing on cheap actions aimed at worker behaviour and awareness. He advised, "Maintain a thorough and painstaking joint program of hazard elimination and of safe and adequate behaviour development" (Heinrich & Blake, 1956, p.22).

Heinrich opened pointing out the problem of possible abuse of *any* opinion about causation by those in power to decide about preventive actions. Instead,

> ...first consideration invariably should be given to the possibilities of correction that lie in guarding, relocating, rearranging, automation or other forms of engineering revision, regardless[13] of whether the primary cause is man or machine.
>
> (Heinrich & Blake, 1956, p.18)

Heinrich then proceeded by explaining that the "often quoted and misquoted ratio" was about primary causation, which should guide corrective action, but should not prohibit giving priority to actions on "subcauses." According to Heinrich, knowledge of "all correctible causes, both primary and secondary, both personal and mechanical" was necessary. Heinrich was sceptical about using the ratio as an excuse for not taking necessary action, and besides, he remarked, "...one who evades responsibility on such a basis would readily find some other available excuse"[14] (Heinrich & Blake, 1956, p.20).

Regardless of debates and reflections like these, the 88:10:2 ratio proved to be resilient and became in one form or another part of the safety curriculum. Half a century after the Heinrich/Blake debate, Manuele remarked, "Of all Heinrich's concepts, his thoughts on accident causation, expressed as the 88-10-2 ratios, have had a significant effect on the practice of safety, and have resulted in the most misdirection" (Manuele, 2011b, p.55).

He repeated Blake's point: the belief in human failure as the main factor in accidents misdirects (intentionally, or not) safety effort towards the sharp end rather than trying to work on organisational or system issues. This means that actions are often sub-optimal, and blame is attributed to the weakest part, getting management "off the hook."

As mentioned above, this was not Heinrich's intention, repeatedly emphasising that changing the environment should be the first approach for corrective action. Also, he was clear that the first responsibility rested with management. Still, many interpreted Heinrich's legacy the way Manuele and Blake cautioned of.

6.2.5 Dismantling the "grand statement"

Just as Manuele, Todd Conklin points at harm done by the 88:10:2 ratio,

> This idea causes much harm to understanding actual catastrophic failure. To be fair, Heinrich did go on to tell employers that the managing of the workplace hazards was more effective than managing worker decisions and behaviours, but after the grand statement that almost all of the problem (88%) was of the workers doing, you can't really blame management for going after worker behaviour.
>
> (Conklin, 2017a, p.61)

This sounds like a plausible assessment and this section offers four elements to defuse the "grand statement."

6.2.5.1 Direct and proximate causes

As Conklin suggests, some of Heinrich's message has gone missing in translation or been drowned out by other parts of the message. One may speculate on some factors. Attributing accidents to behaviour and going for relatively cheap and superficially effective remedies as retraining, reinstructing, and creating a new rule may be an attractive approach for some stakeholders. Another factor may be that most people do not read very well, or selectively.[15] Not only did Heinrich's recommendation for prioritising actions get lost, another element seemed also to be missed. After all, Heinrich spoke of *direct* causes.

This detail is relevant for assessing the value of the ratio. In recent decades, safety theory and practice has emphasised the importance of underlying causes (e.g. Petersen, 1971; Bird & Germain, 1992; Manuele, 2002). Petersen and others said that by focussing on unsafe acts and conditions one ended up dealing with accidents on a symptomatic rather than a causal level. Besides, as we will see, also Heinrich emphasised the need to look further upstream for the reasons of unsafe behaviour and not get stuck on the behaviour itself.

These thoughts about direct cause had been around for a long time. Boyd Fisher, for example, wrote that proximate causes were merely "apparent causes." These were the most easily visible, and thus quickly appointed as "the" cause. However, things were caused by an infinite number of things. For something to happen, many external, underlying, and fundamental issues were necessary and eventually, also the proximate, "Viewing the causation of accidents the other way round, we shall see that none of the fundamental mental causes operates, in itself, to cause an accident" (Fisher, 1922, p.21).

6.2.5.2 Biased data

The first element weakens the relevance of the ratio for contemporary safety thinking. Additionally, one can point towards several methodical problems with the numbers presented by Heinrich.[16] Even though we cannot verify for sure, because the unavailability of Heinrich's data, we can be rather sure that his data was biased. After all, a large part of the material consisted of insurance claims. Almost anyone filing a claim will be inclined to colour the information in one direction or another. This most likely influenced the results of the study. The same applies to the employer records which were used. Blake offered this argument as part of his critique:

> [I]t was gained partly from injury reports made for the purpose of de-termining the compensation due and partly 'from the records of plant owners.' But except where such reports are based on thorough investiga-tion by competent personnel, the information as to cause is limited and frequently unreliable. In other words, the analysts going over the 75,000 accident reports covered by this study had to do a lot of guessing.
>
> (Heinrich & Blake, 1956, p.19)

6.2.5.3 Allowing only one

In contrary to other, similar studies, Heinrich's study was based on the an-alytical choice to allow for *only one* (direct) cause per reviewed case. Where more than one cause was mentioned or possible, the Travelers engineers chose the cause that they considered the most important.

Question is: what makes a cause the "chief" cause? Is it the cause with the largest contribution to causality? But how could one determine this? When all factors are necessary, as Fisher suggested above, how could you select a chief cause? Heinrich did not describe the method followed clearly[17] and he admitted that applying judgement was not without problems, "Personal judgement may lead to error, but it is defensible and necessary and in the majority of cases results in fair conclusions" (HWH1941, p.20). While some cases may have seemed clear-cut, others must have been either biased (the researchers did have a hypothesis in mind that most accidents were attributable to man-failure) or arbitrary.

Before Heinrich published the ratio, several other safety authors had spoken critically about choosing a single cause. Lange offered critique about the habit of some safety professionals to only tabulate one cause per accident in statistics. He thought it "…perfectly clear that such figures are very misleading" (Lange, 1926, p.23) critiquing those who claimed certain percentages about human or machine contributions. Lange suggested that multi-causal representation was the only way to get useful information, "Only by taking all factors into consideration on each accident can we know the relative importance of the main hazard groups" (Lange, 1926, p.73).

DeBlois was very critical of single causation and single responsibility as well:

> …the simplest way out of the tangle of causes and circumstances that are brought to light by thorough accident investigation lies in the selection of a single cause and a single responsibility for each accident, and this greatly simplifies any ensuing statistical procedure, but since accidents occur as the result of combinations of circumstances or causes, the responsibility for all of which can rarely be attributed with any fairness to a single individual, any tendency to fix the blame on one man or regard responsibility as analytically indivisible will ultimately cloud rather than clarify the issue.
>
> (DeBlois, 1926a, p.59, 60)

6.2.5.4 Causes are constructs

As discussed earlier in this chapter, causes should be regarded as constructs since they depend on the analyst's choices. That also means that causes are not objectively countable objects, but products of our minds to make sense of certain situations and circumstances. That makes them subjective and uncountable. Choosing to count them anyway, the result will be a number that is dubious because it depends on the choices made of how to count them.

Heinrich's work even serves as an example for these statements. The bottom categories in the *Origin of Accidents* chart are "Failure to use safe attire or personal protective devices" on the "unsafe act" side and "Unsafely clothed, no goggles, gloves or masks, wearing high heels, etc." on the "unsafe condition" side. The differences between these two appear to be minimal and whether a case was counted in one or the other may have been a matter of personal preference. Even more is Heinrich's changing of "supervisory causes" (HWH1931a) into "unsafe acts of persons" (HWH1941) a perfect illustration of how causes are constructs – it depends on choices what they are called.

With these four elements in mind one must conclude that the 88:10:2 ratio should be regarded as a historical artefact and as Manuele (2011b) suggests, be "dislodged" from the practice of safety.

6.2.6 Another reflection

As argued above, forcing people into an either/or choice of unsafe conditions or unsafe acts is unhelpful. As Blake and others argued, and Heinrich acknowledged, often it is rather "both/and." Acts are done in a context (conditions), acts can create conditions, and likewise conditions can create acts. While Heinrich chose to select only one cause for the *Origin of Accidents* study, he was not binary in his other work. In the examples he included to illustrate better approaches to causal analysis, he mentioned both acts and conditions, and as the next sub-chapter shows, also the domino sequence would present them both together. For good accident prevention, "Knowledge in regard to both points must be had before the remedy can be effectively applied" (HWH1931a, p.44). At one point he even explicitly stated, "...no accident, not even one which falls in the so-called unpreventable group, ever occurs unless a hazard of some kind exists" (HWH1932a, p.31). Also the cause code regarded both factors *and* the underlying reason,

> Under no circumstance is the person doing the analyzing and coding to determine whether the unsafe act or the unsafe condition is the primary cause of the accident. He should code for both factors, as well as for the personal cause if he can determine it from the report. Otherwise there is too much room for bias. And further, if both conditions exist, both need attention and therefore require indicating.
>
> (Kossoris, 1939, p.529)

There is another reflection to make. How does one determine what is "unsafe"? An act is rarely inherently unsafe.[18] The situation it is committed in can make it so. Likewise, a condition is not inherently unsafe, only the presence of a person makes it unsafe for that very person. This makes it even more difficult to separate acts and conditions. This is even so when defining "unsafe" as deviating from some standard, because rules and standards also depend on context. It becomes even more problematic because often the label is applied only after a (near) accident.

Therefore, "unsafe" can be a problematic statement, seen from a theoretical, critical perspective. For practical purposes, however, it can provide a useful shorthand, and most people manage to point out the two in everyday activities as a guideline for what they do and how they do it. As long as one avoids categorising and counting, unsafe acts and conditions can provide a tool for practical safety work. A problem, however, is that many safety programs do exactly that, for example through observation schemes.

6.3 Accidents as processes

DeBlois and Heinrich addressed the imprecise way that many of their contemporaries spoke about causes, accidents, and outcomes. "The words accident and injury are loosely and often incorrectly used. They are by no means synonymous" (DeBlois, 1926a, p.46). To resolve this, they offered definitions, and more importantly, a model that saw accidents as a process, as a sequence of before (causes), during (accident), and after (outcome).

Heinrich's accident sequence developed over a couple of papers. The first time appearing in the *Origin of Accident* papers (HWH1928a, b), "...the terms 'injury' and 'accident' and 'cause' and 'type' are confused. When a person is injured, the sequence of events is as follows, first, the cause, second, the accident, third, the injury" (HWH1928b, p.9). The accompanying chart (Figure 6.1) also pictured kind of a causal chain. However, this was arguably clearer in the revised 1941 version (Figure 6.2) because this presented a continuous logical flow of various causes leading to accidents.

The sequence appeared in passing in the *Foundation of a Major Injury* papers, "the accident and injury are distinct occurrences" (HWH1929a, p.2), Heinrich stated, making this a premise for his triangle. The early sequence was gradually expanded to a sequence of events of which the unsafe acts and unsafe conditions were essential elements.

The 1931 book was the first time he explicitly connected upstream factors to the sequence, stating that "Psychology lies at the root of sequence of accident causes" (HWH1931a, p.127). However, Heinrich did not substantiate this assertion, nor would psychology spring out in the development.

During a presentation at the Fourteenth New York Safety Congress in December 1930, Heinrich spoke about "unsafe practices and conditions." The paper from this was subtitled, "What they are – How to identify and Eliminate them." Here he clearly worked on developing the middle part of the accident sequence, notably the direct causes and their underlying causes. He suggested looking for a "secondary cause and remedy," because correcting these causes would prevent recurrence (HWH1931b, p.10). The paper[19] discussed "unsafe practices" and "unsafe conditions" in more depth than the 1931 book.

The *Safety in the Small Plant* paper showed another suggestion of the importance of underlying causes, further upstream in the sequence, "reasons why man-failure exists, are a valuable part of accident facts" (HWH1932a, p.32). Further development of the accident sequence was found the year after, "...an accident (whether it be preventable or not, from a practical viewpoint) invariably occurs only when one or both of two conditions exist - namely, (1) a physical or mechanical hazard, and (2) the unsafe act of a person" (HWH1933b, p.53). This quote revealed some of the language that accompanied one of Heinrich's best-known metaphors.

6.3.1 The dominos

During an address before the Down River Section of the Detroit Safety Council in November 1934,[20] Heinrich presented the dominos to the world:

> The occurrence of a preventable injury is the natural culmination of a series of events or circumstances. These events invariably occur in a fixed and logical order. One is dependent on another and one follows because of another, thus consisting a chain that may be likened to a row of dominoes placed on end and in such alignment in relation to another that the fall of the first domino may precipitate the fall of the entire row. An accident is merely one link in the chain.
>
> (Travelers, 1934, p.3)[21]

The dominos were one of the first accident models, and possibly the first graphic representation of an accident model.[22] It showed a complete accident sequence in five steps of which the first three represented causes:

1 Social environment and ancestry
2 Fault of person
3 Unsafe act and mechanical and/or physical hazard[23]
4 The accident
5 Injury

The dominos were a strong, appealing visualisation of the accident process. The metaphor was helpful to explain simple linear causation (Figure 6.4). In

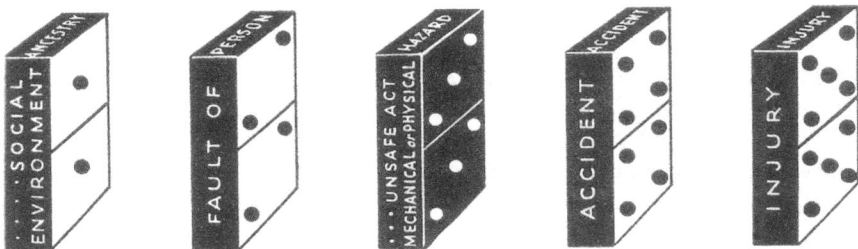

Figure 6.4 The accident sequence.
Source: Accident Sequence Brochure, 1934. Used with permission.

the brochure, and all future editions of *Industrial Accident Prevention*, Heinrich explained the five dominos. The first tile is a bit puzzling (and will return in the next chapter) because Heinrich did not explain where this came from. Previously he had claimed that psychology was at the root of the accident sequence (HWH1931a), but the dominos presented "Ancestry and social environment" as the first domino. The latter could be explained by culture and influence of context, while the former was personal, "...undesirable traits of character may be passed along through inheritance. Environment may develop undesirable traits of character or may interfere with education" (Travelers, 1934, p.2).

The explanation offered for the second tile partly overlapped with the first tile. The label "Fault of person" was one of the things that made the dominos very centred on the human element, unlike the later Bird variation. The second tile aimed at "secondary" causes but included only human secondary causes in the domino sequence. In some of his texts he was clearer than the narrow representation in the visualisation.

The last three tiles had been discussed before. The middle tile pictured direct and proximate causes, which then led to an accident and (possibly) injuries. According to Heinrich, the model enabled that "...accident prevention can be portrayed as a science and as a work that deals with facts and natural phenomenon" (Travelers, 1934, p.3). This may be the first time that Heinrich described it as a Newtonian cause-and-effect sequence.

Compared to the development of Heinrich's other main themes, the dominos are bit of an outlier. One sees a development of the accident sequence through several papers in the early 1930s, until the metaphor and description appear suddenly in a final form in 1934.[24] Then Heinrich did not mention them again until the 1941 book where most of the 1934 text returned and the dominos featured most prominently. They would remain unchanged until Lateiner's and Bird's dominos in the 1950s and 1960s.

There are two exceptions to this in Heinrich's work. In *Basics of Supervision* (HWH1944), he presented a simplified, three-tile domino sequence of "Fault of Person," "Unsafe or Improper Practice or Condition," and "Injury or Production Loss," which he named "Cause-and-effect-sequence" (HWH1944, p.144). This simplification's goal was most likely to provide foremen and supervisors with an even simpler tool for practical safety work. The sequence was expressed in two theorems: "An injury or production loss cannot possibly occur in any industrial operation, at any time or place, unless some person acts unsafely or improperly or unless some unsafe or improper condition exists" (HWH1944, p.125, 126), and "No unsafe or improper practice or condition ever occurred or can ever occur in any industrial operation, at any time or place, unless caused or permitted by the fault of some person" (HWH1944, p.128).

These theorems inspired a presentation of the accident sequence as a set of theorems in the next edition of *Industrial Accident Prevention*:

1   A personal injury only occurs as the result of an accident.
2   An accident occurs only as the result of a personal or mechanical hazard.
3   Personal and mechanical hazards exist only because of the faults of persons.
4   Faults of persons are inherited or acquired by environment (HWH1950a, p.14).

6.3.2 A powerful metaphor

The dominos provided a power metaphor, helpful for intuitive understanding of cause-and-effect, as well of thinking of accident and injury prevention. Safety practitioners quickly started using physical dominos to demonstrate the principle as part of safety instructions. A downside was that the dominos invited to a rather narrow understanding of accident processes, not in the least through the wording. The demonstration by Lieutenant Fred Lippert, director of training of the New York Navy Yard, at the 1942 National Safety Congress serves as an example. The transactions give us possibly the first description of the use of physical dominos, and also the first description of how they can be misinterpreted. After building an actual row of dominos, Lippert explained to the audience,

> First of all, you have the heredity, the environment, the background of the person concerned. Then after that you have the fault of the person himself. In other words, the person is doing something wrong by which he sets the stage a little further for the occurrence of an accident. Then the next step is the actual unsafe act or mechanical or physical hazard which exists, which further sets the stage. This unsafe physical or mechanical condition, or this unsafe act, takes place, and there is then existing what is known as an accident, a fortuitous occurrence, something unplanned for, but it happens nevertheless as the result of these unsafe conditions.
>
> (NSC, 1942a, p.80)

In this interpretation, which was not necessarily what Heinrich intended, it all seems to deal with one and the same person, making the process very compact and linear. This connects to two main points of critique of the dominos: linearity and a focus on direct causes.[25]

6.3.3 Critique 1: linearity

Heinrich's dominos proposed a model of how accidents happened. It "attempted to understand the sequential factors leading to an accident and heralded in what can be termed a period of simple sequential linear accident modelling" (Toft et al., 2012, p.2). One can find relatively early critique of such simple, linear accident investigation techniques, for example a paper, drawing on the work of Jacobs (1961): "The level of causal analysis thus represented usually does not go beyond the identification of the immediate injury-producing factors in the environment. The other causes important in the sequences eventuating in accidents are given less attention" (McFarland, 1963, p.687). Petersen recognised the problem and argued for a less narrow understanding of the dominos, suggesting multiple causation as a way to expand that view (1971). Petersen spoke of root causes, however, and was still in a fairly linear paradigm.

Contemporary critique often points out that linear models are insufficient to explain events and effects in complex, dynamic systems (e.g. Hollnagel, 2004; Dekker, 2011, 2014). In addition, many "new view" authors display a critical stance towards "causes." Hollnagel in particular has some strong arguments against the concept of root causes and the "causality credo" (Hollnagel, 2014a):

> If accidents have explanations, then we should rather try to account for how the accident took place and for what the conditions or events were that led to it. The response should not be to seek out and destroy causes, but to identify the conditions that may lead to accidents and find effective ways of controlling them.
>
> (Hollnagel, 2004, p.29)

Hopkins agrees with this view in this footnote of his *Issues in Safety Science* paper:

> Some authors (e.g. Ladkin, 2001) argue that it is possible to identify a discrete set of factors that together provide a sufficient cause. However, their analysis assumes that all other factors that might affect the outcome remain unchanged. This is a crucial limitation. It may be that the best way to prevent recurrence is not to focus on a discrete set of causes but to identify some background factor that, if changed, would prevent a recurrence.
>
> (Hopkins, 2014, p.12, f.7)

Hopkins mentions that others, like Betrand Russell, even point out that no finite set of causes can ever be regarded as sufficient. One can always add more detail to the description, which reflects the problem of the stop rule in accident analysis. How far back should one go – until Creation or the Big Bang?[26] This discussion is outside the scope of this book. Instead, let us consider why it may have made sense for Heinrich to use a simple, linear model.

6.3.4 Why did linearity and simplicity make sense?

Hollnagel, who frequently refers to the dominos as the prototypical linear accident model, offers on several occasions an explanation why it made sense to propose a simple, linear model of accidents in the early twentieth century (and why this is insufficient to explain accidents now). Industrial work in the 1920s was much more comprehensible and easy to understand – often manual, regular, stable and relatively simple, with few variables, loosely coupled, which were usually linear. This allowed simple models to be sufficient (Hollnagel, 2014a).

Today, one may regard the dominos as a simplification and too linear to reflect reality. Heinrich would partly agree, "The purpose of this discussion is to promote a more complete knowledge of the circumstances, conditions, and events preceding and following an accidental injury" (Travelers, 1934, p.6), but he had also other goals with his model/metaphor: "...to help dispel uncertainty and the ambiguity of terms, and to assist in placing the profession of safety engineering on a sound, practical and more effective basis" (Travelers, 1934, p.6).

Not only were they sufficient to grasp important factors, apparently, linear, sequential models were successful. They provided practical value and guidance to safety practitioners and managers finding measures to prevent accidents. The approach was a vast improvement over common practice at the time, "Heinrich's emphasis on separating the cause of the accident (the hazard) from the cause of the injury encouraged more sophisticated injury investigations" (Aldrich, 1997, p.152).

As such, the accident sequence (and the highly memorable dominos) represented the state of the art of the accident prevention thinking that was around at the time – even though some safety thinkers had more progressive thoughts already at the time.[27] Surely, with what we know today, Heinrich could or should have gone further, but simplicity made it useful, even though reality was much more complex.

This usefulness must be stressed. Heinrich did not merely write for safety experts and accident investigators. His main audience were those who did not work with safety as their main interest. A simple, understandable, memorable, and practical model was what these readers, often managers, needed and wanted, and Heinrich gave to them "one of the most understandable and the clearest theories defining accident processes" (Sabet et al., 2013, p.73). It was understandable because of the "easy visual representation of the 'path' of causal development leading to an accident" (Toft et al., 2012, p.3) and therefore suitable for "those who are not specialists in safety" (Sabet et al., 2013, p.75).[28] The model also guided the determination of solutions. It was a problem-solving tool.[29]

Thinking in linear causal sequences appeals to "common sense" and offers a practical solution to busy managers, like, for example, Toyota's Five

Whys. Actually, the dominos could be seen as an early (and graphically, and metaphorically more appealing) version of Five Whys. The use of such simple models may lead to missing out on nuance and context, but they offer managers and workers a practical tool, making a trade-off between a search for solutions to prevent recurrence of accidents and other objectives. The dominos and similar tools are an ETTO (Hollnagel, 2009) regarding accident analysis. Besides, despite the critique that linear, sequential models are insufficient to deal with today's complexity, Hollnagel concedes, "On the other hand, a closer study of the many cases in the book shows that little has changed in how humans go about accomplishing their work, differences in tasks and technology notwithstanding" (Hollnagel, 2014a, p.36).[30] This suggests that linear models still have a place and application, even in today's complex, connected, and dynamic world.[31]

6.3.5 Critique 2: direct causes

Heinrich described in a series of drawings how the domino sequence worked; when the first domino topples, the others follow. He also showed how one could stop the accident from happening, namely by removing the domino preceding it (Figure 6.5): "In accident prevention the bull's eye of the target is in the middle of the sequence - an unsafe act of a person or a mechanical or physical hazard" (Travelers, 1934, p.1; HWH1941, p.13), and "If one single factor of the entire sequence is to be selected as the most important, it would undoubtedly be the one indicated by the unsafe act of the person or by the existing mechanical hazard" (Travelers, 1934, p.4).

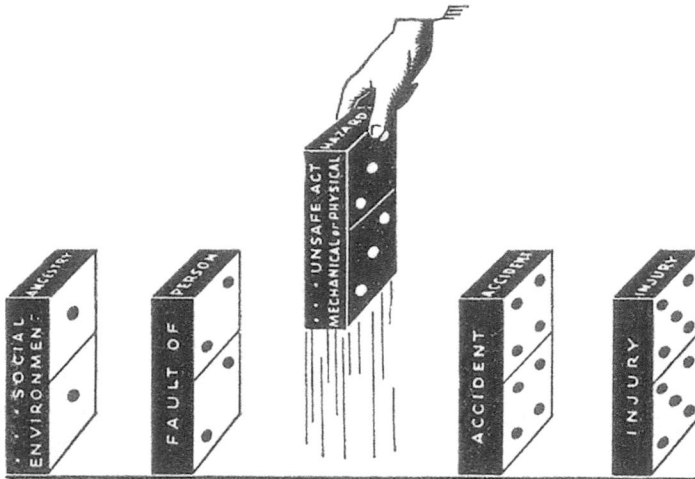

Figure 6.5 Removing the middle domino "prevents the accident."
Source: Accident Sequence Brochure, 1934. Used with permission.

Manuele critiques this focus on direct causes, saying that there are other, more important causal factors to pay attention to.

> Note that the first proximate and most easily prevented cause is to be selected. That concept permeates Heinrich's work. It does not encompass what has been learned subsequently[32] about the complexity of accident causation or that other causal factors may be more significant than the first proximate cause.
>
> (Manuele, 2002, p.22)

Others agree:

> Merely identifying a proximate cause as the 'root cause' may, however, lead to the elimination of symptoms without much impact on the prospect of reducing future accidents (Marais et al., 2004; Leveson, 2004). In order to identify systemic causes, one may need to supplement with models representing alternative mindsets in order to spark the imagination and creativity required to solve the accident risk problem.
>
> (Hovden et al., 2010, p.954)

One wonders whether Heinrich took the bull's eye metaphor too literally. From today's point of view, one would not choose to aim for that tile, except to address an immediate hazard. The question, however, is why this made sense to Heinrich.

6.3.6 Why did focus on direct causes make sense?

Today's safety professionals and incident investigators are taught to concentrate on underlying, "root" causes, and on the context. This was not always so, "It should however be noted that different methods have historically had different foci – for instance, Heinrich (1931) promoted a focus on the *most easily preventable causes* that were *most proximate* to the accident" (Lundberg et al., 2010, p.2133, emphasis added).

This focus may be puzzling and ineffective nowadays. Studying Heinrich's writings, one can identify several reasons why it made sense to him to focus on direct causes. The first suggestion is that he was influenced by the standards of the time. As Heinrich discussed, the Committee on Statistics and Compensation Insurance Cost recommended in 1920, "...the accident should be charged to that condition or circumstance the absence of which would have prevented the accident; but if there be more than one such condition or circumstance, then to the one most easily prevented" (BLS, 1920, p.33).

It made sense that Heinrich chose to define a term in a way he used in his everyday work, following "best practice," especially given the fact that much of his work was a compilation and systemisation of "best" practices within safety from his time. However, Heinrich did not shy away from critique on many other practices, for example pointing out "wrong" assignment of causes by his contemporaries. Additionally, he defined accidents in a quite novel way by classifying them as an accident regardless their consequences. Therefore, pointing towards a standard is probably not enough to explain why this made sense to Heinrich.

Not everyone aimed for direct causes at the time. While Heinrich was strongly influenced by DeBlois regarding causation, he deviated from DeBlois who wrote,

> The proximate cause may or may not be the cause which it is most desirable or efficacious to remove in order that future accidents may be prevented. The only way to find out is to ascertain all the causes and then determine which are remediable and, in particular, what we may term *the principal remediable cause*. A definition of proximate cause is "a cause which directly, or with no mediate agency, produces an effect or a specific result." It bears, therefore, the closest causative relationship to the accident of any of the contributing events or circumstances. The principal remediable cause may be defined as that cause which is most readily and effectively remediable, and the remedy of which will go farthest towards removing the possibility of repetition.
>
> (DeBlois, 1926a, p.48)

DeBlois was not as adamant about the proximate cause as the point of attack as Heinrich. He suggested looking at all the other causes and aim for the "principal remedial cause" which he defined as the cause when removed gave the greatest chance of preventing recurrence. This corresponds to some contemporary definitions of "root cause." In the above quote, DeBlois suggested a more thorough analysis than just aiming for a direct cause. After this, the feasibility and efficacy of actions to "remove" particular causes should be identified. Heinrich did not follow this suggestion, and instead simplified the process by stating that

> It becomes necessary therefore, to draw a line and limit this discussion and, in fact, restrict the definitions... to an extent that will permit ready understanding and practical application of the thoughts involved.
>
> (HWH1928b, p.123)

Heinrich thought once more of his audience and their needs. An important keyword from the above quote is therefore "practical," which was a recurring theme in his work. He did not want to make things unnecessary complicated; "expediency and practicability" (HWH1941, p.105) were guiding principles. Things had to be practicable and within reach of managers, "...in order to avoid impractical and too extensive analysis, by establishing the first, proximate, most readily eliminated cause..." (HWH1931b, p.9).[33] The approach and causal categories recommended by Heinrich were "crude," but good enough to "serve as a practical guide" (HWH1931a, p.268). In a way, he suggested a basic stop rule[34] for accident analysis, namely practicality: "...there is danger of delving deeper than is practical..." (HWH1928a, p.9).

This limited depth in accident analysis would also find its way into the "Heinrich Cause Code." Kossoris wrote in his evaluation,

> The underlying working philosophy of the Heinrich code is characterized by two items: (1) the fact that the classification of accident factors is *not intended to deal with obscure causative factors, or factors too far removed in the accident sequence to be definitely ascertainable*; and (2) the rules for the selection of the accident factors.
>
> (Kossoris, 1939, p.527, emphasis added)

One can link this to another set of explanations that Heinrich gave in his work for his recommendation to focus on direct causes. While he acknowledged, and to some degree discussed, underlying causes, he chose not to spend too much time on them. First, in Heinrich's opinion, industry was not ready to embrace and practice such more advanced approaches. Many did not even understand and practice the simpler basics: "...the exclusion of detailed treatment is well justified because industry in general is not fully conversant with the more simple and direct approach to accident prevention and does not apply it in practice" (HWH1941, p.109,110). Therefore, they first had to understand these simple principles and act on them before being able to proceed.

Second, Heinrich believed strongly in a systematic approach where things were done in order. "It is axiomatic, however, that error is invited if we act on underlying data without first having determined and made use of the more direct facts" (HWH1941, p.40). He has some support there. While McFarland critiqued simple and straightforward investigation techniques and their diminishing returns, he offered a reason that focus on causes in the proximity of the event makes sense: "...effective measures may sometimes be introduced to interrupt a sequence of events prior to complete understanding of the causal chain and the specific details of the etiology" (McFarland, 1963, p.687). One could read this as an argument for interim measures, awaiting something more systematic, but this must not necessarily so, regarding the illustration given. Heinrich acknowledged the need for both long- and short-term actions,

> The accident-prevention task in industry requires both the immediate approach (direct control of personal performance and environment) and the longer-range approach of training and education. In this text emphasis is placed on the immediate approach, because it is a first as well as continuing task, while at the same time it covers the essentials of safety education.
>
> (HWH1950a, p.2)

Third, taking the domino metaphor literally, it makes sense to remove the tile right before the accident to prevent this from toppling. If one would remove the utmost to the left, the middle one can still tip over and trigger the accident. Heinrich suggested this in his graphical explanation. The caption of Figure 6.4 illustrating how the "direct cause" tile is removed told, "The removal of the central factor makes the action of preceding factors ineffective" (HWH1941, p15). This suggested that one possible reason to "aim" for the "middle" domino was that underlying causes may be necessary, yet not sufficient for direct causes to arise. Direct causes may be necessary to trigger the pre-existing conditions.[35]

A focus on direct causes also makes sense from a retrospective standpoint. After all, with the benefit of hindsight,

> ...it appears that any individual accident could have been easily prevented. A slight change in the environment, a mildly different behavioural response, a simple anticipation or an inconspicuous mechanical change is all that would have been necessary to have forestalled almost any given accident. Generally speaking, these changes would have taken little cost or effort. Remedies which would have prevented any individual accident are almost always apparent from even casual investigation.
>
> (Jacobs, 1961, p.329)

Looking back, it is easy to see the one element in a causal chain that should or could have been different and then designate a remedial action to that element. This is an encouraging approach for managers and those wanting to improve safety. Cynically speaking, the remedial action does not even have to be effective because it is very unlikely that the exact same accident happens again anytime soon.

An additional benefit of a focus on direct causes is that it is possible in many cases to gleam "a direct cause" from only very limited information (Jacobs, 1961). This is very appealing to managers who have little time to spare and need to be done with safety issues as quickly as possible – in line with Heinrich's aim to provide "simple and effective methods of procedure" (HWH1931b, p.5).

One may criticise Heinrich today for his limited and possibly biased accident model; however, it had positive effects. Aldrich notes, "result of the new approach was an increasingly thorough investigation of injuries and hazards." Instead of automatically blaming the worker, "as the writings of DeBlois and Heinrich began to influence thinking about safety, investigations became more impartial and correspondingly more valuable" (Aldrich, 1997, p.158).

6.3.7 Evolution

The domino model was highly successful, becoming one of the leading paradigms of accident modelling. It has been called "the mother of all accident models" (Van Alphen et al., 2008, p.25), spawning several post-Heinrich versions. The model was readily adopted and used by others, as we saw with Lieutenant Lippert's demonstration (NSC, 1942a). It also quickly found its way into safety handbooks, as for example Blake's *Industrial Safety* (1945).

Lateiner popularised the model in post-War Europe. In the 1950s, he travelled to many countries, including Germany, the Netherlands, Turkey, and Norway to teach the "Lateiner Method." This involved large dominos to visualise the process. Lateiner was also the first to make changes to the model, renaming[36] the first domino to "Background" and the second to "Defects of Person" (Lateiner, 1961, 1969). Apart from that, he stuck closely to Heinrich's original.

Bird's versions of the dominos would become the best known, after Heinrich's. Bird used them "to represent modern loss control thinking, since they have been used so widely in the past to convey the principles of accident prevention and loss control" (Bird & Loftus, 1976, p.39). He updated Heinrich's accident sequence to reflect the relationship of management with the causes of accidents and their outcomes.

Bird divided his sequence into "pre-contact" (causes), "contact" (the incident), and "post-contact" (outcomes), renaming the dominos "Management/ Lack of control," "Origin(s)/Basic cause(s)," "Symptoms/Immediate cause(s)," "Contact/Incident," and "Loss/People-Property." Earlier versions mostly focussed on the person, being more ambiguous about other factors. Bird's first domino was very explicit about management control – something which Heinrich wrote, but not pictured in the sequence. Control, in Bird's view, included what he saw as the four functions of management: planning, organising, leading, and controlling (Bird & Loftus, 1976). Also the second domino covered a wider scope. Bird pointed at both "Personal Factors," as knowledge, motivation, and physical or mental problems, and "Job Factors," dealing with issues in the work environment, as design, maintenance, or procurement.

Bird's "Loss Causation Model" was adapted several times, including ILCI's/DNV's SCAT (Systematic Cause Analysis Technique) model, its Dutch counterpart SOAT, and Tata Steel's SDO (Sklet, 2002; Van Alphen et al., 2008) as well as the four-tile S137 method commissioned by the Dutch health and safety inspectorate (Roels, 1992). Many others made their own variations on the dominos. Among these Weaver (1971); Adams (1976); and, recently, a reintroduction of "real" causes, "Acts of God (…) aside, human behaviour, irrefutably, will be the proximate cause of any preventable accident" (Difford, 2011, p.81), reducing the sequence to a mere two dominos.

Over the years more complicated sequential models were developed better capturing multi-causality. Examples include Failure Mode and Effects Analysis (FMEA), developed by the US Army in 1949 (Swuste et al., 2019), and Petersen's accident-incident causation model. Influenced by the Ferrell human factors model, this showed an elaborate connection of factors converging into a linear final sequence (HWH1980).

Finally, some claim that Reason's Swiss Cheese Model (1990, 1997) is a direct descendant from Heinrich's dominos (e.g. Leveson, 2011; Blokland & Reniers, 2019; Dekker, 2019). This attribution is most likely based on visual similarities rather than deeper analysis. Coming from entirely different historical and conceptual backgrounds, it is unlikely that Heinrich influenced Reason and nowhere in Reason's work is Heinrich referenced (Larouzee & Le Coze, 2020).

6.4 A more complete accident model

6.4.1 Multiple causes

A point of critique not extensively discussed above, is that Heinrich's accident sequence gives a limited view of causation (Petersen, 1971; Qureshi, 2008). Heinrich focussed on proximate and direct causes. He acknowledged underlying causes but saw no need to always investigate in-depth. That did not mean that he advocated single causation, even though the language and phrasing may suggest this at times. He spoke often about causes in plural. When he used the singular, one might assume it was either a colloquial, everyday speech use of the term (like when something happens and someone asks, "What was the cause?"), or it was in the context of assigning a single "real" cause. The only time he explicitly assigned single causes was during the *Origin of Accidents* study, which was regarding *direct* causes.

Already in the first edition of *Industrial Accident Prevention*, Heinrich showed a surprising width and depth when he emphasised the importance of causal analysis for accident prevention, going beyond direct causes[37]:

> Analysis for cause, in accident prevention, involves a knowledge of many factors that are not commonly taken into account – such, for example, as the intelligence, experience, and ability of the workman and his supervisor and foreman, and their attitude toward personal and group safety; protective clothing, wages, morals and morale, home conditions, light, noise, and ventilation; the recognition by the foremen of their responsibility for accidents; the attitude of the employer; selection, instruction, and assignment of employees to work; routing and internal plant-traffic conditions; housekeeping; plant or operation layout and procedure; machine and tool design; and many conditions and methods of procedure other than the mere provision of guards for machines, physical hazards, and unsafe mechanical exposures, which is too often considered to be the prime essential in the control of accident frequency and severity. Proper evaluation of these various factors is highly important.
>
> (HWH1931a, p.11–12)

Still, at the time, Heinrich's efforts were mainly directed towards emphasising the need to better distinguish between injury, accident, and causes – hence the attention for "real" causes. Later work built on this basis.

Neither did the categories from the early (1928/1931) *Origin of Accidents* chart show a clear distinction between various levels of causation. Elements as instruction, enforcement, acquiring skill, and mental characteristics were more of an underlying nature rather than direct accident causes. Only the 1941 version's caption specified that the chart was about "direct and proximate causes."

Gradually he expanded his initial view, and soon Heinrich discussed deeper levels of causation – "contributory conditions" (HWH1931a, p.50). Although he used a variety of different terms and was not too strict about definitions, one can roughly distinguish between subcauses or reasons, and underlying causes, corresponding to respectively the second and first domino in the sequence.

6.4.2 Reasons and subcauses

Before he introduced the underlying levels in causation through the domino model (Travelers, 1934), Heinrich suggested to look deeper. In the *Unsafe Practices and Conditions* paper, he advised looking for "secondary cause" (HWH1931b, p.10). When one came across an unsafe practice, one should ask "Why does it exist?" and the same had to be done in the case of unsafe conditions. This means that Heinrich advised to go beyond the direct cause and deal with problems both, as Petersen and others put it, symptom level and deeper levels.

The next year, in the *Safety in the Small Plant* paper, Heinrich continued this thought, arguing that "reasons why man-failure exists, are a valuable part of accident facts" (HWH1932a, p.32). He illustrated this with a case where an employee acted unsafely but had not received adequate instruction. Obviously, prevention had to start with instruction, not the unsafe act.

The thinking about "reasons" for direct causes further matured in *Unsafe Habits of Men*. This emphasised, "By far the most sensible thing to do is to find the *reasons why* unsafe acts are committed and then to devise practical action for correction, be it in the engineering-revision or educational field" (HWH1940, p.112).

From the second edition of *Industrial Accident Prevention* onwards, this paper would be the basis for a section discussing a deeper level of causal analysis and suggestions on how to deal with the findings.[38] Heinrich placed great value on what he called "subcauses" or "indirect causes" (Travelers, 1954). He recognised that a focus on direct causes had limitations and therefore, "…if the accident problem does not yield to the more direct approach, as herein advocated, it is advisable to dig a bit deeper in the work of fact finding" (HWH1941, p.110).

Heinrich put the reasons for unsafe acts in four[39] broad categories: "Improper attitude," "Lack of knowledge or skill," "Bodily defects," and "Safe practice difficult or impossible" (HWH1941, p.39). This list was most likely kept short in order to provide readers with something manageable which to select suitable remedial action. These actions came also in a few broad categories. From *Basics of Supervision* onwards he listed them in prioritised order[40] as "Engineering revision"; "Persuasion and appeal"; "Personnel adjustment"; and, only when everything else failed, "Discipline" (HWH1944, p.57, 1950, p.36,37).

Heinrich's domino sequence pictured subcauses as "Fault of Person." In retrospect, one may regard this an unfortunate choice of words. The word "fault" may be associated with blame, which will be discussed in the next chapter. It also biased the domino model towards the human. Many people, as Lieutenant Lippert in the earlier example, interpreted this domino to deal with the person committing the unsafe act. However, Heinrich intended it to be about *any* person which created or permitted the reason for the unsafe act. In his view, when an untrained worker acted unsafely, this was due to the fault of the foreman or manager who had failed to provide proper training.

Following the interpretation of "Fault of Person" dealing with the person who committed the unsafe act, one might comment that the domino sequence failed to give an explanation about the reasons for the mechanical and physical hazards in the third domino. In Heinrich's view, these could also be related to some person at fault, e.g. an action or omission by another worker or a manager. However, he considered finding reasons for unsafe conditions as less important, "...ordinarily it is necessary merely to identify a specific mechanical fault in order that it may be corrected" (HWH1944, p.152). Also, much of his writing was aimed at supervisors and foremen and he saw limitations for them,

> ...the necessity of finding the reasons why they exist is not as important as it is in the case of personal practices. True, they exist only because of the fault of persons, but the person involved is often remote from the condition itself and sometimes not directly subject to the control of the foreman.
>
> (HWH1944, p.151)

Still, in *Industrial Accident Prevention*, he spent a paragraph discussing possible subcauses of unsafe condition, naming various factors on managerial level, design, and education as possible causes on this level (HWH1941, p.104[41]). Even more, in the 1954 brochure these were pointed out as the prevalent subcause, "Chief among these were failures in management and supervision" (Travelers, 1954, p.4).

Heinrich's emphasis on finding reasons provides another suggestion for why he called the unsafe acts and conditions the "Bull's Eye." In the logic of the dominos, the accident would be prevented if the preceding one was re-moved. In order to eliminate this, Heinrich argued that one needed to know the causes for the (direct) causes and deal with these. This suggests that he was not interested as much in direct causes as one superficially might think. He was interested in *preventing* direct causes, in taking out the direct cause, by means of *addressing the subcause*.

6.4.3 Underlying causes

As illustrated by the first domino in the sequence, Heinrich suggested that one might be going even deeper than reasons and subcauses, "...if the accident problem does not yield to the more direct approach, as herein advocated, it is advisable to dig a bit deeper in the work of fact finding" (HWH1941, p.110). Already in the first edition of *Industrial Accident Prevention*, he discussed this, without being very specific:

> Immediate action is necessary, and none could be more effective than to remove the source of danger without delay or to eliminate its harmful possibilities, and then to proceed in the effort to remove the underlying causes.
>
> (HWH1931a, p.141)

The most extensive discussion of underlying causes in this edition is found in the *Safety Psychology* chapter, where he claimed, "Psychology lies at root of sequence of accident causes" (HWH1931a, p.126).

Heinrich called this the "third step in the ideal accident prevention program." After guarding and eliminating of mechanical and physical hazards without causal analysis (step 1) and the selection of remedial actions based on causal analysis of proximate preventable causes (step 2), one could move on to the "selection of remedies based on psychological analysis of underlying causes" (HWH1931a, p.128). While not exclusively, Heinrich aimed here at "basic causes of accident proneness" (HWH1931a, p.134). However, this third step was still out of reach. Because of the major importance that Heinrich perceived, he found it "...one for which large industries and trade associations should now be laying a solid foundation through intensive research" (HWH1931a, p.128).

By the second edition, Heinrich's thinking had evolved, and although the immediate approach had most attention in the book, there was more discussion of underlying causes. A note in the *Reasons for Unsafe Habits of Men* section distinguished between proximate or direct reasons, dealing with the first level of reasons for causes, and the causes for these reasons, which Heinrich labelled "subreasons" or "underlying causes" (HWH1941, p.40).

The example given dealt with inadequate supervision, suggesting issues at managerial level. This indeed, was one important part, "...underlying causes are faults of management and supervision plus the unwise methods, attitudes, and procedures that management and supervision encourage or at least fail to correct" (HWH1941, p.109). Heinrich saw an analogy between managerial causes for accidents and those for poor quality or production. These were similar, listing lacking responsibility, maintenance, lack of resources, being uninformed, and incompetence as managerial causes.

In the section on "Underlying causes," he even described latent conditions at the managerial level that remind of what Reason would write many years later, "The underlying accident causes in management and supervision, such as described, can and often do occasion or permit the existence of improper methods and procedures, these in turn being followed by direct and proximate accident causes" (HWH1941, p.110). This resonates well with Reason's writing how sharp-end practitioners inherit problems higher up in the system: "Human error is a consequence not a cause" (Reason, 1997, p.126).

His taxonomy based on the duality of human and mechanical factors suggests that Heinrich had no definite thoughts yet about what later would be labelled as organisational factors, for example in the MTO-approach that captures man, technical elements, and organisational factors (Sklet, 2002). Organisational factors are found throughout the text, however, mostly attributed to management or the employer. Sometimes in quite advanced ways, as in the example of bus drivers who experienced problems with their time schedule (HWH1941, p.37).

As illustrated by the domino sequence, where no managerial factors are mentioned,[42] there were other types of underlying causes in Heinrich's thinking. "There are also underlying causes of a social, environmental, and inheritable nature which, in this text, are merely alluded to" (HWH1941, p.109). He did not describe this extensively, and so, we must make assumptions about what he meant with "ancestry" – most likely including accident proneness – and "social environment," which may include the modern catchphrase "culture." Many of these factors were outside the workplace or too remote for effective intervention for the organisations where accidents happened. This was another reason for Heinrich to focus on more proximate factors:

> The reason for omitting detailed explanation is that it would require an exposition dealing with psychology, psychiatry, and pathology, social and governmental conditions, slum clearance, and other matters of less direct significance than the proximate causes that are discussed herein.
> (HWH1941, p.109)

In the 1954 brochure, he wrote, "In the background of both direct and indirect causes it became clear that a third set of causes existed. These had to do with the attitudes and characteristics of employees and management" (Travelers, 1954, p.4). This resonates with Heinrich earlier statement about psychology being at the root of the sequence. However, he cautioned about not going too far in the analysis and therefore a stop-rule was advisable, "When tracing the origin of an accident back through the chain of events which preceded and caused it, practicality imposes a limitation else one might find himself thinking of faults in the Garden of Eden" (Travelers, 1954, p.10). Besides, he estimated that about half of the accidents could be prevented without delving this deeply, and industry was not even mastering the simple approach.

6.4.4 Enhancing Heinrich's accident model

In the light of the material presented in this chapter, one must conclude that the dominos represent an incomplete description of Heinrich's thoughts on accident causation, because they do not capture all of the subcauses and underlying causes that he discussed throughout his work. Given the practical aim of most of his work, a complete model most likely never was his intention.

In retrospect, we can combine the various pieces of the puzzle that he scattered throughout his work and construct a more "complete" model of Heinrich's thoughts on causation. By including the underlying causal levels and nuances discussed in this chapter, one can expand the dominos upstream. Additionally, one can expand downstream by including his thoughts on production loss and "near-misses" as described in the accident triangle.

Taking the dominos as a basis, an "enhanced" Heinrich model could be as below (Figure 6.6).

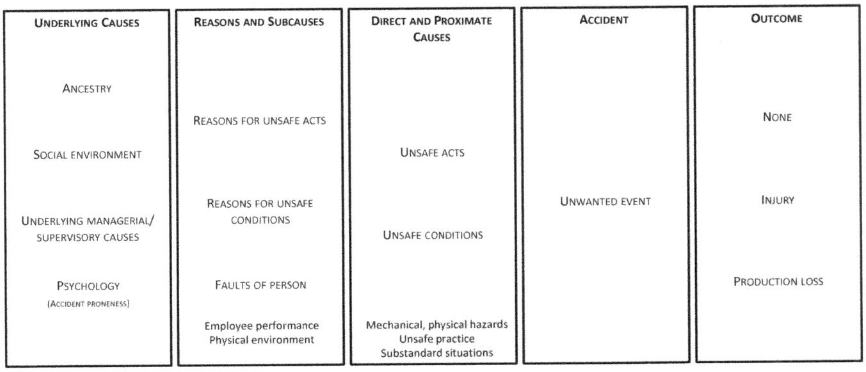

Underlying Causes	Reasons and Subcauses	Direct and Proximate Causes	Accident	Outcome
Ancestry				
	Reasons for unsafe acts			None
Social environment		Unsafe acts		
Underlying managerial/ supervisory causes	Reasons for unsafe conditions	Unsafe conditions	Unwanted event	Injury
Psychology (Accident proneness)	Faults of person			Production loss
	Employee performance Physical environment	Mechanical, physical hazards Unsafe practice Substandard situations		

Figure 6.6 A more complete model of Heinrich's causation.

6.5 Final reflections

This chapter concludes with three reflections regarding causation. First, there is a common belief that Heinrich suggested that unsafe acts are the *root* cause of most accidents. He did nothing of the sort. As discussed in this chapter,

1   The 88:10:2 ratio presented *direct* or *proximate* causes of accidents on which Heinrich chose to focus.
2   Despite this focus on direct causes, Heinrich did not stop at an unsafe act. He emphasised the importance of finding the reason for that unsafe act, which can lie in the organisation, externally, etc.
3   As a first solution, he suggested "engineering revision" and only as the very last resort disciplinary action (which he regarded as a "managerial failure").

Second, while some of Heinrich's language seems rather determined regarding finding and knowing facts and causes, his view was nuanced. As mentioned in the beginning of this chapter, especially his earlier work spoke of choosing or assigning causes, which resonates with a constructivist view. Moreover, although he cautioned not using the term "loosely," Heinrich was well aware of the relative nature of cause and effect, where any factor in a chain of events was both cause and effect:

> All the factors in the chain of events culminating in accidental personal injury, therefore, are 'causes'. For example, environment causes fault of persons. These in turn cause unsafe acts of persons. The unsafe acts cause accidents, and the accidents cause injuries. To go a step further – one that is not necessary to discuss in a text on *accident* prevention – it is clear that the injury also is a cause, inasmuch as injuries result in cost and suffering.
>
> (HWH1941, p.104)

Finally, Heinrich's work shows that his thinking of causes was not limited to the reactive use in accident investigation, "finding" facts and causes after accidents. Knowledge of causes in Heinrich's understanding also included proactive knowledge of causes of accidents and incidents that have not happened yet – what one would commonly describe as risk assessment today.

Notes

1   He did provide working definitions for the most important,

> In this text, except when otherwise expressly stated, the term 'direct and proximate cause' is applied to the accident per se and means an unsafe personal act or an unsafe mechanical condition that results directly in the

accident. The term 'subcause' refers to the personal factor (specific fault of person) that occasioned, effected or permitted the unsafe act or mechanical hazard. The term 'underlying cause' refers to managerial and supervisory faults and to social and environmental conditions that are outside the workplace.

(HWH1941, p.104, 1950, p.116, 1959, p.78)

2 Schreiber also suggested that near accidents were "as instructive as actual accidents" (Schreiber, 1916, p.245), a thought later found in Heinrich's triangle.
3 The final part of this quote reminds very much of Rasmussen's (1990) take on "root cause."
4 In a way, these papers were a call for professionalisation of safety work.
5 It is interesting to see that Heinrich's analysis shown in these examples is considerably richer than the original analysis.
6 A rare exception is found in a reprint of an older text:

> True causes of industrial accidents were not understood, found or recorded. Investigators and statisticians confused the cause of an accident with the *accident* itself, with the *injury*, the *object* or *agency* involved, the *manner of occurrence* and with the *operation* being conducted.
>
> (Travelers, 1954, p.4)

7 Causes to define actions were at the time most likely a huge improvement for structured safety work. It probably has diminishing returns with maturing safety management, "fixed requirements for categorizing and labelling can of course limit rather than empower the actionable intelligence gleaned from such activities" (Dekker, 2017, p.71).
8 Adding: "...which, combined make a third" (Tolman & Kendall, 1913, p.194), namely man and machine together.
9 DeBlois suggested almost ten years before Heinrich to attribute most accidents to supervisors or even top management. In a speech before the Eighth Annual Safety Congress, he cited two axioms of safety, namely (1) that a high percentage of accidents were caused by careless individuals, and (2) that foremen were the logical point of action. However, argued DeBlois, if the foremen were to educate and control employees, did not this mean that a large percentage of accidents should be attributed to them? Or to management who controlled foremen? (DeBlois, 1919). The year before, at the spring meeting of the A.S.M.E., DeBlois did make the distinction between unsafe or structural conditions (10%–20%) and "human defects" or "defects such as lack of proper supervision, discipline, etc. in the organisation" (80%–90%) (DeBlois, 1918, p.22).
10 Based on a November 1932 speech, it was published in *The Travelers Standard*, January 1933.
11 This is interesting in several aspects. First, it predates Bird's way of phrasing things about a decade and a half. Second, it suggests that safety is created (for an important part) by employees, which is a very contemporary view of safety.
12 While both criticized the practice of confusing injury for accident and vice versa, Blake chose to discuss injuries, while Heinrich explicitly chose to speak about accidents, not about their outcomes. This may actually affect their positions because injury without physical hazard is unlikely, while an accident (defined as an unwanted event) without physical hazard is imaginable.
13 This might suggest a departure from Heinrich's assertion that knowledge of cause was necessary for proper action.
14 Consistent with Heinrich's principles of accident prevention which start with the interest and support of top management.

15 Many people take from Heinrich what they want, using his work to reinforce what they already believe. Humans are prone to confirmation bias.

16 Several of these problems one finds even today in similar studies or reports on accident data.

17 He did shed some light on it, e.g. by explaining "Accidents due to mechanical or physical exposures should be assigned to causes in the supervisory group where the foreman had authority to install or maintain guards. In such cases we consider that the chief cause is laxity in supervision" (HWH1928c, p.129).

18 Some BBS authors speak of "at-risk" behaviour (e.g. Geller, 1996).

19 The paper is of special interest because it does contain some definitions (HWH1931b, p.6): Unsafe practice = hazards which are controllable by supervisory methods. Unsafe condition = mechanical hazards, which may be remedied by the installation of guards or by other forms of engineering revision. Cause of the accident = the reason for the existence of these practices and conditions (which means Heinrich aimed beyond the direct causes). Important is that "remedy is always the reverse of the cause."

20 The most common citation error regarding Heinrich is authors referencing to the 1931 edition of *Industrial Accident Prevention* when discussing the domino model. However, this did not appear before 1934, with the accident sequence brochure (Travelers, 1934), and for most readers not before the 1941 edition of *Industrial Accident Prevention*. My thesis explores this further (Busch, 2019b).

21 Heinrich's original speech may be lost. Excerpts have been captured in the 1934 Travelers brochure and in the 1941 book.

22 Sometimes called "Domino Theory" (e.g. Petersen, 1971; Groeneweg, 1992; Kjellén, 2000). This is confusing, because there are several "domino theories" around, e.g. political and for escalating process safety events.

23 This version mentions that the direct cause can be both ("and/or").

24 This development prompts several questions, including where did the metaphor come from? How and why were the various "tiles" (especially the first two) assigned? Why is there an explicit mention of "ancestry," and no other (management) factors that Heinrich does discuss in his writings (and Bird would include some decades later)? Alas, so far, no material has surfaced that can shine a light on the creation of the dominos.

25 Other points of critique, dealing with the lack of multi-causality (e.g. Petersen, 1971; Kjellén, 2000) and the naming of the first two dominos, will be dealt with later in this book.

26 As a quote later in this chapter shows, Heinrich was well aware of this possibly endless chain. He put the start of everything in the Garden of Eden.

27 As for example DeBlois, hinting on complexity, "Many superficially simple industrial accidents arise out of a highly involved network of condition and circumstance" (DeBlois, 1926a, p.47).

28 Another advantage according to the authors is its usefulness to allocate blame. A subject for the next chapter.

29 James Reason made a similar comment, discussing critique of the Swiss Cheese Model:

> Just as there are no agreed definitions and taxonomies of error so there is no single 'right' view of accidents. In our business, the 'truth' is mostly unknowable and takes many forms. In any case, *it is less important than practical utility.*
> (Reason, 2008, p.95, emphasis added)

30 A similar suggestion is made by Hovden et al who conclude that things at the sharp end have not changed that much. They add the nuance, "The traditional

approaches may be good enough; suited to some workplaces but not to others and suited to understanding some accidents but not others" (Hovden et al., 2010, p.954).

31 Note that complexity is not necessarily an inherent property of a situation, it can also be an analytical choice. If you switch on the light, you will most likely not ponder on the complexity of the situation, but turn to a simple, linear, and reductionist way of problem solving: Try the switch again. Check whether the lightbulb is broken. Has the fuse blown out? Is there power in other rooms?

32 This is an interesting piece of critique fuelled by hindsight knowledge. Agreed, Heinrich's premise to focus on causes very close to the accident is very limited and quite contrary to what is taught to safety practitioners and incident investigators during the past decades. However, without access to a time machine or a crystal ball that actually works, it would be hard for him to "encompass what has been learned subsequently."

33 The brochure phrased this as, "Prevention must deal with the immediate conditions and circumstances out of which other similar accidents may be created. Common sense dictates that preventive effort be directed toward the thing most easily and quickly corrected" (Travelers, 1934, p.5). This must have appealed to managers, and others, who wanted solutions, not problems. Also, it echoes the 1920 definition for cause of the Standardization of Industrial Accident Statistics committee.

34 "It is reasonable, further, to proceed toward the correction of the unsafe act by directing attention to the factor immediately preceding it; that is, the reason for its commission, and from them on to dig into the background of the accident *only as far as may be necessary to accomplish results*" (Travelers, 1934, p.5, emphasis added). Heinrich's practice fits well with how Rasmussen later would characterize most root cause analyses: "The stop-rule will now be related to the question of whether an effective cure is known" (Rasmussen, 1990, p.453).

35 As suggested by James Reason (1990, 1997).

36 Lateiner did this during Heinrich's lifetime. Given their close cooperation during those years one expects that Heinrich and Lateiner must have talked about these changes. Alas Heinrich's opinion on the subject did not survive the years.

37 This was in line with the thinking within the Travelers. Already two decades earlier, they presented a remarkably varied view (Travelers, 1913, p.305, 306).

38 An even clearer explanation can be found in Chapter III, *Reasons and Remedies for Unsafe and Improper Practices and Conditions*, of *Basics of Supervision* (HWH1944).

39 In the 1940 paper, Heinrich listed ten reasons. For the 1941 book this was comprised to the four categories listed in the text. These were later renamed to "Improper attitude," "Lack of knowledge or skill," "Physically unsuitability," and "Improper mechanical or physical environment" (HWH1950a, p.36).

40 Earlier, Heinrich presented six categories without clear prioritisation between them: "Education," "Engineering revision," "Placing," "Discipline," "Medical treatment," and "Psychology" (HWH1940, p.117, 1941, p.40).

41 The text on subcauses and underlying causes in the 1941, 1950, and 1959 editions in the *Fact-Finding* chapters are very similar.

42 The domino sequence and the 1941 *Origin of Accidents* chart show a comparable structure. However, they do not match entirely and deviate especially when it comes to underlying factors. The chart lists Management (control) as the most basic cause, while the domino sequence lists social environment and ancestry. Management would only be included as part of the domino sequence in Bird's version about 25 years later.

Chapter 7

The human element

This chapter deals with one of the most debated, and to some degree misunderstood, parts of Heinrich's work. As discussed in the previous chapter, Heinrich attributed the direct cause of most accidents to man-failure. Many interpret this as his focus being on correcting the human, especially their behaviour. That, however, is only part of it, filling the first sub-chapter, dealing with the human as something to control, and why this may have made sense to Heinrich. The chapter looks at Heinrich's view on remedies to deal with the human weaknesses, including a brief discussion of behaviourist approaches.

The "human element" theme envelopes several other subjects – some more debated than others – as accident-proneness, safety psychology, fatigue, and the influence of age on accidents. While Heinrich is known for his attribution of accidents to "unsafe acts of persons," he also acknowledged them as an essential positive factor and advises to draw on their strengths. The chapter, aiming to give a more varied and more complete view of Heinrich's thinking about the human element, ends thus on a positive note.

Heinrich is known for focussing on the human element. Chronologically, however, he spoke about it relatively late. It did not become a main theme before he turned his attention to causation. His first speeches and papers (HWH1923b, 1925) dealt mainly with safety organisation and costs of accidents. Only with the *Origin of Accidents* papers, he started discussing the human element explicitly, suggesting it already in the last *Incidental Costs* paper preceding them,

> Physical hazards are playing an ever-decreasing part in the occurrence of accidents. This is well demonstrated by the many cases in which plants with unguarded machines but with careful workers have fewer accidents than well-guarded plants with careless employees.
>
> (HWH1927c, p.231)

7.1 Humans as an element to control

During the story opening the first edition of *Industrial Accident Prevention*, Heinrich had the works manager saying about their safety problem, "It is almost wholly a case of man-failure, first on the part of the workmen, then also on the part of the foremen. In all fairness, perhaps I should add, on my part as well" (HWH1931a, p.4).

This quote is interesting from several perspectives. First, there is the focus on "man-failure," which is one of the main subjects of this chapter. Second, it foreshadows the various layers of causes, "unsafe acts," "faults of person," and "underlying causes," as discussed in the previous chapter. Third, it emphasises the important role of management, which is the subject of the next chapter.

There are plenty of occasions in Heinrich's writings where he talks about humans as a factor to control. He frequently discussed humans as careless and error-producing beings. One of the clearest examples of the latter is found in his *Accident Cost in the Construction Industry* speech from October 1938. He discussed four cases all containing an "unsafe act" in italics in their description. The general discussion stated, "Violation of simple and elemental safe practice rules occurred in each case" (HWH1938b, p.376), and the major causes of accidents were "continually repeated violations of safe practice rules" (HWH1938b, p.377). He then listed "examples of the hundreds of unsafe practices indulged in by unthinking workmen under the eyes of supervisors" (HWH1938b, p.377).

Sometimes he used quite harsh language, e.g. speaking of "moral delinquents" (HWH1931a, p.41),[1] and in the war-time *Warehouse* paper, he wrote, "Misfits must not be permitted to disrupt the work of good men. Find other work that the misfits can do" (HWH1943c, p.136). Although this quote sounds very negative, the second half had a social element, suggesting fitting work to the worker. It also linked in with Heinrich's humanitarian tendencies: humans often needed to be protected – also against themselves – through supervisory practice and other "remedies."

7.1.1 Why did this make sense?

The previous chapter told that, while moving safety forward in some respects, Heinrich's statements about human failure may have had adverse effects by misdirecting actions. Even though some of today's safety scholars may question the reasoning or the very concept, it made sense for Heinrich pointing towards man-failure as the main cause of most accidents. One may even speculate that it made so much sense to him that this affected his research. The first reason may have been the influence of contemporaries. One may assume that literature such as the Cleveland/Woodhill bus driver study, and Boyd Fisher's writing played a role, although Heinrich does not reference to the latter directly. Another text from Heinrich's bibliography said,

Approximately thirty percent of industrial accidents are preventable by means of safeguards, if properly used and maintained in good condition. On the other hand, at least sixty percent of these accidents can be eliminated by the proper education of employees pertaining to safety; in other words, teaching employees to be cautious and thoughtful at all times, instructing them to think of their own safety and that of their fellow-workmen, and training them to refrain from taking unnecessary risks.

(Cowee, 1916, p.2)

Second, and possibly more important, it may have reflected his own practical experience and/or professional conviction. Over the years, it appears that Heinrich noticed improvement in working situations. Many of the major hazards were dealt with by the organisations The Travelers served, and "physical hazards are becoming less a real factor in safety" (HWH1928a, p.10), he concluded. As equipment and machines improved, failures must be attributed to the human, and alternative remedies needed to be considered[2]:

Accidents caused by unguarded machines are constantly decreasing in number, however, while those resulting from moral or supervisory failure are becoming more common. The recommendations for preventive measures must, therefore, be of a different character.[3]

(HWH1931a, p.259)

This made even more sense because according to Heinrich, machines did no harm out of their own (HWH1933a).

A third reason may have been that insurance inspectors only had a limited time at a plant. Inspecting technical improvements was relatively easy, but regarding education, instruction, and supervision they could only indicate what needed to be done. The actual work had to be done by safety committees and employers. This may have been an important reason to emphasise supervision or humans. These were things that were difficult for insurance companies to steer and inspect efficiently, unlike a physical guard on a machine (Travelers, 1915).

7.1.2 Dealing with humans

A newspaper report from a conference between safety experts and officials of the Department of Labor and Industry tells us,

Mr. Heinrich maintained that not more than 10 per cent of industrial accidents are due to machinery and emphasized the importance of hazards other than those purely mechanical. He showed that, although only one tenth of the accidents were due to machinery, all of the safety workers' energy is being devoted towards that cause. He urged that a proper proportion of this energy be used toward prevention of what he termed the 'abstract' causes of accidents.[4]

It is easy to interpret texts like this, or quotes as "the control of unsafe or improper practices or conditions prevents the occurrence of injuries and production losses" (HWH1944, p.125), as Heinrich seeing humans as the main target for safety interventions. Surely, he discussed a number of remedies that were first and foremost aimed at humans. One of the first things one may think of are safety rules and compliance. Indeed, there are passages in his work where his message appears in a Taylorist command and control manner: follow the rules and do as you are told. On several occasions, Heinrich explicitly stated that "employees are paid to work safely" (HWH1931a, p.77), applying this to all hierarchical levels in the organisation:

> The employees are paid to perform their work in a prescribed way – that is, to stand clear of the hatches while material is being hoisted. The foremen are paid to see that employees carry on their work as instructed. The employer has a right to expect – in fact, to demand – that his wage-earning employees obey the rules. He wants the men to stand clear, and he wants the foremen to see that they do so.
>
> (HWH1931b, p.11)

In Heinrich's view, one of the employer's duties was to determine a safe (which also was the "best") way to do a job, enforcing these safe practices through foremen and supervisors. "One of the most important industrial problems is that of training men in the habits of industry – safe and efficient habits" (HWH1929e, p.23).

7.1.2.1 Education

However, rules and enforcement were only partly an answer. Education and instruction were another essential part. Several of the early safety authors advocated education as a prime approach for safety. Ashe, for example, thought that safe habits could best be established "through educational means" (1917, p.2). DeBlois identified three methods for accident prevention: guarding, engineering revision, and education. However, he emphasised that the latter was "the very basis of all accident prevention" (1926b, p.1) because it taught people about recognising hazards and understanding that accidents were preventable. Similar sentiments were reflected in a yearbook from the National Society for the Study of Education, arguing that accident prevention required a safe environment, supervision, and education, the last being most important as it "furnishes the necessary background for the other two" (Williams & Hillegas, 1926, p.16).

Heinrich agreed that education was important. During his early speech on *The Contractors' and Builders' Safety Problem*, he spoke about the need for education of employees to create "proper attitude towards accident prevention" (HWH1925, p.253) and emphasised safety progress "…has practically all been in the direction of methods and equipment rather than in the education of personnel" (HWH1925, p.254).

However, Heinrich was critical of general education lacking direction. "Now that the unguarded machine as a direct cause of accidents occupies a relatively less important place than formerly, it is commonly believed that education of the workman is a panacea for all accident evils" (HWH1931a, p.12–13). Heinrich noted that much safety education only discussed "matters of common interest to all" (HWH1931a, p.114) which might kindle some kind of enthusiasm for safety in people but did not really help them in the end. The worker was "often left to his own initiative and devices when he returns to work" (HWH1931a, p.114). Instead of believing that general safety education worked, Heinrich was convinced that knowledge of specific causes was necessary for effective accident prevention. "The worker, through knowledge of the job, knows its danger points and must anticipate contact with them, in order to avoid accidents" (HWH1930b, p.145). Therefore, educators should tell workers "what specific dangers in their own line of work should be guarded against and what specific things they themselves may do to avoid injury" (HWH1931a, p.114). This, again, was what was common practice when it came to production and quality, so it should also be applied to safety.

7.1.2.2 Motivation

Another important element in Heinrich's work was motivation, although he rarely used this term and rather spoke of "creating interest." Introducing the subject through a paper (HWH1931c), the second to fourth edition of *Industrial Accident Prevention* all had dedicated – and largely unchanged – chapters on the subject. By "creating interest," he meant to "point out methods whereby safety can be presented so attractively to employees that they will desire" (HWH1931c, p.49), using the metaphor of selling safety to them. Because safety was of personal interest for the employee, humanitarian motives should be used in safety programs that appeal to employees. However, Heinrich cautioned that these might be met by scepticism and cynicism by some employees:

> ...many an employee listens to such an appeal with his 'tongue in cheek' in that he questions to a certain extent the sincerity and the motives of his company executives who are apparently so solicitous of his safety and welfare and of the comfort, security, and happiness of his family.
> (HWH1931c, p.50)

This rings true also today where many safety campaigns under a banner of "safety first" or "zero harm" get the very same reception.

The chapters on "creating interest"[5] serve as good examples of how Heinrich saw practical applications of psychology in safety. Without dwelling on psychological theory, he suggested a list of ten motivating or dominating characteristics of persons which could be used to motivate employees and managers:

1   Self-preservation
2   Personal gain
3   Loyalty
4   Responsibility
5   Pride
6   Conformity
7   Rivalry
8   Leadership
9   Logic
10 Humanity

He explained the application of these characteristics with examples. Heinrich cautioned about one-size-fits all approaches. Persons responded differently to stimulants, so it was necessary to find out what worked – alone or combined – for different persons. This corresponded to his message to look for reasons for problems and only then choosing appropriate and effective action.[6] While far from perfect, Heinrich thought that common-sense psychology would get results. Knowledge of what motivates people, and using this knowledge, increased the probability of success. In his opinion, this approach might have been "the greatest single advance to be made within the next decade, with respect to the safe and efficient production..." (HWH1950b, p.10).

7.1.2.3 Behaviour

Heinrich spoke frequently of wrong practice, people violating rules, and the like. His message that "the human element is chiefly responsible for accidents..." (HWH1931a, p.43) thus seemed to suggest the need to aim on behaviour (mostly through the supervisor), "...more time should be expended upon the elimination of the preponderant supervisory causes of accidents than upon causes that have relatively less bearing upon accident control" (HWH1931a, p.45). This prompts the question whether Heinrich was a behaviourist. Common wisdom suggests this. Several authors (e.g. Swuste et al., 2016; Dekker, 2019) connect Heinrich to behaviourist approaches, sometimes by noting that behaviourism was the dominant psychological school at the time.

It appears, however, that early behaviourists such as Watson and Pavlov did not leave any major marks in safety. Fisher (1922) mentioned Watson's work in his book, but only discussed behaviourism very briefly. This suggests that Heinrich must have heard about Watson's work, but he never referred to Watson. Neither does Heinrich's work show that he drew on Watson's work.[7]

Behaviourists as Watson disregarded the human mind, consciousness, beliefs, and perception. Instead, they found only behaviour relevant for study and intervention – primarily through reinforcement and punishment. People learned through consequences. Behavioural approaches tried to get people to change not through means of motivation or attitude but through behaviour first, with a possibility of attitude change thereafter.

Heinrich did not put a major emphasis on positive or negative incentives. When he spoke about workers having to follow rules it seemed rather to reflect duty ethics or a social contract. Regarding reinforcement, he suggested to favour the positive, "as a rule it is better to stress the positive factors" (HWH1931c, p.51). In his list of remedies Heinrich made clear that discipline (negative incentives) was the very last resort. Although the list of motivators can be (partly) read as a list with stimuli, Heinrich's take on psychology was more often inward-looking and aimed at motivation – "...the real root of the entire subject is the mental or moral attitude – first of the employee and second, of the employer..." (HWH1928a, p.10) – contrary to behaviourists who only studied the visible, outside behaviour.

Chapter 3 discussed Behaviour-Based Safety, a school of safety approaches that often is regarded to be an offspring of Heinrich's work. As argued there, this is an established truth with many nuances and little substance due to entirely different backgrounds. Marsh describes the behaviour-based accident prevention process as "identifying unsafe behaviour, measuring it, feeding it back to the workforce, and facilitating improvement" (Marsh, 2017, p.43). Heinrich did this only to a limited degree, and employed none of the typical behavioural tools, although it is possible to construct links between supervision and observations and various forms of reinforcement.[8] He did not advocate purely behavioural remedies, aimed at correcting the "unsafe acts." Instead he suggested approaches aimed higher up in the hierarchy of control, contradicting much of the critique aimed at his work, such as,

> In practice, safety professionals perceive unsafe act as the 'easiest-to-blame' domino and leave unsafe conditions generally unattended. Out of this narrow focus on unsafe acts, behavioral interventions are commonly designed and implemented, when unsafe conditions demand more attention of the management.
>
> (Khanzode et al., 2012, p.1360)

7.1.3 Causes and actions

In the paper that introduced "creating interest," Heinrich spoke extensively about how motivational efforts are not to be done in isolation. He emphasised the great effect of changing the environment, by making it more ergonomic or user-friendly, changing processes and the like. Educational efforts had their limitations, and "All accident-prevention work *must therefore begin with dangerous physical or mechanical conditions* and must include continuous corrective procedure" (HWH1931c, p.51, emphasis added). One should not stop at "unsafe practices" but investigate the "reasons for their existence" (HWH1931c, p.51) which essentially was asking the question "Why does it make sense to the people on the shop floor to act the way they do."

This may appear at odds with Heinrich's insistence of "man-failure" or "unsafe acts" as the cause of most industrial accidents. However, concluding that human actions are the most frequent (direct) cause for accidents was one thing, dealing with these fallible humans might be another. During the Heinrich/Blake debate, described in the previous chapter, Heinrich stated that knowledge of causes was to guide corrective action, but should not prohibit giving priority to actions on "subcauses." Earlier, he had already cautioned that "…it is unwise to depend automatically and invariably on educational or supervisory methods…" (HWH1940, p.112), and consistently throughout his work, he advocated making workplaces safe as the first priority:

> The most ardent supporters of the belief that man-failure accident causes are predominant are, nevertheless, firmly convinced that mechanical guarding and correction of mechanical and physical hazards is a fundamental and a first requirement of a complete safety program.
> (HWH1941, p.18, 1950, p.16, 1959, p.19)[9]

He found it necessary to emphasise that "This attitude is not at all inconsistent with the emphasis placed herein on the importance of man-failure as a causative factor" (HWH1941, p.18, 1950, p.17, 1959, p.19). Because, even though, "Unsafe acts of persons cause the great majority of industrial injuries, but mechanical hazards are also important, and *their correction should be a first and continuing step* in all industrial operations" (HWH1943c, p.139, emphasis added). This is also clear when reviewing the contents of Heinrich's books where approximately a quarter of the pages was devoted to the subject of machine guarding, along with discussions of engineering revision, illumination, and other subjects.

Numerous examples appear through his work. In his presentation about *The Safety Engineer Aids the Life Underwriter* (HWH1932c), he discussed five examples with focus on redesign of the process and circumstances (and little attention on correcting the person), telling how increased production and safety went together thanks to applied engineering revision. In his 1940s management books he would make similar statements.

This is in line with current-day safety authors.[10] For instance, "There is widespread agreement, however, that the most effective way to deal with hazards is not by altering human behaviour but by redesigning machines and systems of work so as to eliminate the hazards" (Hopkins, 1994, p.30). Apparently, Heinrich knew he was being misunderstood and found it necessary to state,

> When taken out of context, this emphasis on man-failure and its causes may be taken erroneously to belittle the need of mechanical guarding. For this reason, it may be well to state that the guarding of machines and mechanical hazards has been and always should be a fundamental of a complete safety program. Incidentally, guarding and other action of an 'engineering-revision' nature often provide an immediate remedy even for accidents caused chiefly by man-failure.
>
> (HWH1959, p.34)[11]

7.2 Various concepts

Heinrich's work contains several concepts belonging to the human element theme. Some of these have received critique or are debated. Some others, however, are not contended at all and discuss factors that are as relevant today as they were in Heinrich's time. Examples of these include potential risk factors as fatigue and age.

Fatigue was singled out as a "personal subcause" (HWH1941, p.284) with dedicated chapters in the second and third edition. The short chapter(s), mostly a guest contribution by J.R. Garner, MD, briefly discussed its complexity (because of the interaction with other causes and factors) along with problems and possible remedies.

The relation of age to industrial accidents appeared in all editions as a short section of the *Safety Psychology* chapter with more information in the appendices of the first three editions. It was indicated that older employees tended to have lower accident frequencies and less absenteeism. Although reasons were not entirely known, it was assumed that experience and skill coming with age were part of the explanation.

Contrary to some of his contemporaries, such as Lange, Williams, and De-Blois, who had lengthy, dedicated sections about "the new man," Heinrich paid little explicit attention to the education of new employees. However, a brief mention of minors being easily distracted and the need for supervising them appeared in the appendices about the influence of age.

7.2.1 Psychology

The subject of safety psychology had dedicated (but relatively short) chapters in all editions of *Industrial Accident Prevention*.[12] Heinrich lent great importance to the subject, stating among other things, "Psychology lies at the root of the sequence of accident causes" (HWH1931a, p.127). Manuele critiques Heinrich's belief in psychology:

> Although psychology has a place in safety management, the emphasis Heinrich gave to it as being 'a fundamental of great importance in accident causation' was disproportionate, and that overemphasis influenced his work considerably.
>
> (Manuele, 2002, p.77, 78)

Manuele questions especially whether the people who were to apply psychology in their efforts to prevent accidents, notably foremen, were qualified at all. But this critique may have been missing or ignoring a few points.

To some degree, Heinrich's belief in the importance of psychology reflected the thinking of many others of his time.[13] Münsterberg, for example, wrote, "The understanding of psychology is one of the most important roads to success for the modern businessman" (Münsterberg, 1918, p.v). Peter Petersen (1990) notes during the mid-1920s a revival and rediscovery of the ideas of industrial psychology pioneer Münsterberg. Through the work of scholars and authors such as Viteles (appearing in Heinrich's bibliography in 1950 and 1959) and Fisher, his ideas were discussed again.

It is a problem that Heinrich not clearly defined how he understood "safety psychology." At the end of the *Safety Psychology* chapter, he mentioned "fatigue, home conditions, long hours, frequency of accidents after holidays and early in the morning or just before quitting time, age, and other bodily and mental conditions" as some examples of "psychological accident factors" (HWH1931a, p.142). Many of these factors may not strike today's reader as dealing with psychology as such but may be counted within the field of human factors.

From Heinrich's work, we can deduct some subjects he connected to psychology, giving a better understanding of what he meant. In the first edition, he started by suggesting that safety psychology could be a further development connected to underlying causes. Through the discussion of several cases, the text presented the underlying assumption that there was "something" inside the human that could be explained and fixed. This "something" was beyond reach at the moment but might be accessible in the future. It was still largely a "virgin field" (HWH1931a, p.140), but he saw great gain, "...investigations in this direction are quite as likely to lead to reward in ultimate returns as are expenditures for fundamental research in chemistry" (HWH1931a, p.129).[14]

Some of Heinrich's thinking went into the direction of accident-proneness, although the "basic causes" from studies as the *Safety on the "El"* report pointed towards a variety of causes, medical, attitude, and training, several of which were non-psychological. Still, a lengthy section of the *Safety*

Psychology chapter was dedicated to discussions of accident-proneness with references to the work of British researchers such as Farmer and Chalmers, and Greenwood and Woods.[15]

So, Heinrich's first understanding of safety psychology was about mechanisms and factors that might give information about underlying causes of accidents. Knowledge about these could aid future safety work, given scientific progress. There was also another understanding of psychology, namely as a practical tool to assist accident prevention. This was already being applied and sufficiently understood for all practical means, "a great deal of psychology is an inherent part of properly conducted safety engineering and is applied daily under the guise of better understood terms" (HWH1931a, p.140).

This understanding of psychology was clearest when Heinrich described how managers and safety practitioners dealt with humans by using what he in his later writings called "lay psychology" (HWH1944, 1949, 1969). This approach is found especially in the chapters on "creating interest" and in his management books.

Managers and safety engineers needed not become psychologists; it was okay to apply heuristics. "There is such a thing as a rule-of-thumb, shop-type, homely yet practical and fairly effective brand of psychology which, in some degree, is possessed by most everybody" (HWH1950b, p.8). People learned to apply this through experience:

> ...the average person learns and practices a form of psychology, not specially studied, not professional, not always too good, but never the less fairly effective... a shop-type brand of psychology that permits him to deal with the human element in accident causation.
>
> (HWH1950b, p.8)

Heinrich suggested four main categories of preventing and correcting acts and conditions: engineering revision, instruction and appeal, personnel adjustment, and discipline. All of these[16] required some form of simple psychology, emphasising the importance of psychology for safety engineers. Especially in his later work, Heinrich spoke frequently of how traditional safety engineering and psychology needed to supplement each other,

> Psychology and engineering must work together in the industries because the application of mental processes is in, and is strongly influenced by, a mechanical environment. Therefore, the first requisite of the accident preventionist is in engineering. Directly coupled with this qualification is that of the ability to understand and influence persons.
>
> (HWH1950b, p.8)

Even more strongly was the argument from the opening of the 1959 *Safety Psychology* chapter,

> Opportunity exists for professional psychology to be introduced much more completely into the field of practical accident prevention. Evidence accumulates daily indicating the need for accident-control measures based on better knowledge of human behaviour as it is affected by the traits, feelings, processes, and attributes, collectively, of the mind.
>
> (HWH1959, p.170)

The unpublished paper about *Psychology and the Safety Engineer* from which Heinrich quoted in this chapter was a strong appeal to cooperation and for professionalisation:

> Safety engineers agree that psychology has a place of considerable importance in accident prevention. They are not psychologists, but their profession demands that they deal with persons. Clearly this is required because the great majority of accidents are caused by the unsafe acts of persons.
>
> The psychologist could be of tremendous assistance in corrective or preventive action if he would step into the picture and suggest some way whereby the safety engineer may determine why persons act unsafely and what is practical to do about it. An approach of this nature might well lead eventually to the more general acceptance by management of the idea that an industrial psychologist should be included as a member of the plant staff just as a physician is already so included.
>
> (HWH1959, p.171)

With the benefit of hindsight, one might interpret this as an early call for the introduction of human factors into the safety profession.

7.2.2 Accident proneness

This was one of the subjects taking some space in the *Safety Psychology* chapters. Heinrich's stance towards the subject is somewhat unclear. On the one hand, he mentioned it in all editions of *Industrial Accident Prevention*,[17] citing literature on the subject, and one could argue that the first domino with "ancestry" suggested something similar. On the other hand, it appears that little practical remedy came from the subject and he did not pursue the subject as he did with others.

Neither did Heinrich provide a clear definition of what he meant by "accident prone," which is rather typical for much of the literature on the subject

(Visser et al., 2007). Roughly speaking, one can distinguish three different interpretations of accident proneness:

1   Persons with specific personal traits (personality characteristics, or mental/cognitive deficiencies) that cause them having above average number of accidents.
2   Persons who have repeatedly accidents and incidents – on average more than others.
3   Persons suffering from certain factors (mental, physical, or otherwise) that temporary cause them having above average number of accidents.

The most common understanding of accident proneness is perhaps the first alternative where researchers tried to find distinguishing characteristics of people that made them more susceptible to having accidents than others. Some of his work suggests Heinrich thinking of this alternative. However, he thought in broader lines,

> It is known that some individuals are more prone to accidents than others and that under apparently ideally safe conditions certain persons who are considered unusually careful, thorough, conscientious, skilled, physically fit, and experienced nevertheless receive serious injuries.
> (HWH1931a, p.129)

Also, the literature that Heinrich cited was ambivalent. The work for the Boston Elevated Railway took its point of departure in *accident repeaters*, setting out to discover relevant factors. The preliminary study by Slocombe and Bingham suggested links between safe and economic operation, and concluded that those who experienced accidents more frequently were either older drivers with high blood pressure, younger with little experience, or those who had more "delinquencies" in their personal file. These findings left little room for interpretations towards personality factors.

This did not discourage the Railway. Determined to find a "cure" for the "disease" of accident proneness, they looked into the causes of accidents, including their locations, and especially the operating habits of those who had frequent accidents. A series of tests was devised, with close observations by instructors and supervisors. The study eventually concluded that the vast majority (about 75%) of operators had a low frequency of accidents, the second largest group (about 20%) improved their performance thanks to special attention and instruction while the third group to the researcher's surprise shifted from high to low frequency (and back) due to various reasons like health, family circumstances, or a shift in type of rolling stock or line of work. Only a minimal group was suspected to be unfit for the work at all.

The study provided little support for accident proneness as a personality or psychological disorder. It rather suggested that some had a higher frequency due to various circumstances. This supported mostly the third alternative, as other authors suggested. Many of the early safety authors singled out new employees as in a risk group, some pointed towards non-English speaking workers[18] or specific circumstances (fatigue, working hours, environment). Selection, being fit for the job, and training of workers were ways to deal with this and "If the same men are frequently injured, put them where they cannot get hurt" (Cowee, 1916, p.368). This suggestion would be repeated by Heinrich in his "misfits" quote and by Vernon (1935).

The fourth edition of *Industrial Accident Prevention* nuanced the subject accident proneness strongly, although there was some reference to studies dealing with relationships between accident frequency and "safety attitude" (of workers and supervisors) and (in a study about car drivers) of personal traits as impulsiveness. However, the text appeared to define accident proneness mostly in terms of "accident repeaters" (HWH1959, p.176) and concluded that they, "although contributing a troublesome problem, are a relatively small group" (HWH1959s, p.176),[19] suggesting a somewhat different direction:

> This supports the conclusion that the majority of workers having no previous accident history. Thus, the benefits of the psychological approach will derive from influencing the greater number of non-repeater workers, but with special consideration of the relatively small group of accident prones.
>
> (HWH1959, p.176)

Still, the first alternative was not entirely rejected, "It may be that one day we shall be able to determine with reasonable accuracy, those characteristics or qualities of persons which, unless corrected, will eventually lead to unsafe practice, accidents and injuries" (Travelers, 1954, p.10).

Over half a century later, one may conclude that research still has proven to be unsuccessful in finding characteristics to identify accident prone persons. Consensus seems to be that there is a huge number of characteristics or traits that vary from situation to situation. One researcher commented two decades after Heinrich's hope for the future,

> ...attempts to reduce accident frequency by eliminating from risk those who have a high susceptibility are unlikely to be effective. The reduction which can be achieved in this way represents only a small fraction of the total. Attempts to design safer man-machine systems are likely to be of considerably more value.
>
> (Cameron, 1973, p.49)

7.2.3 Ancestry

The element of Heinrich's work that has possibly drawn the strongest reactions, is his labelling of the first domino as "ancestry." It has led to speculations and people calling him a bigot, a racist because he put ethnicity at the root of the sequence, and a supporter of eugenics (e.g. Marsden, 2017; La Duke, 2018a, b).

These claims find little support in Heinrich's writings. Apart from the *Bedford Stone Club* speech (HWH1923b), containing a mildly sexist joke about cavemen practices of finding their wives and a mildly racist comment about "yellow-skinned Japanese," Heinrich's writing is largely clear of sexist or racist comments. He wrote mostly about men, which reflects the era rather than any personal attitudes. On a few occasions, he coloured his stories by having his characters talk with accents. This rather speaks of his talent as a storyteller than that one suspects any racist background.

Eugenics, the set of beliefs and practices that aims at improving the genetic quality of a human population, which many regarded as an accepted science during the first part of last century, does not appear once as a solution for accident problems. Besides, the purely genetics-based eugenics would contradict the other element of the first domino, "social environment," which was about social and cultural factors affecting the person. "The community establishes and maintains the environment of the worker. The environment affects the worker's attitude toward safety and toward his employer" (HWH1934, p.116).

From today's point of view, pointing out "ancestry" as part of the causal chain is not deemed acceptable. And yet it is also intriguing. Because where did it come from? What did it mean? Heinrich hardly explained. The subject remained in the background because of his focus on proximate causes, and the fact that the factors in the first domino were hardly useable for practicable accident prevention.

Heinrich mentioned "ancestry" on several occasions, using terms as "unalterable native endowment" (HWH1931a, p.134), "heredity" (HWH1931a, p.40), "inheritable faults" (HWH1941, p.104), or speaking about a "long list of related causes, which in some cases reach back many generations[20]" (HWH1931a, p.41). Nowhere did he explain where it came from, neither did he point out specifically what he meant.[21] In the accident sequence it was barely explained, "Recklessness, stubbornness, avariciousness and other undesirable traits of character may be passed along through inheritance" (Travelers, 1934, p.2). Neither are there any examples apart from a mention in passing of "family records" which might "justify the belief that reckless tendencies had been inherited" (Travelers, 1934, p.4).[22]

Some of the literature in Heinrich's bibliography provides suggestions as to why Heinrich may have found "ancestry" a valid, yet in practice useless, causal category. Lange spoke little of accident proneness, except for specific groups as non-English speaking employees. This section may have influenced Heinrich in his idea about "ancestry," making it another example of Heinrich presenting the safety knowledge of the time.

> Some interesting figures have been compiled on the non-English speaking workmen. Their efficiency is from ten to forty percent below that of men who understand English. They are injured three times to every two

injuries of their English-speaking fellow workmen, and the injuries are more severe. From the foregoing it is evident you will have to pay special attention to these men.

(Lange, 1926, p.167)

Fisher's (1922) extensive discussion of "predisposition" may also have led Heinrich into this direction, as may DeBlois speaking of "high injury rates among the illiterate, non-English-speaking classes" (DeBlois, 1926a, p.229), supporting this with statistical information. Although he added a nuance because these people would be employed in the "least desirable" work-places, he stated,

...the greater liability of foreigners to accidental injury is generally ad-mitted as well as their relatively low productive efficiency. How much of this is caused by true ignorance and how much is attributable to defects in the system by which foreign labor is put to work and supervised, it is impossible to say.

(DeBlois, 1926a, p.229)

In a paper, DeBlois also mentioned "habits of thought and action that are of racial origin and persistence" (HWH1926b, p.1) that needed to be broken down through safety education.

While we may frown upon attributing accident causes to "ancestry" and "social environment," one has to consider these within their context (but not necessarily condone them). At the same time, one may wonder whether there is really much difference between Heinrich's attribution and blaming accidents on "weak safety culture" as has become common practice in re-cent decades (Guldenmund, 2010). With a decreasing impact of traditional family situations and increasing influence from organisations on their em-ployees, one wonders whether organisational culture has taken over the role in Heinrich's first domino.

7.2.4 Human error?

Modern safety authors frequently associate Heinrich with the term "human error," as in, for example, "Heinrich, well known for his theories regarding human error" (Blokland & Reniers, 2019, p.254), "...Heinrich, who declared that 88% of all accidents are the result of human error" (Heragthy et al., 2018, p.3), or suggesting that the "concept of human error became part of the safety lore" with Heinrich (Besnard & Hollnagel, 2012, p.16).

These authors impose a contemporary interpretation on Heinrich's asser-tion. Heinrich mentioned commonly and frequently terms as "man-failure," "unsafe act," or "unsafe practice." These can be related to error, but Hein-rich did not use the term "human error" until very late in his work and even then, only rarely. One of these occasions was when he spoke about the early days of accident prevention, "Activity centred about mechanical hazards to the neglect of human error" (HWH1956, p.38).

Occasionally, Heinrich did speak of error, but mostly in a somewhat different context, for example, "Accidents are, in fact, errors. They violate the principles of nature and likewise of economy. They are concrete and specific evidence that something or someone has failed" (HWH1931c, p.55, 56). The most intriguing mention, however, is found in the first edition of *Industrial Accident Prevention* where one finds a short section titled "Errors and Accidents" in the *Safety Psychology* chapter. The first paragraph appeared exclusively here, referring to another study by Heinrich which apparently never saw the light of day, confirming his interest in the matter:

> The author is now making a comprehensive study of the relation between error-repeaters and accident-prones or accident-repeaters. The data are as yet inconclusive, but it is becoming clear that a relation does exist and that the application of a true remedy to the employee who is prone to have accidents improves the error record of the same individual. This is but another way of saying that safety and efficiency go hand in hand.
>
> (HWH1931a, p.137)

7.2.5 Carelessness

During the early days of safety, it was very common to attribute many accidents to carelessness – and in many ways this still is the case today. The title of a June 1916 article in *Safety Engineer* is quite illustrative: *The ABC of "Safety First": Always Be Careful* (Koenig, 1916). Another stark example is Van Schaack's book on safeguarding which opens with "ignorance" and "carelessness" before moving on to clothing and environmental factors as illumination and defect machinery.

> It is astonishing how ignorant many workmen are, not only of mechanical arrangements in use in factories, but of palpably dangerous general conditions and the necessity of exercising ordinary care to avoid being hurt. It is equally astonishing how large a part native curiosity plays in the causing of accidents.
>
> (Van Schaack, 1911, p.8)

According to Van Schaack, this could be limited by the employer by making sure to employ qualified personnel, by setting safety rules, putting up warning signs, and by having active foremen and supervisors. A manual for workers said,

> Carelessness is one of the principal causes of accidents. Employees should therefore exercise reasonable caution in all ways and at all times.

Scuffling, 'fooling', running around the plant, and voluntary exposure to unnecessary risk, should be avoided, and all the employees should work together to suppress activities of this kind. It is usually the younger and more thoughtless ones who are responsible for the trifling, and for the practical jokes and 'horse play'; and the older employees should exert their influence to curb these spirits and maintain order and decorum. Careless and thoughtless acts often result in injury to other persons, as well as those who are immediately responsible for them. The natural and unavoidable dangers of the work are great enough, and they should never be wilfully or heedlessly increased.

(Travelers, 1914, p.9)

When Eastman started her research for the *Pittsburgh Survey*, she was told that "95 per cent of the accidents are due to carelessness" (Eastman, 1910, p.84). As discussed in the previous chapter, she had a very critical stance towards this belief, calling it "a heap of unreasoned conviction" (Eastman, 1910, p.85), dismissing it with statistical arguments. Some of the early safety authors were careful about carelessness. Beyer, for example, reflected on the subject of "fault,"

...very few accidents occur where there is not a certain element of fault or carelessness on the part of someone, and the injured man is usually more or less to blame; such carelessness, however, is of an inevitable sort, and might be compared to upsetting a cup of tea at the breakfast table. This involves carelessness, but such carelessness is an unavoidable characteristic of our human make-up. Where men are working around machinery to which the human body offers practically no resistance, a slip or mismove, insignificant in itself, may prove disastrous.

(Beyer, 1916, p.2)

Others were more normative. Lange thought carelessness was not to be tolerated (Lange, 1926, p.32), even suggesting an "unsafe practice committee" whose tasks included sanctions. He also had a quite harsh section on discipline and the obligation to follow rules ending in "YOU DO YOUR PART MR WORKMAN" (Lange, 1926, p.179, CAPS in original).

While Heinrich did not use the term "human error" often, like many of his peers, he used the term "carelessness" several times throughout his work. At one point, he even somewhat sloppily stated – without the nuance that he often applied –

Carelessness in its many forms – inattention, poor judgement, slipshod handling of machine tools and materials; carelessness in giving, understanding and carrying out orders; and violation of safe-practice rules – is responsible for 90 per cent of all industrial injuries.

(HWH1930b, p.150)

However, one might suspect that this was a slip of the pen, because on other occasions he was clear that "carelessness" was not a "real cause." In the notes to the first *Origins of Accidents* chart, he noted, "Terms such as 'carelessness', ...have deliberately been omitted because of ambiguity" (HWH1931a, p.47). Another example can be found in the notes of a meeting with engineers, where Heinrich explained to them how to deal with an "employer who hides under the term carelessness" (HWH1948, p.2). Blake agreed on the critical stance towards "carelessness" and included in his book a sub-chapter titled, "Carelessness not a cause of accidents."

> Instead it is an alibi for industrial executives, foremen and others who unthinkingly are placing the blame for accidents on the workers who are injured. It serves as a boomerang, too, for its use condemns the person who uses it; it is an unthinking admission on his part that he is making little or no intelligent effort to control the actions of workers.
>
> (Blake, 1945, p.47)

Carefulness plays a role in safety, an important role, even. Most readers will have heard their mothers cautioning them to be careful, and no-one phrased it better than John Adams saying, "risk perceived is risk acted upon" (Adams, 1992, p.30), indicating the relative and dynamic nature of risk. Once noticed, humans approach risk differently than they did before they notice. However, carefulness is not the whole story, and only takes you so far, in part because of the inherent limitations of humans.

7.2.6 Shame and blame

From attributing accidents to "man-failure," "faults of person," or "carelessness," it is a small step, willingly or not, to blame someone for what happened. Therefore, Heinrich is frequently associated with the practice of "shaming and blaming" the worker at the sharp end after an accident or mistake. This makes sense, given Heinrich's conclusion that most accidents are caused by human failure. However, attributing a cause is not necessarily the same as blaming, even though one superficially may be led to believe that it is the same. This belief is even enforced by Western culture which regards people as being responsible for their own acts, and therefore "cause" is often associated with "guilt" or "blame" – at least partly (Guarnieri, 2012).

Literature suggests various drivers for accident analysis. Rasmussen mentioned analysis for explanation as to what happened, for allocation of responsibility, and for system improvement (Rasmussen, 1990). An investigation usually cannot cover all of them. Heinrich's prime driver for the investigation of accidents was accident prevention. It is in the title of his book, and clearly stressed on various occasions: the knowledge of causes was necessary to apply practical preventive and corrective remedies. Still, as the next chapter will show, he was also interested in responsibility, and cautioned, "We need to differentiate between blame (guilt) and responsibility. A foreman need not hold himself morally guilty when an employee in his charge is injured, but he cannot and should not evade responsibility under any circumstances whatsoever" (HWH1929e, p.52).

In allocating responsibility, Heinrich did not place the main part with the workers, but first with employers, managers, and supervisors. If he was blaming, one would expect him to rather be blaming them. However, when speaking about "Faults of Person," which often were causes on a supervisory level, he emphasised,

> In making use of this word 'fault' no implication is intended that 'blame' or 'guilt' or any other undesirable quality invariably attaches to the faulty person involved. As can be seen from the list, many of the so-called faults are quite natural and 'forgivable'.
>
> (HWH1944, p.158)

Another reason for associating Heinrich with shaming and blaming might be his at times highly normative and judgemental language. When he talked about "carelessness," "moral delinquents" (HWH1928a, p.10), "moral causes," and "moral fault" (HWH1931a, p.42), or even concluded that the "moral attitude of the employer may not have been ideal" (HWH1931a, p.40), it is hard to understand these terms otherwise than blame, even though he may not have intended it as such. On a few occasions he even left no doubt, saying that accidents are a result of "...stupidity and lack of skill, and becomes something to be ashamed of and to be avoided" (HWH1931c, p.51).

Finally, it is worth noting Heinrich's attitude with regard to dealing with "violations" and sanctions. While Heinrich sometimes saw accidents as "evidence of a bungled job" (HWH1938b, p.5), he urged against the negative, blame and shame-oriented approach, "don't hold up a workman to ridicule by using the 'horrible example'" (HWH1938b, p.5). Heinrich did mention both commendation and ridicule as motivational factors but was quick to point out ridicule's negative side-effects and cautioned to use it by exception (HWH1941, p.355). Likewise, he emphasised that disciplinary approaches were the very last resort for managers (HWH1944, 1949), in part because they indicated managerial failure.

7.3 Humans as an asset

Although Heinrich is best known for his message of controlling human action, his work shows that he did not see humans solely as a problem of an otherwise perfect system, but also as an essential factor in creating and improving safety. For many this may be a surprising side of Heinrich. Already in his very first documented safety speech, he emphasised the importance of doing safety *with* the workers, not to them,

> A safety organisation which is not of, by, and for the men cannot very well be successful. Employees should not be forced but rather should be educated to the point where they will take the initiative themselves in safety matters.
>
> (HWH1923b, p.7)

Involving workers in accident prevention was important. He recognised that "most workers take pride in their skill and ability" (HWH1931c, p.49) and by getting workers interested in safety, they might be engaged in the "development of ideas that would lead to safer procedure" (HWH1931a, p.120). Even though human caused accidents, Heinrich also saw them as an essential positive factor and advised to draw on their strengths, "...in many cases safety may be promoted by employing the ingenuity and mechanical genius of workmen in the average plant, in the revision of process and procedure" (HWH1931a, p.246).

In the *It's Up to the Foreman!* paper (HWH1938a), Heinrich handed the reader five methods for safety intervention, most of which were surprisingly participative: "Arrange for the individual to play a definite part in safety work," "Seek the advice of workers on safety problems," and "Appeal to skill."

Even though he spoke much of the guiding, controlling, and enforcing role of foremen, stressing the importance of safety rules and compliance, Heinrich also acknowledged the importance of autonomy: "men do most frequently and freely the things they want to do" (HWH1938a, p.4).[23] He concluded, if "workers themselves are so well-informed and safety conscious" then there is no need for "urging to work safely" (HWH1938a, p.4). This connected in a way to the opening part of his 1940 paper, where he preferred "...capable, and experienced men who work under unsafe conditions" (HWH1940, p.112) over "...incapable, and inexperienced men who work under safe conditions" (HWH1940, p.112).[24] This resonates well with "new view" thinking of resilient people who have the capability to handle variations in uncertain environments.

In his paper about *Warehouse Safety Problems*, Heinrich again emphasised the importance of worker involvement and using their ingenuity and knowledge, "Remember that the men under you are part of your organisation and want to know what is going on. You will find they will have ideas. Keep them informed" (HWH1943c, p.136).

The most spectacular example of Heinrich portrayal of humans as an asset to create safety, however, is found in the transactions of the National Safety Congress where Heinrich presented the *Origin of Accidents* chart.

He saw the employee as a "root-remedy" and employer-executive as a "secondary root-remedy" because they were capable of controlling causes:

> ...I have indicated in the lower rectangle that the employee is a root-remedy – as of course he is. An employee need not go to work in a physically unsafe plant, and if he is the proper type, he may safely perform his work even under unsafe conditions and without proper supervision.
> (HWH1928c, p.129)

Again, this could be regarded as prototypical Safety-II, or Resilience Engineering (Hollnagel, 2014a). Many years later, he would say something similar in the opening sentence of another National Safety Congress speech. This once more sounds like "new view" in that it pictures humans as both the source for accidents and of safety: "In all truth it may be asserted that the human element, basically, is both the cause and the control of all industrial accidents" (HWH1950b, p.7).

7.3.1 The reward of merit

One of Heinrich's most interesting pieces of work, totally different from most of his other writings, is *Reward of Merit*, based on his address at the April 1930 Sixth Annual Eastern Safety Conference. It may also be his most self-help-oriented work because there is a strong suggestion of how to become successful (and safe) through a list of principles (e.g. alertness, knowledge, initiative, responsibility, foresight, concentration, goal oriented) that together or situational will be beneficial. The engaged, energetic style is striking, most likely this was a powerful speech.

The paper is interesting in several ways. First, the opening, with a discussion of "outworn beliefs," myths, and superstitions, contrasting these to beliefs supported by facts (i.e. "science"). Then the emphasis of safety as a positive thing and how Heinrich stresses that risk is necessary: "The average person, however, must work to support existence and therefore cannot avoid participating in many things that include an element of danger to himself. In fact, a normal person is socially inclined and wants to 'take part'" (HWH1930b, p.146). Especially the last part is interesting, characterising humans as risk seeking creatures. With good reason, because: "...safety and success, are founded upon the same factors, that pioneering and daring go hand in hand with safety, when properly carried on" (HWH1930b, p.151).

With some minor adjustments large parts of the paper are translatable to 2020 and new approaches to safety. Parallels can be seen in the call for evidence-based approaches and taking as a premise that the sources of safety and success are identical. One even finds three of the four "cornerstones of resilience" (Hollnagel et al., 2011): anticipation, response, and monitoring.

Heinrich spoke of individual resilience, however, about the ability to handle hazards and variability. Through this focus on the ideal of an individual who can handle things (and in passing also a brief discussion of organising for success), there is some straying into safety as a (moral) choice. The paper did contain some sections that speak about carelessness, but the main message was elsewhere. Safety depends on action, Heinrich argued; it has to be created actively. The closing section was another call for seeing safety as part of normal work:

> Accident-prevention work has for too long been relegated to a position similar to that of semi-charitable institution. It has been considered as something that is deserving, that should be supported, and that would receive attention sometime - probably as soon as some of the more important moral-uplift and foreign-relief-work questions had been disposed of. Such an attitude is distinctly harmful to society, to industry, and to the individual.
>
> (HWH1930b, p.150–151)

Unlike other papers from the period, this did not have major impact on *Industrial Accident Prevention*, but one can see similarities between the paper and Heinrich's mid-1940s books for supervisors. To put his thoughts about the human element in the right context, it is important to take note of his writing on management which would dominate his work from the 1940s onwards. That is the subject of the next chapter.

Notes

1 However, even worse were some of the newspaper pieces following the *Origin of Accidents* study, reporting that "… a man who is mentally disturbed is as much of a danger to himself and fellow workers as if he were physically disabled" (e.g. Reading Times, 15 November 1928; Muncie Express, 20 November 1928).
2 This was in no way a new idea. Donkin discusses French scholar Charles Dupin who preceded Heinrich by a century. Dupin wrote in 1829, "We have been very much occupied in perfecting the machines and the tools which the worker uses in the economic arts. We have hardly attempted to improve the worker himself." Interestingly, just as Heinrich would, also Dupin noted positive characteristics of humans as well, emphasising their "immeasurable advantage of being an instrument who observes and corrects himself, a self-stopping motor which functions with the motivation of its own intelligence and which perfects itself by thinking not less than by work itself" (Dupin, 1829, in Donkin, 2001, p.141).
3 One can interpret this passage as a recognition of risk homeostasis where "safer" machines lead to "unsafer" handling. However, it is more likely that Heinrich not spoke of newly introduced hazards, but rather of previously unaddressed ones.
4 Harrisburg Telegraph, 26 July 1933.
5 The 1944 book and the paper on the "human element" (HWH1950b) contain similar discussions.

6 The 1950 paper is quite interesting here, both because it shows that Heinrich understood hindsight bias, and was able to handle critique. After giving an example of applying a motivating characteristic (appealing to someone's ambition of leadership), he said,

> A friendly critic to whom this story was told said, 'I see nothing unusual or particularly psychological about it. It looks to me like a simple case of bribing a workman with a half-promise of promotion. The supervisor, knowing so well that Tom Burke was susceptible to the lure of leadership, should have had common sense enough to try that approach in the first place.' This criticism is highly illuminating. It illustrates vividly the very circumstances that point to the need of expressing psychology in shop language. Indeed, the critic is correct in his conclusion that the case required merely a simple common-sense approach. However, its simplicity is evident only now that all the facts are known.
>
> (HWH1950b, p.10)

The case is also a good demonstration of how Heinrich was able to take abstract and theoretical concepts and translate them into understandable language and practical tools.

7 Possibly because he did not see a practical application to accident prevention of what Watson did. As described in Chapter 3, only Skinner's research and its successful application in other domains triggered safety practitioners to explore the possible application of behavioural approaches in safety.

8 One may say that there are strong similarities between Heinrich's looking for reasons for unsafe behaviour and BBS's looking for antecedents and consequences, but that similarity applies to any method looking for underlying causes/factors.

9 Interestingly, this passage preceded the discussion of the 88:10:2 ratio in the referenced editions of *Industrial Accident Prevention*.

10 Also, other sources from Heinrich's era suggest he was not alone in his thinking, advocating the same prioritisation when selecting remedies:

> The first rule of safeguarding is this: If a certain kind of accident can be prevented either by a mechanical guard or by carefulness on the part of the employee, always provide the guard.
>
> (Williams, 1927, p.90)

A war-time *Handbook of Industrial Safety Standards* to which Heinrich may have contributed said,

> Safety organisation is essential to successful accident prevention, but no matter to what extent we are able to provide 'safe men', we cannot afford to neglect the safeguarding of their working conditions. Such measures protect against those moments of distraction, thoughtlessness and fatigue to which all are subject at times. They are the best guarantee against accidents to the newly employed, who are otherwise more prone to accidental injury than are seasoned workers. Physical safeguards, moreover, are evidence of the sincerity of the employer's statement to his employees and to others that he believes in maintaining a safe establishment.
>
> (National Conservation Bureau, 1942, p.6)

11 Similar quotes are found elsewhere, e.g. (HWH1940, p.112, 1941, p.35).

12 The chapter moved around in the book. For the second and third edition it was placed only *after* the long chapter on machine guarding. The1941 version was almost identical with the original version of the chapter, while the main change

in 1950 was that the three lines of development for accident prevention was expanded to five lines of development, giving prevention based on subcauses and underlying causes an explicit place. The fourth edition of *Industrial Accident Prevention* provided significant changes to the *Safety Psychology* chapter. It was moved back to the first half of the book, directly after the chapter on *Creating and Maintaining Interest* and while parts of the original text returned, the text first drew together various psychological strands from other chapters in the book. Then it quoted a lengthy section of a hitherto unpublished 1954 paper arguing for the need of psychology aiding safety engineering, notably though making psychology usable and understandable, through "conversion of the applicable principles of psychology to lay language" (HWH1959, p.171).

13 DeBlois (1926a) drew strongly on the work of Boyd Fisher (1922) and his main categories of "mental causes of accidents," spiced up with many quotes from other relevant authors. Interesting is that little of what is discussed would be counted as psychology these days, as most deals with what is now understood as ergonomics. One also finds a lengthy section on illumination, something that other authors discussed in terms of conditions – which only illustrates how arbitrary the division between men and conditions can be...

14 Some went considerably further in their thinking of safety psychology. For example, Dr. Meyer Brown, lieutenant in the U.S. Naval Reserve Medical Corps and associate professor in nervous and mental diseases at Northwestern Medical School, at the *Fundamental Causes of Accidents* session of the 1942 National Safety Congress,

> I believe the psychiatrist can render a service to industry in this way. Our biggest job is education; to make foremen, plant managers, safety engineers realize and understand more fully these mental misfits, to be a little more familiar with psychological quirks in normal people, to realize that a man's performance on the job depends not only upon the kind of man he is, but the kind of place he is working in and the kind of home he comes from. I think in some cases psychiatric consultation may prevent a serious accident if a patient is seen early. Some of you may have come across accidents of a very serious nature committed by a man who several months later was in an institution for the insane. A trained observer could have detected that case very early.
>
> (NSC, 1942a, p.84)

15 Unlike most other chapters, the *Safety Psychology* chapters had relatively many quotes and footnotes.

16 Heinrich claimed with the exception of engineering revision. However, one might say that also engineering and process revision requires insight in how and why humans behave, and thus psychology.

17 The first two editions had an appendix on "Causes of Accident Proneness" drawing on a study from 1923 to 1928 on the Woodhill Division of the Cleveland Railway company. This study was very similar to the Boston Elevated Railway study.

18 Interestingly, DeBlois (1926a) nuanced the possibility of accident proneness by suggesting that these people are often employed in higher risk occupations or companies without proper provisions.

19 Simonds and Grimaldi cite a study concluding similarly, "...the accident repeater is not a significant factor in the safety problem" (Simonds & Grimaldi, 1956, p.389).

20 As a word of caution. The passage does not state clearly what exactly reaches back generations. It may be about mental or physical characteristics that a person inherits, but just as well about "latent condition" as may be the case in the fall of a tower mentioned further down on the page.

21 In a talk about *Home Safety*, Heinrich may have been hinting at the first domino saying, "Like charity, safety begins at home" (HWH1951, p.8) because there was a "paramount influence of safe home life in the accident occurrence in the shops and on the highways" (HWH1951, p.8).

22 During Heinrich's lifetime, Lateiner changed the description of the first domino from "Social Environment and Ancestry" into the more neutral "Background." No material survived to shine light on the reasons for this, or Heinrich's thoughts about the change. Neither does Lateiner's change illuminate its meaning because Lateiner does not describe the first domino in any elaborate way either. At one occasion he illustrates it with an example of a "sick child at home" (Lateiner, 1961, p.261) providing a distraction for the worker. Bird would move even further from Heinrich's first domino by attributing it to "Management" (Bird & Loftus, 1976). Interestingly, Weaver's dominos (1971, p.24) showed "heredity" instead of "ancestry."

23 Similarly, he discussed foremen as generally being competent and coming to work to do a good job: "...foremen in general know their stuff - sense their responsibilities - are eager to do their part and are already doing a grand good job..." (HWH1942c, p.191, 192).

24 This text was subsequently included in the second to fourth edition.

Chapter 8

The role of management

Heinrich is widely known for his focus on workers' behaviour; however, (top) managers were his primary audience (HWH1931a, 1941)[1] – one of many things he agreed on with DeBlois. Acknowledgements that Heinrich "also" paid attention to management often focus on his writing about foremen/supervisors (Swuste et al., 2013). Indeed, these middle/low managers got much of his attention. Stating that this was the only group managers that had Heinrich's attention means grossly underestimating his scope. This chapter aims to give a fuller picture, discussing subsequently, the role of foremen and supervisors, the role of executives and the employer, responsibility (especially of management), accident prevention as part of everyday business, and safety management, including Heinrich's Axioms.

After the second edition of *Industrial Accident Prevention*, Heinrich concentrated on management with a series of papers and books, mostly aimed at supervisors. This was partly a contribution to the war efforts, since supervisors were seen as key persons in production efforts. There was another reason, however. According to Jesse Bird, management was Heinrich's favourite subject (Bird, 1976). One sees the theme throughout his entire work. Already Heinrich's speech at the Bedford Stone Club (HWH1923b) dealt with safety management and organisation, and his last published work, the posthumous book with Lateiner (1969), was a reworked version of his supervision books.

It is important to understand that the roles of top management and foremen in the late nineteenth century and early twentieth century differed from today. At the time, foremen hired workers, directed the work, and they were also responsible for safety. Employers, factory owners, and top managers did not involve themselves in these matters. Only later came the realisation that top managers and employers did have a significant responsibility and role in safety. DuPont pioneered this view. They and others as U.S. Steel provided an example for other organisations to follow (Aldrich, 1997; Swuste et al., 2019).

Aldrich speaks of a "shift in perspective from work accidents as routine matters of individual carelessness to the modern view that accidents reflect management failure" (Aldrich, 1997, p.2). Heinrich's work, emphasising that management had the best opportunity to improve things, fit in this shift and helped further the modern point of view. This view was shared by people outside of safety. Lansburgh opened his influential book,[2] *Industrial Management*, by stating,

> Management, the unseen force which drives all that is physical within a factory, is by far the most important factor of the present industrial age. Machinery and materials may be put to work, workers may labor; but without adequate management to organize and consolidate them into a profitable, co-ordinate whole, to distribute the results of their work effectively, and to govern their operations during performance, this performance may become so uneconomic as to cease entirely.
>
> (Lansburgh, 1928, p.1)

Today, focus on management is possibly stronger than before as evidenced by a continuous stream of management literature. It is also reflected in management standards (e.g. ISO 9001 and 45001), where management responsibility is an essential element, and in regulations as the health and safety at work acts of many countries that foremost stress the duties and responsibilities of employers over those of employees.

Before turning to higher management, let us start with what at the time was generally regarded as the "key people" in work situations.

8.1 Foremen and supervisors

8.1.1 Contemporary thinking

The first group of management that had Heinrich's interest were what we today would call middle managers. They were commonly referred to as foremen in the early twentieth century, and later supervisors or superintendents. They were in close contact with the day-to-day work and employees and were regarded as being in the best position to anticipate, observe, and correct "unsafe behaviour" and physical hazards. Focus on foremen was a constant theme in early safety literature and is found in all the early safety handbooks and many articles. "The foreman is generally the responsible head for the workers immediately under him, for he sees that new workmen are instructed in the use of machines to which they are assigned" (Tolman & Kendall, 1913, p.28).

Foremen were the intermediate between top-management and the workers:

> The employer is not always in close personal contact with his men. In such cases, he must rely on his superintendent and foreman to keep in

close touch with the employees, so as to give them the proper instruction and advice as to their safety welfare.

(Beagle, 1917, p.136)

This gave them a key position in both production and accident prevention. One finds this in many texts. Foremen held "the strategic point in the warfare against accidents" (Tolman & Kendall, 1913, p.28), they were "the real keystones of 'Safety First'" (Bowie, 1916, p.238) and they were to be used "as centers from which to radiate safety ideas" (Ashe, 1917, p.26). The foreman was seen as the "key man." "He is the top sergeant of industry. To the men, the foreman represents the company, and their morale is largely determined by his treatment of them" (Williams, 1927, p.20).[3]

For workers, the foreman was the visible representation of management. "Foremen can do much towards securing the interest and support of employees. They can do more than anyone else in the company to make the safety organisation a success" (Cowee, 1916, p.9). While other authors would argue that the role of top management is even more important, all agreed that foremen should lead by example, demonstrating the importance of safety.

If foremen do not display real, continuous, intelligent interest in applying safety methods in the same spirit in which they exhibit their technical abilities, and in that manner lend the influence of their personalities, almost complete failure will be the result.

(Bowie, 1916, p.238)

DeBlois (1926a) was among those writing about the role of top management, but he also emphasised the important role of direct line managers, especially in controlling the "human factor." Guarding, design, and education created safety progress, but "...the results of engineering revision, whether general or specific are worthless unless followed by supervision, and by continued supervision" (DeBlois, 1919, p.199).

The foreman, in other words, has every facility and every liberty in instituting accident prevention work and in using discipline if necessary, to make working conditions safe. There is every reason, therefore, why foremen should feel the responsibility for such accidents as occur in their departments.

(Ashe, 1917, p.26)

Not only safety literature emphasised the role of foremen. Lansburgh and Spriegel introduced a chapter on foremen in the reworked edition of *Industrial Management* (1940), titled "The Foreman – A Representative of Both Men and Management." Discussing the roles from a management perspective this matched well with the safety perspective. According to Lansburgh and Spriegel, the foreman's position was leading a unit which interacts with other units and upon which other units depend, just as his depends on others. Foremen had responsibility towards three main groups: management, employees under and materials, environment and equipment. This included taking care of workers, as well as machines, resources, efficiency, quantity and quality of production. They emphasised the importance of a balanced relationship (recognising that foremen might be biased towards management, or otherwise), being fair and impartial, but establishing a good tone with workers.

There was also some dedicated literature aimed at foremen, including an early Travelers Insurance brochure.[4] Under the caption *Relation of the foreman to the employee*, it said "The success of any accident-prevention campaign will depend largely upon the nature of the relation between the foremen and the employees over whom they are in authority" (Travelers, 1918, p.5). Arguing that foremen should have genuine interest in the workers' welfare and treat them fair and friendly but keep their distance and become not one of them. If they were the "right type" and set a good example, then workers would "usually be ready and willing to follow his instructions in spirit as well as in letter; and this makes discipline easier to maintain" (Travelers, 1918, p.5).

8.1.2 Heinrich and foremen

Heinrich followed the contemporary thinking. Already the *Bedford Stone Club* speech mentioned the "vital importance in safety work of the foreman" because this was "the man who knows both ends of the game" and in this key position[5] "let him practice and encourage safety and the rest will be more easily converted" (HWH1923b, p.8). In the early versions of *Origin of Accidents* (HWH1928a, b, 1931a), Heinrich attributed 88% of accidents to causes of a supervisory nature, meaning that in his view foremen could control most of the accidents.

After Heinrich had published his three best-known ratios and started discussing causation it was time for the next main theme. *A Message to Foremen* (HWH1929d, e, 1930a) was the first paper dedicated to foremen, appearing in several versions. From the language used, one senses that this was a subject very close to Heinrich's heart. The paper pointed out qualities of good foremen[6]: "Common-sense logic, initiative, and acceptance of responsibility..." (HWH1929e, p.23), speaking about leadership, taking responsibility, and empowerment.

Over the next years, he occasionally returned to the subject, for example by laying out supervisory methods in *Industrial Accident Prevention* and discussing their roles in accident prevention, "A superintendent or foreman is above all a handler of men. His chief function is to issue orders and enforce them" (HWH1931d, p.102).

While the subject never had been really absent, *It's up to the Foreman!* (HWH1938a) was for the first time in eight years that Heinrich dedicated a paper entirely to this subject.[7] One might say that it marked the beginning of a period where writing for managers was dominating his work. In part it was a call for greater safety awareness and Heinrich thought the foreman/ supervisor was in the best position to intervene. Among others by setting an example and "walking the talk": "actions may speak louder than words" (HWH1938a, p.6).

When the United States entered the Second World War, foremen were a point of focus for the Department of Labor. "Without the active support of supervisors, the most elaborately conceived and carefully worked out program of safety is doomed to failure" (Ainsworth et al., 1944, p.9). Heinrich put his interest in management to good use, doing a series of speeches and papers on the role of the foreman.

> It's a foreman's job, so far as war production is concerned. He is the man who knows that motions are purposeful and useful – what motions are unsafe, delaying and wasteful. He is the man who understands best that there is but one right way – one best set of production motions to do a given job and who knows that way to be the safest and quickest way.
>
> (HWH1942a, p.5, 1942b, p.175)

Heinrich also wrote a book dedicated to first-line managers. This first appeared as a thin booklet, *The Supervisor's Safety Manual* (HWH1943d) and the year after as the book *Basics of Supervision* (HWH1944). The latter was presented in short form in a three-part series of articles in *The Industrial Supervisor* (HWH1945a, b, c). These books and papers were intended as an aid to foremen and supervisors in professionalising their work. Theirs was a major responsibility, "It is a big job. None but a well-qualified person can do it and *he needs help*" (HWH1945a, p.6).

8.1.3 Professionalisation

Long before he wrote his management books, Heinrich advocated the need to support first-line managers. "Unfortunately, in the average case, he enters upon this phase of work uninstructed in the art of executive leadership..." (HWH1929e, p.23), he wrote most likely from first-hand experience. In February and April of that year he had lectured at the Foremen Safety Training courses in St. Louis organised by the Safety Council.[8]

> Industrial foremen are getting 'fed up' on repeated assertions that they are responsible for the safety of their men and that they should accept responsibility and act on it in an authoritative way. They are fed up and

a bit irritated, not at all because they disagree, but because they feel that a burden is being shoved on their shoulders without helpful suggestions as to how best the job can be done.

(HWH1942c, p.191)

This interest returned throughout his career. He saw the need to educate and train supervisors on their own level, teaching them "how," especially because of the lack of (in his opinion) suitable education, which was in stark contrast to their importance: "...it is no less paradoxical that supervision is not being taught, either to practicing or prospective first-line leaders..." (HWH1950a, p.369). He emphasised the need to educate them in their own language, "The basics of supervision should be expressed for their benefit in simple understandable language" (HWH1945b, p.12). Supervisors needed practical aid to be able cope with their problems:

> What he needs to know about business economics and business management, labor relations, psychology, etc. is what can be boiled down out of such subjects into short, crisp bits of information which he can put immediately to good use in solving his everyday problems.
>
> (HWH1945c, p.14)

He also suggested to facilitate experience transfer between foremen instead of preaching safety to foremen – in Heinrich's opinion that had been done too often already. He proposed organising a "round table" where only foremen participated and discussed practical problems and solutions. His expectations were high: "...from what I have learned by inquiry, an honest-to-goodness, down-to-earth meeting would take place between the very men on whom we are depending to get results..." (HWH1942c, p.194).

Heinrich's contribution to the professionalisation of foremen was a structured approach to what he saw as their task, which he at one point defined as "...the art of controlling the performance of persons and, to the extent of authority vested in the supervisor, controlling also the conditions of their environment. Emphasis is placed on control" (HWH1950a, p.372).

Basics of Supervision introduced a new metaphor (Figure 8.1), "The chain of supervision links men, machines and materials with safe production" (HWH1945a, p.6).

The chain consisted of four Groups of four Factors each[9]:

I Attitude

1 Willingness to lead and direct
2 Recognition of responsibility to management
3 Recognition of responsibility to employees
4 Sense of fair play

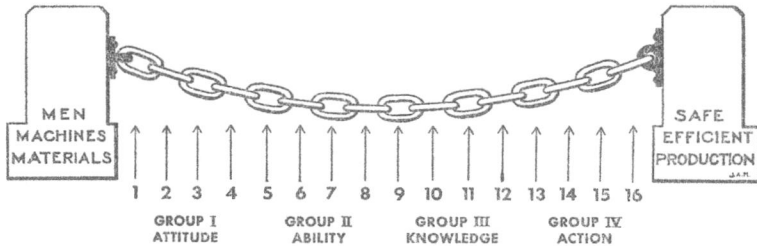

Plate 4: The Supervisory Chain and its Sixteen Links.

Figure 8.1 The chain of supervision (HWH1944, p.13).

II Ability

    5   Planning and organisation
    6   Training and instruction
    7   Observation and analysis
    8   Persuasion, convincing, and command

III Knowledge

    9   Company policy
   10 Job, unsafe, and improper conditions
   11 Worker performance, unsafe, and improper practices
   12 Control methods

IV Action

   13 Training and instructing
   14 Giving orders
   15 Checking worker performance
   16 Prevention and correction

Heinrich, being a man of practice and action, found Group IV the most important. He described the first three groups as "static" because these are things that a supervisor should "have," "be," or "know." The fourth group was the "the dynamic expression of the first three" (HWH1945b, p.12).[10] The "action" group of "training and instruction, giving orders, checking compliance and prevention and correction" (HWH1945b, p.13) may be regarded as kind of a PDCA-cycle (Deming, 1986).

The 16th factor, Prevention and correction, received most attention in the books. He suggested a hierarchy of supervisory action: "engineering revision, persuasion and appeal, personal adjustment and discipline" (HWH1945b, p.13). Worth noting is again that Heinrich first suggested to address the situation and not went for the worker at the sharp end. He explained this in terms that stand up well even today:

> Engineering revision often is a remedy for problems arising out of unsafe and improper practices of persons when these practices are committed because it is difficult, impossible, awkward, uncomfortable, inconvenient, embarrassing, etc. for the person to perform safely and properly. The method also applies when the unsafe and improper practice is so attractive, enticing, convenient, etc., as to make it preferable.
>
> (HWH1945b, p.13)

Discipline, on the other hand, defined as "persuasive or penalty methods of enforcing rules" (HWH1945b, p.14), should be used "only as a last resort when all other methods fail," and then only with "greatest of discretion and caution" (HWH1945b, p.14). Even stronger, "the resort to discipline is an admission of failure" (HWH1945b, p.14).

In *Formula for Supervision*, he expanded the basics, i.e. the four groups of four factors with a "four step formula" to follow in the solving of supervisory problems. This simple formula related well to Heinrich's earlier principles of scientific accident prevention. Again, this provided an easy to remember set of steps to support everyday work.

> First - Identify the problem.
> Second - Find and verify the probable reason.
> Third - Select the remedy.
> Fourth - Apply the remedy.
>
> (HWH1949, p.21)

Heinrich emphasised the dynamic parts of a supervisor's job, illustrated by the description of a case describing all the things that occupy a supervisor's ordinary workday, including many distractions and short-term problem-solving activities and adaptations/adjustments. A supervisor's day was messy and complex: "The foreman doesn't have ideal conditions to deal with. He is an exceptionally busy person if he is really on the job" (HWH1945c, p.14). This suggested that Heinrich understood well the difference between plan and action, or between work-as-imagined and work-as-done (Hollnagel, 2014a).

8.1.4 Supervisors and investigation

Despite the fact that Heinrich emphasised "accident-cause analysis by a trained safety engineer" (HWH1930c, p.1122) as one of the two essential elements for accident prevention, he advocated accident investigation by foremen. In Heinrich's view, supervisors were in the best position to investigate because of their proximity and knowledge of what went on. Also, they were responsible for workplace safety.

> The person who should be best qualified to find the direct and proximate facts of individual accident occurrence is the person, usually the supervisor or foreman, who is in direct charge of the injured person. This individual is not only best qualified but has the best opportunity as well. Moreover, he should be personally interested in events that result in the injury of workers under his control. In addition, he is the man upon whom management must rely to interpret and enforce such corrective measures as are devised to prevent other similar accidents. The foreman, therefore, from every point of view, is the person who should find and record the major facts (proximate causes and subcauses) of accident occurrence.
>
> (HWH1941, p.111)

This, again, was in line with the thinking of the time. Most safety authors assumed that supervisors were closest to the work and that they knew best about the details of what had happened. Already Tolman (1913) suggested that the foreman should be the one to investigate. Of the early safety authors, only Lange was sceptical of having foremen leading accident investigation, "Foremen's report on accidents should not be taken as final when it is possible to avoid doing so, for the reason that foremen sometimes try to shield themselves by giving an incorrect version of how the accident occurred" (Lange, 1926, p.57).

This critique is also heard today. Fred Manuele suggested conflicting interests: "All such personnel will be averse to declaring their own shortcomings. Similarly, it is not surprising that supervisors and managers are reluctant to report deficiencies in the management systems that are the responsibility of their superiors" (Manuele, 2014b, p.35) Additionally, Manuele thought that supervisors were not qualified because they lacked appropriate training.

Heinrich had apparently thought about this. At the 1942 National Safety Congress he replied to a question whether safety education should be extended to from safety engineers to all in supervisory positions:

> I certainly know of nothing more important in carrying on the job of industrial safety than that of creating in the mind of a foreman the knowledge, the belief, that he is, first of all, responsible; secondly, that he has the best opportunity in the world to do something in regard to safety in his department; and thirdly, that he should go out and do it, *especially that part of it which is involved in the investigation of the cause of the accidents which occur in his department.*
>
> (NSC, 1942b, p.109, emphasis added)

Documents as the cause code (ASA, 1941) and the BLS's *Accident Records Manual* (Kossoris, 1944) were intended to aid and guide their work. These documents pointed to the foreman as first analyst and intended helping him doing the job by structuring the process and reminding him of elements to assess. Also, Heinrich's management books spent much time on discussing how to analyse problems and find suitable solutions, for example through the "four step formula."

8.1.5 Post-Heinrich

Heinrich's work about foremen fit well with the thinking about their role in industrial enterprises of the early and mid-twentieth century. Afterwards, and even today many regard direct managers as the key to safety and health work in organisations. Unsurprisingly, those drawing heavily on Heinrich's work followed the same thinking. Alfred Lateiner worked as a management consultant and his book *The Techniques of Supervision* (1954) went through numerous reprints and was published in the updated version until at least 1988. Bird and Loftus discussed the principle of point of control as one of the principles underlying loss control management, "The greatest potential for control tends to exist at the point where the action takes place" (Bird & Loftus, 1976, p.63).

The National Safety Council published a *Supervisors Safety Manual*. Since its first edition in 1956 it went through continuous updates with as most recent the eleventh edition from 2018. The second edition (1961) was subtitled "Better production through accident prevention." The book aimed to assist training of the supervisor, teaching him the fundamentals of accident prevention and basic human relationship techniques. It said in its Preface,

> The supervisor is the key figure in any occupational safety program. The best way to promote hazard-free operations and to arouse safety consciousness in the individual worker is through the supervisor, who is management's direct link to the working force.
>
> (NSC, 1961, p.iii)

Supervision and safety are still relevant focal points for safety authors (even those considered as "new view" safety), as illustrated by recent titles as *Selling Safety: Lessons from a Former Front-Line Supervisor* and *LEAD Safety*.

Over the years several authors have offered critique, for example remarking that foremanship as described by Heinrich does not fit in today's working relations, a subject that will not further discussed at this point. Another point raised by several authors is that one must look further than the direct manager. "Appropriate decisions and effective line managers are prerequisites for successful production, but they are not themselves sufficient"

(Reason, 1990, p.201). Regarding the view of foremen as "key men," Dan Petersen wrote,

> In a way, yes, he is. However, although the supervisor is the key to safety, management has a firm hold on the key chain. It is only when management takes the key in hand and does something with it that they key becomes useful.
>
> (Petersen, 1971, p.17)

This resonated well with what DeBlois remarked four and a half decades before, introducing the next main object of interest from this theme,

> Many have said that the foremen are the keymen in industrial accident prevention. Converting each foreman to safety may perhaps unlock the doors leading to the rank-and-file, but the master key to the entire establishment is the executive head.
>
> (DeBlois, 1926a, p.40)

8.2 Top management

Stereotypical beliefs tell often that employers and top managers show little interest in occupational safety and health. After all, safety and health measures cost money and may hinder production. Interestingly, Aldrich remarks, "...when safety came to the workplace, it came from the top down, not the bottom up" (1997, p.91). He suggests that many workers saw accidents and bad conditions as belonging to the jobs. However, employers and plant owners at some point realised that providing safer and healthier workplaces was in their best interest, contributing favourably to their reputation, improving internal relationships, the humanitarian[11] thing to do, and creating financial benefits through saving costs. The example set by U.S. Steel quickly became the model to follow for other companies (Aldrich, 1997).

Heinrich's main audience were top management. He addressed much of his work to them – despite he did not write about them at length. He was influenced in his thinking by others as DeBlois, who stressed the importance of management – unsurprisingly regarding DuPont's emphasis on the role of management in safety. DeBlois indicated top management as the most important factor of "controlling the destiny of industrial accident prevention" (DeBlois, 1926a, p.vii). Heinrich followed this thinking by placing executive interest and action at the start if his principles of accident prevention. If top management did not want to work on improvement, progress would be difficult, as apparently demonstrated by organisations familiar to the authors, "The safety movement has not progressed faster toward fruition because the industrial executives, especially those of the medium-size and smaller companies, have not learned this lesson" (DeBlois, 1926b, p.3).

DeBlois (1926a) saw repetition of similar accident scenarios in a company as an indication of management not taking safety seriously enough. One finds this link to mismanagement with many later safety authors, such as Petersen (1971, 2001), Turner (1994), and Hopkins (2001, 2008). Although Heinrich did not write about mismanagement, he was clear about the fact that top managers were those who actually have the power and opportunity to make safety happen.

> Compensation laws place the burden of accident responsibility upon the employer of labor, and educational methods reach the employee mainly through the employer. The guarding of machinery must be accomplished by the employer or with his full cooperation. The elimination of unsafe methods and processes is under the control of the employer. And in fact, the entire structure of the accident-prevention program rests upon employer-executive participation and action.
>
> (HWH1927c, p.221)

The Travelers Insurance also propagated this view. Jesse Bird's biography mentions a March 1927 article from Travelers' *Protection* magazine saying,

> Special stress was laid on the fact that it is the human element on a risk that counts. If the man at the top can be convinced of the value of safety engineering and safety organization work, the men under him will quickly swing into line. It may take years for an idea to work up from the foreman or superintendent to the top; it only takes months for that same idea to permeate the entire organization if it starts at the top and works down.

Again, one finds this thinking early on in Heinrich's work and once more the *Bedford Stone Club* talk is a good example. Here, the first of ten points for a "live effective safety organisation" mentioned is top management,

> There must be someone executive in the company, preferably the highest in authority, the President or General Manager, who must himself take an active interest in the safety organisation and he must let it be known that he is so interested.
>
> (HWH1923b, p.5)

This was followed up by emphasising the example set by foremen and executives. They, as he wrote two years later, "...personally, must recognize accident prevention as part of your routine work" (HWH1925, p.256).

One finds attention for top management throughout *Industrial Accident Prevention*. The story opening the book sees the executive vice president taking firm control of the accident problem, setting the course and deciding the action. The book contained an entire chapter on executive enforcement. Heinrich realised that this probably was not the typical reaction of top managers, suggesting "...a perfectly natural tendency on the part of business executives to give first consideration to pressing matters of profit and loss and only second thought, at best, to their moral obligation to insure safe working conditions for their employees" (HWH1931a, p.38, 39). Safety second, rather than safety first. However, while workers and foremen might make mistakes and disregard instructions, *the responsibility lies, first of all with the employer*" (HWH1931b, p.11).

Heinrich thought it important that safety was seen as a management activity, which in his opinion "cannot be legislated nor otherwise forced upon the employer" (HWH1932a, p.31).[12] There were better ways to get executive interest. However, "Seldom has prevention been portrayed from the viewpoint of the business executive" (HWH1933b, p.56). He suggested that earlier proponents of safety made worthwhile efforts but did not align their interests with those of management. In order to merge it "practicably and effectively with managerial routine" (HWH1933b, p.56), he offered his systematic, "scientific" approach. Interestingly, he saw actually an advantage for small plant executives. Because they were necessarily closer to the sharp end, it might be easier for them to get first-hand information, and correct problems.

8.2.1 Planning

Heinrich spoke regularly of activities necessary to create a safe workplace, including engineering and education. Planning was a subject he rarely discussed, even though planning arguably can be essential for safety. In *Planning the Day's Work* (HWH1932d),[13] Heinrich connected planning to the efforts of learning from accidents, facts, and proper action. Because it treats a rare subject, it is an interesting piece in Heinrich's management writing, revamping the "scientific accident-prevention principles" into "planned executive procedure." Unlike most other papers, it was not included in any of his books.

> Let executives and supervisors accept these fundamentals (1) the will to achieve, (2) the determination of accident facts, and (3) action based on the facts; let them recognize the striking analogy between control of production and control of accidents; let them proceed on the basis of planned executive work; and industrial-accident prevention will no longer be the mystifying problem it now appears to be.
>
> (HWH1932d, p.77)

He was also aware of the limitations of planning: "Few persons are privileged to plan the day's work and to carry it to completion without interruption or interference from outside sources and routine tasks" (HWH1932d, p.72). Heinrich's description of managers whose plans are ruined by distractions and interruptions is applicable even in today's situation. It also connects well to Heinrich's writing elsewhere about dynamic elements of managerial work.

Probably to reinforce his message, Heinrich used "axiomatic" statements as "There are no exceptions to the rule that planned work is more productive of results than work which has its origin in emergencies and troubles" (HWH1932d, p.73) and hinted at a Taylorist division between them who do as they are told and "executives who are paid to do their own thinking" (HWH1932d, p.72).

8.3 Responsibility

Responsibility is closely related to the subject of (top) management. Heinrich was not shy to emphasise management's responsibility and even turned this into one of his Axioms, "Management has the best *opportunity* and *ability* to prevent accident occurrence, and therefore should assume the *responsibility*" (HWH1941, p.12). Fixing responsibility was a "fundamental of organisation" (Lansburgh, 1928, p.42) in industrial management. Many other authors wrote about the subject,

> ...effective accident prevention requires that the general management shall place responsibility for safety in each operating unit, squarely upon the executive of that unit. The manager of the plant or other operation – no one else – must be held responsible for safety...
>
> (Williams, 1927, p.5)

However, few spoke of it as often as Heinrich. It was a recurring subject in many publications, connecting it to various other aspects. As with other subjects, already the *Bedford Stone Club* speech emphasised responsibility:

> As long as you are an employer or as long as you are in charge of men and direct their work you cannot escape a moral responsibility for what they do, particularly when it has been pointed out to you that there is an easy way to prevent injuries, namely by organized safety.
>
> (HWH1923b, p.3)

According to this quote, knowledge came with responsibility. The same applied to authority and leadership: "acceptance of responsibility is a vital factor in industrial leadership" (HWH1929e, p.23). Quoting management consultant Henry Gantt, Heinrich wrote "The authority to issue and order involves the responsibility to see that it is properly executed" (HWH1929e, p.23). He emphasised this with a chapter about follow-up (HWH1931a).

On many occasions, Heinrich connected responsibility with initiative and action. "The initiative and the chief burden of activity in accident prevention rest upon the employer..." (HWH1931a, p.126), because "the responsibility lies, first of all, with the employer" (HWH1931b, p.11). Heinrich emphasised that management had the main part of responsibility, because they had "... the best opportunity to prevent accidents and should logically take the initiative" (HWH1934, p.116).

Sometimes his writing extended to personal responsibility, "It takes courage, sometimes, to warn fellow employees of dangerous practices and to make suggestions for the improvement of unsafe conditions" (HWH1930b, p.143). And he recognised how fear of accountability could prevent people from taking responsibility: "...every man who tries may fail: every man who leads must accept the consequence" (HWH1929e, p.23).

On many occasions, Heinrich connected taking responsibility to success: "Dodging responsibility or 'passing the buck' is practically an unknown quality in the make-up of successful men" (HWH1929e, p.23). This was illustrated by the examples in the *Message to the Foremen* papers. Here he suggested a positive side of risk taking, "...in accepting responsibility we at least expose ourselves to the probability of winning sometimes..." (HWH1929e, p.23). Shortly after he acclaimed, "responsibility is undeniably necessary to success" (HWH1930b, p.144).

In the *Safety Wins a Place in the Sun* paper, Heinrich thought the most important factor to "progress of startling significance" (HWH1934, p.112) was "employers in rapidly increasing numbers are themselves actively engaging in safety work" (HWH1934, p.112). This he illustrated with several examples where employers and executives "...in increasing numbers in the last several years have taken decisive and constructive steps toward the inclusion of practical safety work as an integral and worthwhile part of the executive and managerial program" (HWH1934, p.114) and that "recognition of responsibility for safety is likewise admitted" (HWH1934, p.115).

Almost always he wrote in a positive sense about responsibility, although he emphasised possible consequences of *not* taking responsibility as when writing about the need to make safety a normal part of business: "Management cannot escape this responsibility, and until it does accept it and initiate appropriate produce, industry must continue to be harassed by burdensome accident costs" (HWH1933b, p.54). Discussing the *Contractors' and Builders' Safety Problem*, he even portrayed "irresponsible contractors" as the largest problem. These were those contractors who lacked competence, resources, and care for safety, "cutting corners in an unwise attempt to save time and money" (HWH1925, p.250) in an attempt to underbid others and secure the work for themselves. Heinrich's description of these organisations with no interest in safety reminds of what Hudson might characterise as a "pathological" organisation (Hudson, 2007).

The second to fourth edition of *Industrial Accident Prevention* contained a dedicated section, dealing with the responsibilities of various players. Once more, Heinrich stressed here the responsibility of management over that of foremen, safety engineers, and employees. This was emphasised on various occasions:

> ...it is true that the employee is responsible to a certain extent. The employer, however, must accept the lion's share, because in the final analysis it is he who creates the working conditions, offers employment, selects and assigns workers, and initiates the entire series of events included in industrial operations. Any disclaimer, by management, of responsibility for accident occurrence is specious. It also retards business recovery, inasmuch as it leaves and unnecessary expense problem unsolved.
>
> (HWH1933b, p.54, 55)

The *Place in the Sun* paper contained a lengthy discussion of the responsibility of the various actors – managers, supervisors, employees, community, and government – "...responsibility for safety is practically universal...," followed by the appeal "All the more reason, then, to quit arguing about it and get to work" (HWH1934, p.116).

8.4 Integrating safety

> ...safety is an integral part of management and not simply something to be taken care of by a special department charged with the administration of safety features only.
>
> (Boston Elevated, 1930, p.1)

An important element of the early safety movement were safety committees, typically including workers as well as management representatives, that played an important role in creating safer workplaces. The first modern safety organisation was developed at the South Chicago works of Illinois Steel around 1907. They organised safety committees on various organisational levels whose members were paid employees that inspected the works and made recommendations. They also were authorised to stop work and discipline workers. This model was soon adopted by U.S. Steel and soon after by other companies (Aldrich, 1997).

Many safety authors during the first decades of the twentieth century spent much attention discussing safety organisation and committees. In effect this meant creating a safety organisation in parallel to the normal organisation. While safety committees were generally seen as a good thing, especially when employers and top management participated, they also had drawbacks:

> A committee is at best weak in execution. It has the advantage of bringing together the viewpoints of the group and its joint judgement is normally better than that of any individual in the group; but prompt,

effective, and orderly execution depends upon the placing of authority and responsibility in the hands of one person and the faithful discharge by him of that responsibility...

(Blake, 1945, p.268)

Heinrich chose another route. The story opening the first edition of *Industrial Accident Prevention* was not about a safety committee discussing the firm's accident problems. Instead it told the story from a management meeting and how to solve the issue from a management perspective. Heinrich spent relatively little space on the common form of safety organisation, instead focussing on integrating safety in everyday work and management.

In many of his papers, Heinrich emphasised that management, production, efficiency, quality assurance, and safety build on the same principles and go perfectly together. He saw integration of safety and production as a main responsibility of management. There was an analogy between production and accident prevention: "it is therefore not only possible to prevent the great majority of accidents, but also to do it by *the exercise of the very methods* that make for *economy, greater production and greater profits*" (HWH1928b, p.130).

Again, this was not a new way of thinking. Several others had preceded Heinrich, or had similar thoughts, although few may have delivered the message as frequently and consistently. In a speech on supervision, DeBlois sketched two possible routes for safety management to follow: "Safety may be made a part of the business of production, operation or whatever the function of the business may be; or it may be put on as a sort of a "sideshow," endorsed by the management" (DeBlois, 1919, p.201). The latter option, delegated safety to the safety engineer. Thereby it would always be regarded as of secondary importance. Choosing the former route, however, would lead to safety becoming "an integral part of the work, and the conflict that otherwise arises between production and safety is largely avoided" (DeBlois, 1919, p.201). The Committee that conducted a study on safety and production recommended, "...the same executive direction and control be given to decreasing industrial accidents as is given to increasing productivity" (AEC, 1928, p.35).

Heinrich made this one of his Axioms: "The methods of most value in *accident prevention* are analogous with the methods required for the control of the quality, cost, and quantity of *production*" (HWH1941, p.13). He had been discussing this for many years. Referring once more to the *Bedford Stone Club* speech, he mentioned as the sixth of ten elements of effective safety organisation, "Sixth, all employees must be taught to recognize safety as an element in their routine work" (HWH1923b, p.5). And managers of building firms and contractors he advised to "Give accident prevention its due share of attention *in all of your activities*..." (HWH1925, p.253), and "...the systematic and orderly merging of accident-prevention work with construction supervisory routine, on a basis of recorded facts, in a more regular and effective manner" (HWH1931a, p.60).

One way for Heinrich to convince managers that "...accident control is just as vitally a part of the operating program as is quality and volume of production..." (HWH1931a, p.62, 63) was that he demystified safety management for them, by arguing that the principle approaches for the management of production and safety were alike, that it could be done without much effort and that it was something they already mastered, "Accident prevention may therefore be merged practicably and smoothly with routine manufacturing and maintenance work and be carried on by the same executive and supervisory staff, by like methods, and with little additional expense" (HWH1931d, p.99).

Additionally, Heinrich made a lot of effort to explain it to managers in language they understood, with examples they could relate to, offering them practical tools as easy-to-remember lists such as the principles of accident prevention. As an aside, Heinrich hardly ever mentioned the well-known slogan of the early safety movement: "Safety First." It appeared in passing in some of his earliest writing and after that he probably realised that the literal meaning of the slogan conflicted with his message of integration.

In retrospect, some suggest that following the advice of integrating safety into the normal business may be one explanation for the progress of occupational safety and working conditions, "While workmen's compensation acts helped to bring forth the modern safety movement, the effectiveness of safety work resulted from its institutionalization within the management structure of large firms" (Aldrich, 1997, p.275).

However, many organisations still run safety rather as, in DeBlois's words, a "side-show" with a major administrative machinery parallel to ordinary management (e.g. Dekker, 2018) and an almost century-old message apparently needs reiteration, "Safety professionals that align safety objectives and activities with other organizational strategies, targets and business processes are effective at stewarding and sustainably improving safety" (Provan et al., 2017, p.110).

8.5 Safety management

8.5.1 Early developments

As we saw in this chapter, Heinrich spoke about safety management and organisation early on in his career and during the late 1920s and the 1930s he developed tools and principles, but safety management was explicitly introduced in the second edition of *Industrial Accident Prevention* with a short chapter titled *Safety Organisation*. On a mere seven and a half pages, he discussed safety organisation and committees with the various roles and responsibilities, and also included other provisions. He emphasised,

"It includes the interest, support, direction, and participation of the higher executives, the application of effective procedures and the provision of adequate first-aid, medical, and hospital facilities" (HWH1941, p.314).

In line with his practical orientation, Heinrich found the basic ("inescapable") elements of organised safety to be *activities* rather than organisational, administrative, and bureaucratic functions: inspection and surveys, accident analysis, and corrective action regarding the findings (HWH1941, p.315).

8.5.2 Axioms

Heinrich's first step towards a more complete structure for Safety Management was providing ten Axioms. Axioms are "self-evident truths," statements that are the basis of something, needing no further explanation. "The actual work of accident prevention depends, first of all, upon the recognition and knowledge of the fundamental truths involved..." (HWH1926, p.257) he said early on, but the Axioms of Safety did not appear until the second edition. They are a central part of Heinrich's work from that moment on and provide the framework for the 1941 book, and later editions. Several authors later suggested alternate axioms (e.g. Petersen, 2001a) or commented on them.

Heinrich's earlier work brought the "principles of scientific accident-prevention" that can be regarded as predecessors of the Axioms. These principles changed somewhat between papers, but by and large they said:

1　Executive interest, support, and action.
2　Knowledge of accident facts.
3　Appropriate and effective action based on these facts.

The *Bedford Stone Club Safety Talk* (HWH1923b) contained ten "fundamental principles" for successful safety organisation, some of which can be linked to the later axioms while others were more oriented on practical safety management. Heinrich was not the first to suggest Safety Axioms. In the August 1914 issue of *Safety Engineering*, one finds six safety axioms that were printed on the pay envelopes of the Pierce Arrow Motor Car company:

1　Make safety your first thought.
2　Accidents are someone's fault, don't let it be yours.
3　Find danger points and report them to the supervisor.
4　Preventing accidents makes you someone's benefactor.
5　Lost time is waste if accident is preventable.
6　Your duty is preventing that waste by being careful (Anonymous, 1914, p.146).[14]

Although some subjects may be common, this early list differs strongly from Heinrich's. One important difference is the audience. While the above list addresses workers, Heinrich's dealt with general management and were most of all aimed at managers and safety professionals. Many people – especially managers – are fond of lists, evident truths, and the like because they clarify, they create order, they invite to action, and they give a sense of control. Therefore, the Axioms were yet another way in which Heinrich managed to anchor his message, making it memorisable, decluttering some of the safety work and giving them something that made sense and was practicable.

The original version of the Axioms said:

1 The occurrence of an injury invariably results from a completed sequence of factors – one factor being the accident itself.
2 An accident can only occur when preceded by or accompanied and directly caused by one or both of two circumstances – the unsafe act of a person and the existence of a mechanical or physical hazard.
3 The unsafe acts of persons are responsible for the majority of accidents.
4 The unsafe act of a person does not invariably result immediately in an accident and an injury, nor does the single exposure of a person to a mechanical or physical hazard always result in accident and injury.
5 The motives or reasons that permit the occurrence of unsafe acts of persons provide a guide to the selection of appropriate corrective measures.
6 The severity of an injury is largely fortuitous – the *occurrence* of the accident that results in the injury is largely preventable.
7 The methods of most value in *accident prevention* are analogous with the methods required for the control of the quality, cost, and quantity of *production*.
8 Management has the best *opportunity* and *ability* to prevent accident occurrence; and therefore, should assume the *responsibility*.
9 The foreman is the key man in industrial accident prevention.
10 The *direct* costs of injury, as commonly measured by compensation and liability claims and by medical and hospital expense, are accompanied by *incidental* or *indirect* costs, which the employer must pay (HWH1941, p.12, 13).

These axioms were changed for the following editions:

1 The occurrence of an injury invariably results from a completed sequence of factors – the last one of these being the accident itself. The accident in turn is invariably caused or permitted by the unsafe act of a person and/or a mechanical or physical hazard.
2 The unsafe acts of persons are responsible for a majority of accidents.

3   The person who suffers a disabling injury caused by an unsafe act, in the average case has had over 300 narrow escapes from serious injury as a result of committing the very same unsafe act. Likewise, persons are exposed to *mechanical* hazards hundreds of times before they suffer injury.

4   The *severity* of an injury is largely fortuitous – the *occurrence* of the *accident* that results in injury is largely preventable.

5   The four basic *motives* or *reasons* for the occurrence of unsafe acts provide a guide to the selection of appropriate corrective measures.

6   Four basic methods are available for preventing accidents: *engineering revisions, persuasion and appeal, personnel adjustment,* and *discipline.*

7   Methods of most value in *accident prevention* are analogous with the methods required for the control of the quality, cost, and quantity of *production.*

8   Management has the best opportunity and ability to initiate the work of prevention; therefore, it should assume the responsibility.

9   The supervisor or foreman is the key man in industrial accident prevention. His application of the art of supervision to the control of worker performance is the factor of greatest influence in successful accident prevention. It can be expressed and taught as a simple four-step formula.

10   The humanitarian incentive for preventing accidental injury is supplemented by two powerful economic factors: (1) the safe establishment is efficient productively and the unsafe establishment is inefficient; (2) the direct employer cost of industrial injuries for compensation claims and for medical treatment is but *one-fifth* of the total cost which the employer must pay (HWH1950a, p.10, 11, 1959, p.13, 14).

Significant changes were:

• The first axiom combined the first two from 1941.

• The third axiom has changed significantly, referring now also to the ratio. One could argue that it emphasised human behaviour, although the second part still spoke of "hundreds."

• The fifth and sixth axiom were influenced by the *Formula* book, mentioning a set number of reasons and methods for prevention.

• The sixth axiom was new, describing four basic methods for prevention.

• The ninth axiom was expanded, referring to the formula for supervision.

• The tenth axiom was also expanded, mentioning humanitarian motives, linking safety and efficiency and suggesting a ratio for indirect costs.

A point of critique is that assuming self-evident truths may match badly with a "scientific approach," especially when some of the Axioms – as for example the second about unsafe acts of persons being the cause of the majority of accidents – are debatable. However, even though they may not really be axiomatic, Dekker (2019) suggests that they *became* axiomatic over time.

8.5.3 Structured management tools

During the Second World War, as part of their education of supervisors and engineers, Heinrich and Granniss had worked with a tool that later was included in what today would be called Travelers' quality management system, and the later editions of *Industrial Accident Prevention*. *The Hazard Thru Track* (Travelers, 1945) was a schematic help for simplified accident prevention, combining different "principles" with (elements of) Heinrich's accident sequence. The "new" principles resembled the steps in the formula for supervision that Heinrich presented some years later:

1   Know about hazards
2   Find and name them
3   Select a remedy
4   Apply the remedy

The contents of the steps were in accordance with what Heinrich proposed in *Basics of Supervision*. Hazards were defined as personal (unsafe acts) or mechanical (unsafe conditions) and in the second step, the safety engineer was urged to find reasons for their existence. In the third step, remedies were found in four groups with a suggested hierarchy:

1   Engineering revision
2   Persuasion, appeal, instruction
3   Personal adjustment
4   Discipline as last resort.

Besides being a helpful instrument, this flowchart was also meant as a means of professionalisation of safety engineers: "It is when obstacles to accomplishment are encountered that the ability of the safety engineer is taxed the most. He must have ingenuity enough to overcome obstacles" (Travelers, 1945).

Heinrich and Granniss were also involved in another project related to safety management. The creation of a *Form for Use in Self-Appraisal of Industrial Plants* had already been suggested in June 1939 by the ASME. A draft was presented the next year, but due to war priorities little was done in the years after. However, some of the safety documentation provided by the Department of Labor during the War did include short self-appraisal checklists. By the end of the War, in July 1947, a final draft of the self-appraisal form was finished and approved as ASME Standard.

This tool was intended as "a uniform method of measuring the degree of progress made by an individual plant in the development and application of safety and accident prevention measures" (ASME, 1947, p.4) with subjects covering among others planning, construction, safeguarding, supervision, housekeeping, education, code compliance, management control, and leadership. It was believed that

> the periodic use of such a uniform method of self-appraisal would also stimulate efforts leading to accident prevention on a national scale since it would provide a means for comparison of the degree of progress made in a given individual plant with that made in other plants.
>
> (ASME, 1947, p.4)

The checklist provided a maximum score per item of which the assessor could "claim" a certain credit. The various items were clustered into three groups: physical/engineering activities, educational activities, and management control. Interestingly both management control and educational activities had twice the weight of physical/engineering, emphasising the importance of management activities. The ASSE self-assessment tool can be seen as a direct predecessor to the International Safety Rating System, which was created by Frank Bird Jr. in 1978 and is sold by DNV GL until today.

8.5.4 Another metaphor

In the first editions of *Industrial Accident Prevention* safety management did not appear as a subject of its own. By the third edition, however, it was labelled as such and accompanied by the appearance of another visually appealing model. His final metaphor, a five-step ladder was the visual representation of a basic safety management system. It is interesting that Heinrich opened the 1950 edition of *Industrial Accident Prevention* with this picture, stressing safety management as an important subject.[15]

One could view the safety ladder as a further development of his principles of "scientific" accident prevention. The ladder stood on a robust foundation, representing the basic philosophy of accident occurrence and prevention. This basic philosophy was detailed in Chapter 2 of the book – presented in the previous versions and relatively unchanged – and summed up by the ten Axioms. The foundation also showed a grand vision for safety management; based on attitude, ability, and knowledge there was a desire to serve industry, the country, and humanity.

Five steps went from this foundation upwards, representing the basic safety organisation and management and then essentially the principles of accident prevention: gathering facts, analysing the information, selection of remedy, and application of remedy. In part, the approach compares well to later management system structures based on the Plan-Do-Check/Study-Act cycle, although one might say that the feedback and improve-loop from the PDCA-cycle was less clearly expressed by Heinrich (Figure 8.2).

The metaphor also illustrated the fact that safety, efficiency, and production went together and are not separate objectives – the five steps led up to them both.

It is somewhat surprising that Heinrich did not dedicate any speeches or papers to safety management and the ladder (unless these are yet to be discovered), and they only appeared in the third and fourth edition of *Industrial Accident Prevention*. However, around the same time, Heinrich wrote

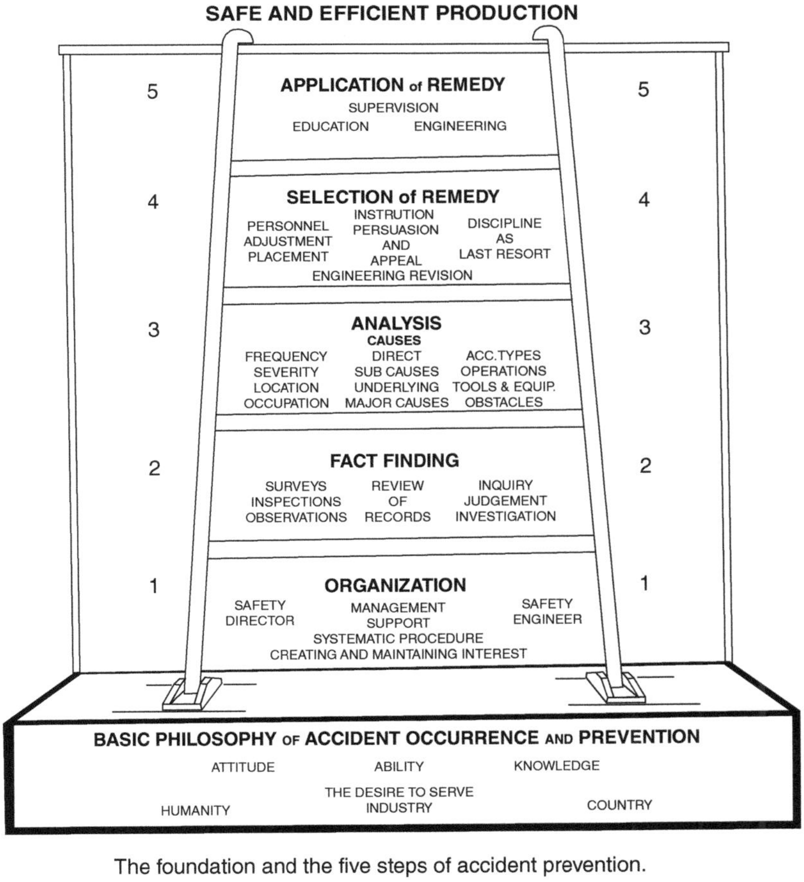

SAFE AND EFFICIENT PRODUCTION

| 5 | **APPLICATION of REMEDY**
SUPERVISION
EDUCATION   ENGINEERING | 5 |

| 4 | **SELECTION of REMEDY**
INSTRUCTION
PERSONNEL   PERSUASION   DISCIPLINE
ADJUSTMENT   AND   AS
PLACEMENT   APPEAL   LAST RESORT
ENGINEERING REVISION | 4 |

| 3 | **ANALYSIS**
CAUSES
FREQUENCY   DIRECT   ACC.TYPES
SEVERITY   SUB CAUSES   OPERATIONS
LOCATION   UNDERLYING TOOLS & EQUIP.
OCCUPATION MAJOR CAUSES OBSTACLES | 3 |

| 2 | **FACT FINDING**
SURVEYS   REVIEW   INQUIRY
INSPECTIONS   OF   JUDGEMENT
OBSERVATIONS RECORDS   INVESTIGATION | 2 |

| 1 | **ORGANIZATION**
SAFETY   MANAGEMENT   SAFETY
DIRECTOR   SUPPORT   ENGINEER
SYSTEMATIC PROCEDURE
CREATING AND MAINTAINING INTEREST | 1 |

BASIC PHILOSOPHY of ACCIDENT OCCURRENCE and PREVENTION
ATTITUDE   ABILITY   KNOWLEDGE
THE DESIRE TO SERVE
HUMANITY   INDUSTRY   COUNTRY

The foundation and the five steps of accident prevention.

Figure 8.2 Heinrich's safety ladder (HWH1950a, p.7).

unpublished speeches on the subject of *Safety Organization – A Function of Management*. He had a surprisingly modern definition of safety organisation (or management): "...a planned, continuing and systematic procedure designed to coordinate and make effective all activity in the prevention of avoidable accidents. Emphasis in this definition is placed on 'procedure' and on 'activity'" (HWH1952a, p.1).

He addressed the failure of safety professionals to "sell" managers in an appealing way on their responsibilities. Instead of emphasising compliance, they should show the benefits – "the great opportunities for efficiency in production that lies [in safety organisation]" (HWH1952a, p.1). He shied not away of criticising some of his peers, "Isn't it quite likely that the profession is so impressed with 'self-evident truths' that it becomes impatient with the folks who don't see it the same way and so fail to take the indicated corrective action?" (HWH1952a, p.1). Even stronger, "...there is too much 'inbreeding' in the safety engineering profession - that we live too much *to ourselves*" (HWH1952a, p.3).

He also spoke in critical terms about management education, "...some, but not all college business administration courses include a bit of safety. It isn't enough. Further, the new courses do not benefit the executives who are already on the job" (HWH1952a, p.2). As a consequence of these factors he found that, "accident prevention exists at best as a step-child" (HWH1952a, p.2). Therefore Heinrich named a number of points of what the safety profession can do to convince management and actually involve management in the work. Because, "management not only has the responsibility for initiating and carrying on safety organisation, but also has the best opportunity to do so. In fact, no one else has either the authority or means to do so" (HWH1952a, p.3).

According to Heinrich, safety organisation consisted of "certain inescapables" which one will find as requirements in today's safety management systems:

1   Management approves, initiates, supports, and controls.
2   First aid, authorised medical, and hospital care.
3   Appointment of one person to direct the work in detail.
4   Committees and sub-committees as necessary.
5   Periodic survey of operations and equipment.
6   Selection, instruction, training, and supervision of personnel.
7   Recording and investigation of accidents, including corrective action.

For the "systematic, continuing and orderly procedure in accident prevention," Heinrich suggested his principles/four step formula and emphasised, "The work of accident prevention cannot make progress without Safety Organisation. Safety Organisation in turn is ineffective until management accepts its responsibility" (HWH1952a, p.5).

8.5.5 Misconceptions

Finishing this chapter, there are some misconceptions regarding Heinrich's approach of (safety) management that deserve to be addressed.

8.5.5.1 Lagging indicators

Long suggests that Heinrich promoted the use of injuries and other lagging indicators as a measure of safety (Long, 2014a, 2015, 2017c, d, 2018a), making statements as "...apparently, injury rates are a measure of safety, thanks Mr. Heinrich" (Long, 2017d). While Heinrich may have contributed to a much wider acceptance of accident prevention as something that management has to pay attention to, the use of lagging indicators can hardly be attributed to him. Safety has been measured in terms of accidents and outcomes since its earliest days (e.g. Eastman, 1910; Beyer, 1916; Ashe, 1917; DeBlois, 1926a), and some – including Heinrich – even rather talked about "accident prevention" than "safety" (Beyer, 1916; Ashe, 1917).

So, it is easy to debunk the attribution to Heinrich. One can strengthen the argument against this attribution further, however. First, Heinrich was very clear about the difference between accidents and outcomes, stressing the difference of the various phases of an accident many times in his work. He placed little value on the consequences since these were rather random, as expressed in his Sixth Axiom.

Second, at times Heinrich was outspoken and critical on the use of lagging indicators. "Since rates must always be predicated on past experience, they are always behind..." (HWH1925, p.258), he said already early in his career. In his discussion of no-injury contests, Heinrich explained with examples how LTIs were a weak measure of safety: "The occurrence of lost-time accidents or serious injuries, in itself, does not always provide an accurate measure of the conditions out of which they arise" (HWH1931a, p.123). He also was very clear on the subject in the chapter on Accident Statistics: "For the average plant, a record of lost-time injuries only is insufficient" (HWH1941, p.342). Instead, Heinrich advocated including other – minor – events as part of the statistics and suggested to pay attention to behaviour and conditions *before* something happened.

If anything, Heinrich contributed to nudge safety management into a *less* reactive direction through the triangle, by suggesting paying attention to accidents (events, conditions, acts) rather than injuries.

8.5.5.2 Safety bureaucracy

In many organisations "record-keeping and incident analysis" has become "a bureaucratic initiative" (Dekker, 2018, p.82). Some authors attribute this safety bureaucracy to Heinrich's influence, to a lesser – "classical bureaucratic approach to safety management" (Kjellén, 2000, p.13) – or greater degree, "...in tribute to my many safety colleagues who waste countless hours each week reporting on things that don't matter, tabulating data of irrelevance and maintaining worship to Heinrich and Bird by statistics..." (Long, 2018b).

While Heinrich on several occasions promoted the improvement of incident statistics (HWH1932c, 1935) and even dedicated full chapters to the subject in *Industrial Accident Prevention*, he did not intend them as a bureaucratic enterprise. Instead, he saw them as a way to support and improve accident prevention by presenting organisations with more and richer information to base their decisions on and direct their accident prevention work.

> The value of statistics lies in the conclusions that may be drawn from them and upon which corrective action may be based. Statistics provide a valuable clue to the causes of future accidents. ... with the conclusions drawn from statistics as a guide, a study of present conditions may be profitably made and a final decision as to the method of existing known accident causes more readily reached.
>
> (HWH1931a, p.263, 264)

Even more, Heinrich was a practical man who was no fan of unnecessary paperwork and unpractical solutions. "Practicability and common sense must prevail in safety as in other things..." (HWH1941, p.139), he wrote. In his *Safety in the Small Plant* speech (HWH1932a), he was quite clear about this. The second paragraph discussed how small plant owners might be reluctant to engage in safety because it looks overwhelming with procedures, records and "red tape."[16] Heinrich even went one step further later in the speech: "The accident preventionists, moreover, because they fail generally to emphasise simplicity and principles, share responsibility for this situation..." (HWH1932a, p.31). This is reminiscent of *The Safety Anarchist* (Dekker, 2018) and the "safetycrats" of Marriott's book (2018). Heinrich recognised how safety practitioners can be their own worst enemy. Early on in his career, he cautioned,

> Don't let 'Safety' drift into a cut and dried spineless matter of form proposition. Sometimes good businessmen, in all effort to systematize their organisation eliminate from it all personality and they lose by doing so as for instance when using stereotype form letters for correspondence.
>
> (HWH1923b, p.7)

This he would keep up until the end, showing himself to be an anti-bureaucrat and man of action. Committees, engineers, records, and forms were important, but "These are important factors in some degree at least, but they constitute only the mechanism or the vehicle whereby *activities* are made possible" (HWH1952a, p.1). Heinrich thought that "...the measure of a real Safety Organisation lies in its activities" (HWH1952a, p.1).

Notes

1 A critic may comment that this makes perfect sense. After all, managers like hearing that the worker's behaviour should change (suggested by many authors, e.g. Smith, 2011). This means ignoring a major part of Heinrich's arguments, however.

2 This book saw five editions between 1923 and 1955. The third edition was co-authored by William Robert Spriegel who became the main author when Lansburgh passed away in 1942. Just as Heinrich's *Industrial Accident Prevention* represented the current status in professional safety knowledge at the time, Lansburgh's book mainly served to provide a solid overview of contemporary management science. It even included a chapter on Industrial Safety, which in the second edition mainly was based on Volume 123 of *The Annals of the American Academy of Political and Social Science* which was a special on the subject.

3 However, Williams did not spend more than a mere five pages on the subject.

4 This was addressed to foremen but was "intended for the use of all persons holding positions of authority, the word 'foremen' being used throughout the text merely for the sake of avoiding a useless and tiresome repetition of titles, where persons in authority are mentioned" (Travelers, 1918, p.3). Heinrich, however, clearly differentiated in his work between top and higher management, usually indicated with words like "executive" or "superintendent," and first-line management, supervisors, and "foremen."

5 Besides calling foremen the "key men" (e.g. HWH1931a, p.96), Heinrich also called them "hub of the masses" (HWH1929e, p.23). It was also Heinrich's ninth axiom (HWH1941, 1950a).

6 It is interesting that Heinrich discussed how the promotion of good craftsmen to foremen does not provide good leaders (HWH1931a, p.290). A phenomenon only all too know from today's organisations.

7 To my knowledge, this is the first time Heinrich published in a magazine, directly aimed at foremen. The paper is quite different from several others in the period that are more or less "variations on a theme."

8 *The Tribune*, 27 February 1929; Our Mountain Home, 17 April 1929.

9 In *Formula for Supervision* (HWH1949) he actually called them "links." Quoted are the descriptions from *Basics of Supervision*. Heinrich changed some in *Formula for Supervision*. Ten changed into "Knowledge of right and wrong job conditions", and eleven changed into "Knowledge of human engineering and of right and wrong employee practices." Most likely to downplay the safety part in favour of a more general (production) orientation. It is puzzling, however, as why Heinrich chose to use a more abstract and difficult term like "human engineering."

10 One could comment here that neither attitude, ability, nor knowledge are truly "static" because they will evolve based on experience and learning. Heinrich seemed to realise this, labelling them as "inactive" in *Formula for Supervision* (HWH1949).

11 Will delivers some interesting critique in his thesis, starting from a Marxist view of exploitation of workers by capitalists, and Marx's disbelief that the situation for workers would improve with economic development. Marx's perspective stands in contrast with that of safety authors praising the progress in safety and working conditions which in large part (according to Will) is attributed to humanitarian motives. Will does not believe such claims are valid, "It is important to note that those who have defended management attitudes do not cite any reliable opinion surveys as grounds for their claims; instead, they usually quote statements made by various corporate executives in addresses to safety

conferences" (Will, 1979, p.14). He thinks that these quotes cannot be accepted as reliable evidence about general management attitudes. This point of view is supportable. However, the repeating of these statements may have been a tool of persuasion by the early safety authors, creating a reality where things can become true if you just say them often enough...

12 Employers can probably be forced into action, but it is interesting to see that "modern" initiatives including ISO standards, safety culture schemes and culture-oriented regulations try to exactly do that – often with an "appearance-based" result.

13 Unlike most other publications in *The Standard*, no specific date this speech was delivered is mentioned. This might suggest that this was a generic speech prepared to be held (possibly by others) on suitable occasions "for the purpose of emphasizing the need of planned accident-prevention work" (HWH1932d, p.72).

14 Some readers may find these axioms embarrassing, but most likely each of them has one or more comparable versions in today's safety discourse.

15 The emphasis on safety management was also shown through the inclusion of the chapter on safety organisation early in the book. This in contrast to the claim of a more behavioural approach in later work (Dekker, 2019). Interestingly, there is no mention of behaviour or "unsafe acts" to be found in the figure. It also indicates a richer set of causes than one would gleam from the dominos alone.

16 The exact same argument returned also in the *Sand and Gravel Industry* paper later that year (HWH1932b).

Chapter 9

The triangle

This chapter deals with Heinrich's most famous and most debated concept, known as "Heinrich's Pyramid," "Safety Triangle," "Iceberg," "Heinrich's Law," or simply "Accident Ratio." Heinrich would not recognise many of the contemporary names. Initially, the graphic representation was not triangular, and Heinrich never drew it as a 3D pyramid shape.[1]

This chapter presents what Heinrich wrote about the triangle and whether we can reconstruct how he reached his conclusions. It systematically reviews ways to read the triangle and discusses its underlying principles. Then, it turns to various interpretations and critique. The chapter closes with some reflections on the triangle. As most ideas, the triangle developed over time. The first section traces its evolution.

9.1 Origin and development

9.1.1 Conception

Many important discoveries are made by coincidence rather than design. This also applies to the "triangle" which was a by-product of the research that produced the 1:4 ratio for indirect costs. The principle was mentioned for the first time in the November 1927 paper.[2] As part of the conclusions, it said:

> For every accident involving compensation or professional medical cost, there are approximately thirty minor accidents that receive merely first-aid treatment…" and "For every actual injury that occurs there are several 'near accidents' resulting in property damage, spoilage, lost time and other costs, difficult to calculate with exactness.
>
> (HWH1927c, p.230–231)

These passages are interesting in several aspects. First, Heinrich presented them mainly as observations although he already suggested causal connections.[3] Second, they show that he had not yet adopted his specific definition of accidents. Soon after, he categorised near-miss events as accidents and rejected "minor accidents." Third, he mentioned explicitly that the ratio was an approximation. Fourth, the second quote suggests a non-specified one-to-many ratio for property damage, predating Bird's work by almost four decades.

The observation stuck. When Heinrich presented his *Origin of Accidents* paper in mid-1928, the ratio was further developed and now carried a lesson:

> We have found that for every single injury involving compensation or professional medical cost, there are at least 30 injuries of a minor nature wherein no such cost is incurred and that are not recorded or tabulated. We find also that the same causes predominate in this group as in the major accidents. We find that each minor cause[4] is potentially serious. and
>
> The cause in each instance is the same, yet we ordinarily record only the serious case. We unnecessarily limit our exposure (and consequently the value of our conclusions as to cause) when we select only compensable, lost time, or major cases for analysis.
>
> (HWH1928a, p.11–12)

Here, Heinrich attached a causal statement to the observation, suggesting that "bigger data" (Swuste, 2016) contributed to better accident prevention. According to Heinrich, the Travelers had reacted successfully upon this lesson, "...results show conclusively that concentration upon the reduction of minor-accident frequency automatically takes care of the severe accidents as well" (HWH1928a, p.12).[5]

9.1.2 A ziggurat

By the end of the year, the idea had matured and was presented at the Twelfth New York Industrial Safety Congress in December. *The Foundation of a Major Injury*[6] (HWH1929a, b) for the first time presented the finished concept. Much of the paper re-appeared in the 1931 book. One of the new key elements was an expansion:

> ...for every mishap resulting in an injury, there are many other accidents in industry which cause no injuries whatever. The investigation has enabled us to establish the conservative ratio of 10-to-1, between no-injury or potential-injury accidents and those causing injuries.
>
> (HWH1929a, p.1)

This was a significant insight based on the realisation that events without outcomes hold the potential to cause injury. Therefore they "...present to the capable supervisor who is intelligent enough to take advantage of it, a splendid opportunity to anticipate and prevent actual injuries" (HWH1929a, p.4). This was strengthened by the fact that there was a much greater number of such low-outcome accidents:

> In view of these facts, it should be obvious that present day accident-prevention work is misdirected, because it is based largely upon the analysis of one major injury – the 29 minor injuries are recorded (but seldom analysed) and the 300 other occurrences are, to a greater extend ignored.
>
> (HWH1929a, p.5)

The paper argued forcefully that a focus on events with major outcomes was limiting. They were rare, so they were a weak measure of a safety program. Neither were they a good guide for improvement. Key to the idea was the distinction between accident and injury (a point made before *Origin of Accidents*), which was an argument not being led by (random) outcomes, but instead concentrating on the events.

The *Foundation* paper included a graphic representation of the idea, although the "blocky" first version was more a ziggurat than a pyramid (Figure 9.1). This early version mixed two approaches:

1   Accidents per outcome ratio as shown in the "ziggurat."
2   Causal codes per outcome ratio as shown in the table.

The top part was straightforward, sorting accidents with major and minor injuries as the first two levels, using the "conservative estimate" of 1:10 for injuries to no-injuries to create the third level.[7] Underneath, a table presented the result from the analysis of 50,000 accidents. All accidents were assigned *one* causal code.[8] Because the 1:1:1 relationship between event, outcome and causal code, the resulting ratios for outcomes and causes were the same. More surprising, however, is that also each causal code had a distribution of consequences of approximately 1:29:300.[9]

The text below the table claimed showing that the causes of major and minor injuries are the same. That was too bold a statement. While the table presented an outcome ratio for the assigned causal codes, which is the same as the outcome ratio for all accidents, this does merely suggest that the average case would have that distribution. However, it does not show common causes as such. For that, one would have to look at scenarios and within accident types – as Heinrich did in the various cases discussed in the paper. Apparently, Heinrich realised this and altered the text for future work.[10]

Heinrich presented the concept as "a splendid opportunity to discover and correct accident-producing conditions long before an injury actually happens" (HWH1929b, p.10). This positive approach is generally forgotten by the critics who focus on the numbers, the research behind it or

THE FOUNDATION
OF A MAJOR INJURY
■

I MAJOR INJURY

29 MINOR INJURIES

300 NO INJURY ACCIDENTS

THE RATIOS GRAPHICALLY PORTRAYED ABOVE----1----29----300------
WHEN EXPRESSED IN PERCENT OF <u>ALL</u> ACCIDENTS SHOW THAT------
00.3% OF ALL ACCIDENTS PRODUCE MAJOR INJURES-------------------
08.8% OF ALL ACCIDENTS PRODUCE MINOR INJURES-------------------
90.9% OF ALL ACCIDENTS PRODUCE NO INJURES-----------------------

THE TOTAL OF 330 ACCIDENTS AS SHOWN BY THE TABLE BELOW
ALL HAVE THE SAME CAUSE.
SINCE IT IS TRUE THAT THE ONE MAJOR INJURY MAY RESULT FROM
THE VERY FIRST ACCIDENT OR FROM THE LAST OR FROM ANY OF
THOSE THAT INTERVENE, THE OBVIOUS REMEDY IS TO ATTACK

<u>ALL</u> ACCIDENTS.

ANALYSIS OF
50,000 INJURIES BY CAUSE

REAL CAUSES OF ACCIDENTS	50,000 INJURIES		500,000 ACCIDENTS NO INJURY	%
	MAJOR INJURY	MINOR INJURY		
FAULTY INSTRUCTION	491	14509	150000	30
INATTENTION	375	10625	110000	22
UNSAFE PRACTICE	230	6770	70000	14
POOR DISCIPLINE	202	5798	60000	12
INABILITY OF EMPLOYEE	134	3866	40000	8
PHYSICAL UNFITNESS	49	1451	15000	3
MENTAL UNFITNESS	17	483	5000	1
MECHANICAL HAZARDS	168	4832	50000	10
	1666	48334	500000	100

NOTE - THIS TABLE SHOWS THAT THE CAUSES OF MINOR INJURIES AND
OF NO INJURY ACCIDENTS ARE SUBSTANTIALLY THE SAME AS THE
CAUSES OF MAJOR INJURIES. BY REMOVING THE REAL CAUSES OF
EITHER MINOR INJURY OR NO INJURY ACCIDENTS WE THEREBY
PREVENT MAJOR INJURIES AS WELL.

THE TRAVELERS INSURANCE CO.

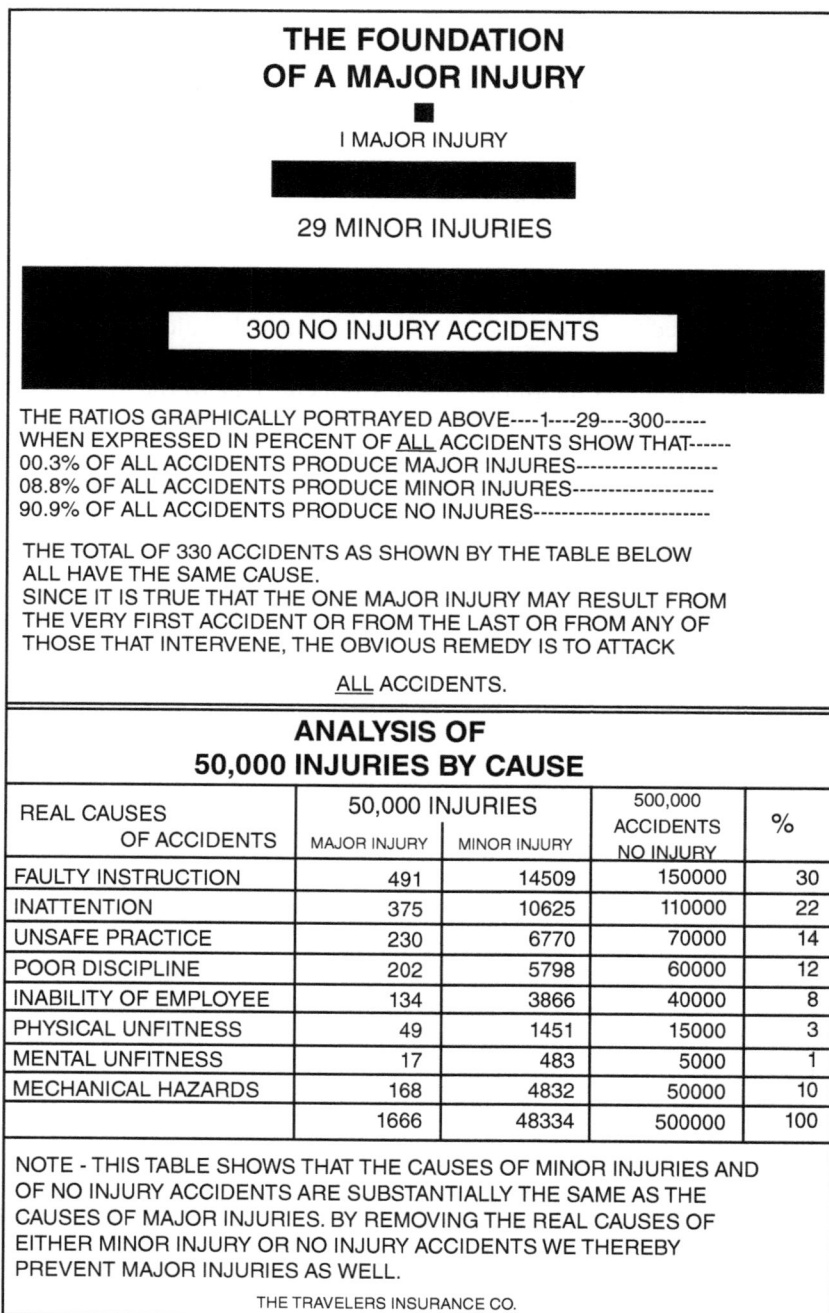

Figure 9.1 Foundations chart.
Source: Travelers Standard XVII, January 1929. Used with permission.

confounded applications. It also echoed the concept of risk, without using that word: "The importance of any individual accident in prevention work lies in the potential power to create injury and not in the fact that it actually does or does not so result" (HWH1929b, p.10). Overall, it was a significant nudge of accident prevention into a more proactive direction.[11]

9.1.3 Researching and building

Initially, the realisation that there were many minor injuries for every major one sprung from the research of 10,000 cases taken randomly from the Travelers files (HWH1927c). It then developed further into a 1:29 ratio during the *Origin of Accidents* study of 75,000 cases. Heinrich called the ratio an "incidental" (HWH1929a, p.4) to that study, which he decided to follow up. However limited, Heinrich revealed some of the research forming the triangle's base.

9.1.3.1 Classification and sorting

The first step in building the triangle was categorising the data. As part of the *Origin of Accidents* study, 75,000 cases were assigned one causal category according to Heinrich's scheme as described in Chapter 6. In order to construct the triangle, it was also necessary to distinguish between accidents with major and minor outcomes.

Heinrich used the following definitions: *Major Injuries* were all cases reported to insurance carriers or state compensation commissioner. *Minor Injuries* included all other injuries, mostly labelled as first-aid cases. *No-Injury Accidents* were all other unplanned events having the probability of causing personal injury or property damage. On several occasions Heinrich specified that the latter were "narrow escapes" (HWH1950a, p.24).

In many later/contemporary representations, the top level of the triangle is presented as dealing with fatalities. That was not Heinrich's interpretation, however.[12] His definition allowed including minor injuries in the "major" category. Expressing this caveat, he indicated that the "real" major to minor injury ratio might be even larger.

The *Foundation* paper mentioned the analysis of 50,000 cases, which were sorted by the above consequence categories. Of these, 1,666 were found in the major category, and 48,334 were found in the minor injury category, forming the two top levels of the triangle.

9.1.3.2 Building the lower level

Critique of the triangle (Manuele, 2002, 2011b, 2014a) frequently suggests that it was impossible for Heinrich to know the number of no-injury accidents. This, he admitted himself, "There are few existing data on minor injuries – to say nothing of no-injury accidents…" (HWH1929a, p.4). Manuele even questions whether Heinrich based his research on anything at all.

However, Heinrich gave some clues how his analysis of 50,000 accidents[13] causing injury established a "conservative ratio of 10 to 1, between no-injury or potential-injury accidents and those causing injuries" (HWH1929a, p.1). In the text, Heinrich described various cases with ratios varying from 1:3,000 to so-called 1:1 (first event causes injury).[14] Heinrich's team used these to estimate an average for the injuries to no-injuries ratio.[15]

Some judgement was involved. Heinrich discussed the selection of cases to be included and excluded, "each injury was included where there was any substantial indication whatever of the existence or lack of prior accidents" (HWH1929a, p.7). However, they excluded most of the accidents "caused" by machines because it would inflate the no-injury group too much. This was illustrated by an example of a mechanical defect that had existed over several years and each rotation of the machine presented a near accident (millions of them), until finally all factors came together producing an injury. While one could interpret this as another sign of Heinrich's focus on behaviour, it is more likely that Heinrich chose to present what he regarded as a conservative ratio, rather than one that would be less believable – although he acknowledged that the ratio could have been much higher and was never fully knowable.

> The number of such no-injury or potential-injury accidents in comparison to actual injuries has always been a nebulous quantity, and it probably will never be known exactly. Nevertheless, Travelers engineers have arrived at a minimum which in itself is so high that it proves conclusively the necessity for supervisory control enforced through adequate executive participation.
>
> (HWH1929a, p.4)

The bottom level of the triangle was merely an estimate that, however imprecise, was considered convincing enough to act upon.

9.1.4 Earlier ideas

Heinrich was probably the first to put a number on the concept, but the idea of reacting on minor events was found earlier. Richardson wrote a short article in the September 1916 issue of *Safety Engineering* where he already discussed some of the elements that Heinrich popularised through the triangle. "There is probably no source of information which can be of greater use... than the lessons from the observation or study of the causes of minor or "near" accidents," he wrote. "...each minor or 'near' accident contains just as important a lesson as may be drawn from those having more serious results." Like Heinrich over a decade later, Richardson had strong focus on behaviour:

> Usually the bad practice has continued until disregard for the danger caused the carelessness which resulted in the accident, or else

carelessness was the general habit. It is therefore safe to assert that in a large number of cases the occurrence of serious accidents might have been foreseen had the occurrence of the minor or "near" accident received consideration.

(Richardson, 1916, p.189)

Like Heinrich, he thought this was easy[16]: "To learn the lesson requires only the powers of observation and a realization of the possible consequences." Richardson's closing paragraph fits with the triangle or pyramid metaphor, saying that it would be a good thing for those working on safety "...*to start at the bottom*[17] in the effort to eliminate the minor accidents. The more serious ones will then be fewer in number." This final sentence even implied a common cause between serious and minor accidents.

DeBlois suggested the existence of common causes, speaking of a contributory cause in a group of accidents which may be wiser to attack than a great variety of proximate causes (DeBlois, 1926a, p.52). And while it is doubtful that it influenced Heinrich's development of the ratio, Williams presented a case illustrating the idea very well:

Consider likewise the case of a labourer pushing a truck load of material through a shop. A piece of material falls off - because it was poorly piled, or because the truck was not of the right sort, or because the floor was rough, or the lighting poor, or the labourer untrained or careless. This happens a hundred times, always causing some loss of time, not only for the trucker but for others behind him, perhaps for the machine tender waiting for the material. The hundredth time, the piece of material falls on the labourer's foot, and he is injured. This accidental injury calls attention to the fact that something is wrong with the truck, the material, or the man.

(Williams, 1927, p.143)

9.1.5 1931

Heinrich mentioned the triangle concept in his *Message to the Foremen* papers (HWH1929d, e) and the *Foreman's Responsibility for Accidents* talk (HWH1930a), emphasising the opportunity to act before harm occurs. There were no further developments in these sources. Still, over the years, Heinrich would keep tinkering with the concept as evidenced by changes in the text and explanations. The first appeared in the 1931 edition of *Industrial Accident Prevention*. Based on the papers, this book presented the concept to a larger audience. Heinrich made some minor changes to the original chart and text, which he placed in the chapter *Selection and Application of Remedy* under the header of *Accidents - Not Injuries Point of Attack*. Strikingly, he removed any mention of the 50,000 cases study, and the link to the *Origin of Accidents* research.[18] The other main difference is the text in the chart where he toned down the causal claim, presenting now a simple description, "This table summarizes the analyses of 50,000 injuries substantiates the assertion that there are 29 minor injuries and 300 no-injury accidents for each major injury, and that this ratio applies to each one of the causes listed" (HWH1931a, p.91). In all later editions of the book, Heinrich would omit the table with causes.

At this point, it is worth mentioning that he apparently understood randomness and probability, since he explicitly mentions as part of the chart (and would do so for future updates) that, "Since it is true that the one major injury may result from the very first accident or from the last or from any of those that intervene, the obvious remedy is to attack all accidents" (HWH1931a, p.91). For Heinrich, this was an additional reason to act on the minor events. After all, the very first event might result in a serious injury.

9.1.6 The triangle

The second edition of *Industrial Accident Prevention* made some changes to the concept. Manuele, and others echoing him, criticise these "unexplained changes," but most likely they are intended as clarifications of the concept – as Heinrich suggested in the preface to the second edition.

The most striking change is the principle's presentation, shaped as a triangle (Figure 9.2). The new graphic had many lines converging from the bottom of the triangle towards the top. It is hard to say whether there is a deeper thought behind these lines, or if it is just a matter of aesthetics. One might be reading too much into the matter, but one could argue that these lines symbolise the various "unit groups" of "similar accidents" (per scenario) in the total data set.[19]

Apart from some streamlining, the changes in the text were limited, but quite significant, indicating that the unit group of 330 accidents were "similar" and "of the same kind" (HWH1941, p.26). As discussed further below, this is an important addition regarding the understanding of "unit groups" and common causation.

Another significant innovation was in the text accompanying the picture, namely the memorable one-liner, "Moral – Prevent the accidents and the injuries will take care of themselves" (HWH1941, p.27). As any one-liner, superficial interpretation without knowledge of underlying thinking may lead to wrong applications. Here may be one seed for the misguided belief that focussing on small accidents will prevent major disasters.

9.1.7 Variations

During the 1940s, the concept was featured in Heinrich's management books. These are interesting because they feature different and novel graphic

THE FOUNDATION
OF A MAJOR INJURY

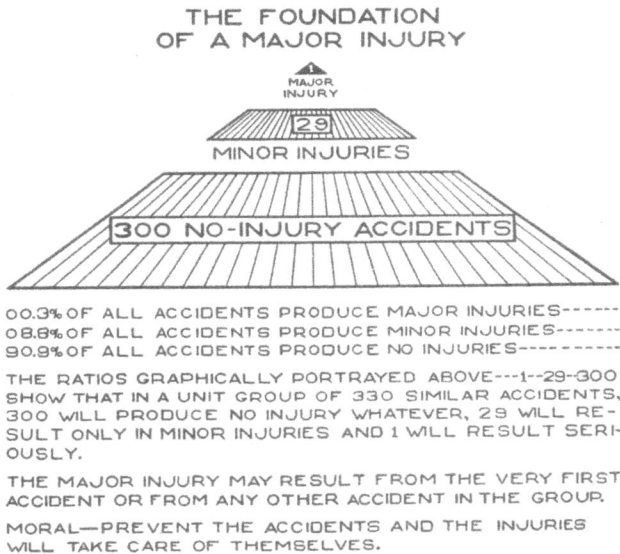

MAJOR
INJURY

29
MINOR INJURIES

300 NO-INJURY ACCIDENTS

00.3% OF ALL ACCIDENTS PRODUCE MAJOR INJURIES------
08.8% OF ALL ACCIDENTS PRODUCE MINOR INJURIES-------
90.9% OF ALL ACCIDENTS PRODUCE NO INJURIES----------

THE RATIOS GRAPHICALLY PORTRAYED ABOVE---1--29-300
SHOW THAT IN A UNIT GROUP OF 330 SIMILAR ACCIDENTS,
300 WILL PRODUCE NO INJURY WHATEVER, 29 WILL RE-
SULT ONLY IN MINOR INJURIES AND 1 WILL RESULT SERI-
OUSLY.

THE MAJOR INJURY MAY RESULT FROM THE VERY FIRST
ACCIDENT OR FROM ANY OTHER ACCIDENT IN THE GROUP.

MORAL—PREVENT THE ACCIDENTS AND THE INJURIES
WILL TAKE CARE OF THEMSELVES.

Figure 9.2 1941 version of the triangle (HWH1941, p.27).

representations, as well as showing some additional thoughts around the concept. First, there was the representation in the appendices of *Supervisor's Safety Manual* (HWH1943d). This was relabelled as the *Background of a Lost Time Injury*, probably to appeal more to the language of its audience. The picture was a drawing of an endless row of workers lifting an object with a bend back, symbolising that this "wrong" way of doing the job in the end may lead to a serious injury and loss of valuable time. The text concentrated on behaviour, mentioning a *violation* of a safe practice rule (Figure 9.3).

The second appearance of the concept was in the *Basics of Supervision* (HWH1944). This once more showed a blocky figure, instead of the triangle, possibly because the book was published by a different publisher (Figure 9.4). As in the 1943 book, the text emphasised behaviour and opportunity. Heinrich mentioned two morals: "Moral One – Why gamble, do the job the safe way always. Moral Two – Look at the hundreds of opportunities that a foreman has to stop unsafe practices before injuries occur" (HWH1944, p.35).

The text suggested extending the foundation of no-injury accidents with thousands of unsafe practices that lead to neither accidents, nor consequences. Another addition to the concept was the inclusion of production losses. Although Heinrich had no data available for a ratio, he suggested that the principle also applied to production losses and supported this by several cases. The conclusion was the same as for injuries, "Whatever the

The Background of a Lost Time Injury

1 SERIOUS OR LOST TIME INJURY
29 MINOR INJURIES
300 No INJURY

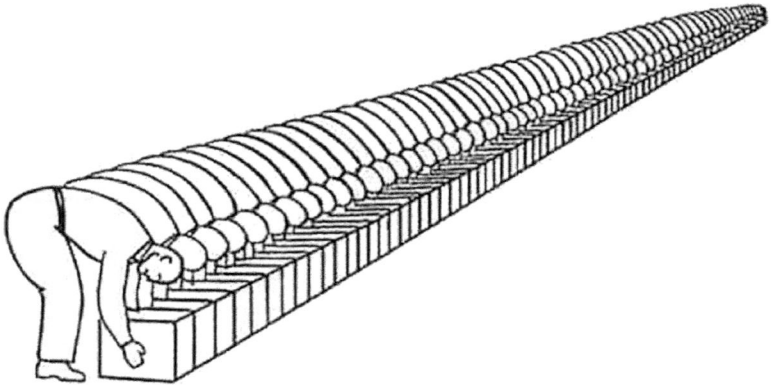

THE BACKGROUND
OF A LOST TIME INJURY
IS THE FOREMAN'S OPPORTUNITY

Figure 9.3 A novel depiction of the 1:29:300 ratio (HWH1943d, p.23).

ratio is, it permits the foreman to take preventive or corrective action before real trouble occurs" (HWH1944, p.37).

9.1.8 1950

When Heinrich published the third edition of *Industrial Accident Prevention*, he used the 1941 figure and text as the basis, but again made some changes. The first main difference was the caption of the figure, addressing misunderstandings, "The 300 accidents shown in the lower block are not merely unsafe practices. They all are falls or other accidents which resulted in narrow escapes from injury" (HWH1950a, p.24).

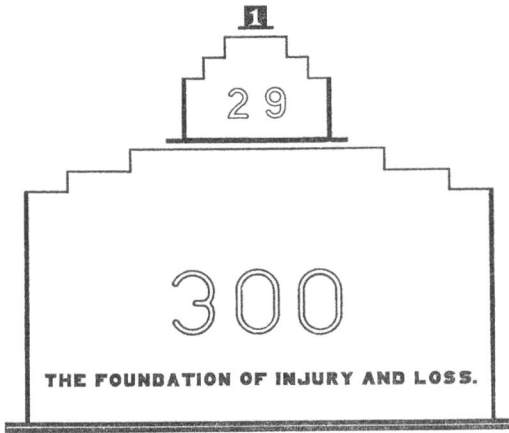

Figure 9.4 The graphic representation from *Basics of Supervision* (HWH1944, p.36).

Another significant addition was made in the text introducing the concept. The 330 accidents were now "of the same kind and involving the same person" (HWH1950a, p.24). This may strike the reader as a peculiar statement, and drew some critique, because how could Heinrich possibly have data on accidents occurring to the same person? However, keeping Heinrich's definition of accident in mind (including near-misses) and taken as an assumption that the addition is meant for the purpose of clarification, it can be interpreted in two ways:

1   In the literal sense, it may point towards the empirical cases offered to illustrate the triangle. These, e.g. the worker crossing rail tracks many times before being hit by a train, do often involve the same person.
2   Alternatively, it was intended to stress the principle of similarity in a statistical thought experiment of 330 repeated events.

Heinrich toned down the common causality, by eliminating a whole paragraph discussing the "predominant causes" which had been in all previous versions. Instead, he bolstered the conclusion of this sub-chapter, "Equally evident is the conclusion that the *unsafe practices and conditions* which result *in neither major nor minor injuries* should be corrected *before* injuries result" (HWH1950a, p.31), adding a new ending that possibly fit better with the safety management focus of this edition, showing Heinrich, the management advisor, giving suggestions how to practically and efficiently deal with problems.

In the average case it is not necessary that a permanent system be established for a complete cause analysis of first-aid accidents. However, it is extremely valuable to do a thorough job at periodic intervals or where there are indications that changes have occurred.

(HWH1950a, p.32)

9.1.9 The triangle expanded

Heinrich's final version of the concept brought some major changes. While the section on the triangle in previous editions was headlined, "Accidents, not injuries point of attack," there was a new header: "The 300-29-1 Ratio Spells Opportunity." A lengthy paragraph emphasised this, "Viewed as an aid in accident prevention this ratio is significant because it vividly emphasises preventive opportunity" (HWH1959, p.26). Heinrich addressed some misunderstanding, not discussing these in detail. His comments suggested that people either thought that the ratio was stable and universal or that a long series of near-misses was necessary before one needed to act. "Misunderstanding and misquotation of this ratio compel the author to reiterate the fact that this ratio is an average. Sometimes a minor or major injury occurs the very first time…" and therefore, "They should and can be controlled long before one of the 300 no-injury accidents ultimately causes an injury" (HWH1959, p.26).

As he had suggested in 1944 Heinrich expanded the triangle downwards, "Another significant point in this ratio is sometimes overlooked. It indicates still greater opportunity" (HWH1959, p.27) because the basis contained only near-misses, and there was an underlying base of hundreds or thousands of unsafe acts or exposure to hazards. "Underlying and causing all accidents, including those resulting in no injury or in either minor or major injury, there is an unknown number of unsafe practices or conditions, often running into the thousands" (HWH1929, p.27). A newly drawn triangle expanded by a block of an unknown quantity of unsafe acts and conditions illustrated this (Figure 9.5).

The text of the figure was based on previous versions, and expanded with the explicit statement, "These ratios only apply to the average case." This extension also led to an extension of the moral:

Moral 1. Prevent the accidents and there can be no injuries.
Moral 2. Prevent the unsafe practices and unsafe conditions and there can be neither accidents nor injuries.

(HWH1959, p.27)

The basis of the text was still the original paper, and a continuation of the text from 1950, but with some significant changes throughout the text. Noteworthy is a section on near accidents, suggesting that this is a wrong term, because it really is about no-injury accidents. There was also a paragraph connecting the triangle and the dominos, indicating the third domino as the

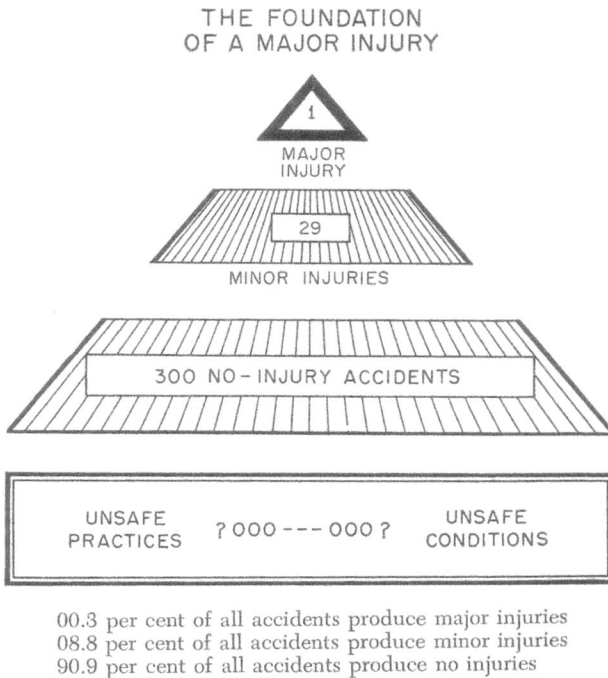

THE FOUNDATION
OF A MAJOR INJURY

1
MAJOR
INJURY

29

MINOR INJURIES

300 NO – INJURY ACCIDENTS

UNSAFE   ? 000 --- 000 ?   UNSAFE
PRACTICES   CONDITIONS

00.3 per cent of all accidents produce major injuries
08.8 per cent of all accidents produce minor injuries
90.9 per cent of all accidents produce no injuries

Figure 9.5 The expanded 1959 triangle (HWH1959, p.27).

foundation of unsafe acts and conditions, the fourth domino as the lower level of the triangle, and the fifth domino as the two upper levels of minor and major injuries.

9.1.10 Post-Heinrich triangles

The triangle concept was highly successful and was adapted by many others. First among these was Heinrich's collaborator, Lateiner, who used the original ratio, stressing the opportunity for prevention and introducing the iceberg metaphor.

The accident problem is like an iceberg with only one eleventh of its mass visible. The invisible base – made up of no-injury accidents – is ten times greater. We ordinarily look at an accident critically only when it produces an injury. In brief, for every accident we report, investigate, analyse, and record there are 10 that we usually ignore.

(Lateiner, 1958, p.14)

The first, and probably best-known post-Heinrich ratio study was the 1969 study done by Bird. Earlier, he already presented the results of a study done at Lukens Steel. Based on over 90,000 accidents in a seven-year period, he found a ratio of 1:100:500 in which the bottom layer consisted of property damage incidents instead of near-misses. The book also featured a cartoon illustrating the concept that an event can lead to different outcomes (Bird & Germain, 1966).

Bird's 1969 ratio study covered over 1.7 million accident reports from almost 300 companies across the USA. This gave a new ratio, 1:10:30:600, adding property damage as the new third level with no consequence/near accidents as the basis (Bird & Loftus, 1976; Bird & Germain, 1992). According to Bird, property damage was another predictor of major injuries, and this became one of the important pillars for his Loss Control programs.

> It isn't really important what the ratio is for a specific operation; the important point that can generally be assumed is that many more property damage and near-miss accidents occur than those terminating in loss. The value of this truth is two-fold. There are many opportunities occurring that do not result in injury but that possess an enormous potential for predictive injury or damage preventive efforts, as the case may be, and the enormous cost reduction potential is there to be tapped.
>
> (Bird & Loftus, 1976, p.107, 108)

During the 1970s there were several ratio studies, commonly accompanied by triangle-styled graphics, including Fletcher's 1972 ratio of 1:19:175, the Tye-Pearson ratio, established on behalf of the British Safety Council in 1974–1975,[20] resulting in a 1:3:50:80:400 ratio and the 1984 Procter & Gamble Port Ivory study, producing a 1:39:292:730 ratio (Fulwiler, 2002).

The 1980 edition of *Industrial Accident Prevention* did not offer a new triangle. The book presented the 1959 text discussing Bird's work before Petersen and Roos critiqued some misdirected applications:

> ...we indicated that there had been considerable confusion surrounding the triangle concept, the relationship between severe- and minor-injury causes. And there has been. The difficulty stems from the use of statistics. Our original data of 1-29-300 were based on "accidents of the same kind and involving the same person." The figures are averages of masses of people and all kinds of different accident causes and types. It does not mean that these ratios apply to all situations. It does not mean, for instance, that there would be the same ratio for an office worker and for a steel erector. It might mean that they could be averaged to this (or a similar) ratio; but certainly neither of these extremes would fit the ratio.
>
> (HWH1980, p.64)

The UK's Health and Safety Executive presented several triangles during the next decade. Studies in five organisations in the oil, food, construction, health, and transport sectors produced a 1:7:189 ratio (HSE, 1997). Two years later, the HSE (1999) presented another accident triangle based on another data set, showing a 1:207:1402:2754 ratio. This report mentions individual case studies with wide variations between industries.

Several authors, notably from the field of BBS, followed the example of Heinrich and Granniss in the 1959 edition of *Industrial Accident Prevention* adding behaviour at the bottom of the triangle (Krause et al., 1990; Geller, 1996; Fulwiler, 2002; McSween & Moran, 2017). Oil company Conoco Philips did another ratio study in 2003[21] which expanded the triangle downwards, "for each single fatality there must be assumed a hidden bottom end of the pyramid of 300,000 at-risk behaviors" (Freibott, 2012, p.260).

Even in recent years, studies presented new triangles, or created them for a specific industry (e.g. Collins, 2010; 2011; Prem et al., 2010; Walker, 2017; Dunlap et al., 2019).

9.1.10.1 Serious injuries and fatalities

At some point, several authors (Manuele, 2008; Krause & Murray, 2012) commented that while there had been great reductions in workplace injuries, the number of serious injuries and fatal incidents (SIF) showed a much slower decline. The conclusion was therefore that the premise of looking at the entire triangle (which according to them was what Heinrich had said) was wrong: "...the premises on which the pyramids, the triangles or the specific ratios (e.g., the 300-29-1 ratios) were built are not valid requires a major concept change – and the data show this is necessary" (Manuele, 2008, p.34).

Several studies followed (Martin & Black, 2015; Inouye, 2018). These suggested that incidents with the potential for serious injuries and fatalities have different root causes and contextual factors leading up to them. Therefore, it was necessary to focus on SIF-precursors instead of any minor incident as slip, trips, and falls.

Many of the SIF papers and tools still present the triangle, but single out a certain area of the triangle indicating that there are some minor events that have the actual potential to create something major, while others do not have this potential. The message of SIF-proponents is to concentrate on those which have the potential of major harm.

These findings were presented as shocking new discoveries and the new way forward. On a critical note, one might comment that it was odd that these insights came from people who had been promoting and selling a particular form of Heinrich's triangle along with behavioural programs for years. Second, one might note this was not new. Petersen and Roos added in the fifth edition of *Industrial Accident Prevention*, "Safety workers for years have been attacking frequency in the belief that severity would be reduced

as a by-product. As a result, our frequency rates nationwide have been reduced much more than have our severity rates" (HWH1980, p.64). Numerous other authors, including Bamber (discussed in Hubbard & Neil, 1985),
Hale (2002), and Heinrich himself,[22] mentioned the need to focus on specific
scenarios in order to implement effective measures.

> The conclusion is that clearly articulated and understood scenarios
> must drive prevention activities. We should discriminate between the
> scenarios that can lead to major disaster and those which can never
> get further than minor inconvenience." adding "If we tackle minor in-
> jury scenarios it should be because minor injuries are painful and costly
> enough to prevent in their own right, not because we believe the actions
> might control major hazards.
>
> (Hale, 2002, p.40)

9.1.10.2 Other domains

Heinrich's work was mainly within occupational safety, and the triangle's
background is there – even though some of Heinrich's examples cross over
in other domains. Over the years the idea was adopted in many other domains, including aviation (Ranter, 2002; Nazeri & Donohue, 2008; Rieder &
Bepperling, 2011; Walker, 2017; Nascimento et al., 2013), the maritime sector (Anderson, 2003; Bhattacharya, 2009; ICS, 2013; Andersen, 2018),
sports (Hawkins & Fuller, 1996, 1998), air pollution (Environmental Protection Agency, n.d.), security, quality, corporate governance, finance and
risk management (Oktem, 2002; Oktem et al., 2010),[23] compliance (State of
Queensland, 2018), and bicycling (Hamann & Peek, 2017).

Swedish traffic researchers presented triangles with clear similarities to
Heinrich's and Bird's work, but without referencing them. Hydén (1987) suggested that "the processes that result in near-accidents or traffic conflicts
have much in common with the processes preceding actual collisions – only
the outcome is different." His traffic safety pyramid had a tiny tip with accidents of varying seriousness (fatality, injuries, and damage). Lower levels
showed more near accidents, even more "slight conflicts," a greater number
of "potential conflicts," and an enormous basis of "undisturbed passages" –
things going mostly right. Svensson and Hydén (2005) even widened the
concept by including "normal behaviour," reminiscent of "new view" safety
approaches.

For many years, the triangle has seen applications in process industry
(e.g. Jones, Kirchsteiger & Bjerke, 1999; Phimister et al., 2003). On many
occasions, however, based on wrong premises. Investigation of major accidents as Texas City and Deepwater Horizon have documented problems of
mixing indicators for occupational and process safety. Several authors (e.g.
Hopkins, 2008) have suggested the development of separate applications.

There appears to be widespread use of the triangle in the health domain. Taxis et al. (2005) discuss several applications in patient safety, along doing some research into the matter themselves. Headlining, "What you see is only the tip of the iceberg," the Centers for Disease Control and Prevention (2014) used both an iceberg and a consequence pyramid on their World Polio Day infographic, and even the World Health Organization (2010) used an injury pyramid as a graphic representation of the demand on the health sector caused by injuries, while various health agencies used pyramids during the Covid-19 crisis (Folkehelseinstituttet, 2020).

9.2 Studying the triangle

Before investigating several interpretations, it is useful to go systematically through different ways to read it (some not mutually exclusive) and some principles underpinning the triangle.

9.2.1 Ways to read the triangle

9.2.1.1 A whole

The first way to read the triangle is seeing it as the gathering of all accidents and incidents of an organisation, sector, or country with their associated outcomes and presenting them in one triangle. When someone calculates a "Heinrich ratio" (e.g. Hawkins & Fuller, 1996; Johnson, 2003) of something, or when companies, government agencies, or regulators (e.g. HSE, 1996, 1999) present accident data in triangle form, they adopt this view.

This is essentially the approach followed by Bird who took a certain amount of accident data and determined the ratios between the various outcome categories. The usefulness is limited. It gives an average of outcomes within the data set. Manuele remarks,

> ...ratios can be produced from most statistics on accident occurrences. And they may have value in their own settings. However, in a constantly changing world, those ratios may not have permanency, nor are they universally applicable in all work environments.
>
> (Manuele, 2002, p.40)

So, besides giving a snapshot of a that specific data set, and having the opportunity of comparing for example a development over time, this view means little but for statistical purpose. For preventive use, one would need some more, "...while the ratio results are interesting in their own right, they are not related to causal patterns, they are merely descriptive of the inter-relationship between frequency and severity" (Davies et al., 2003, p.45).

More problematic, this view can be misguiding. Hopkins discusses the misconception of seeing occupational safety injuries as an indication for the management of major accident hazards/process safety. Hopkins calls the triangle a "theoretical support" for this thinking that was a relevant factor in the Texas City disaster, because it led management to believing that they controlled all safety hazards. "The single-triangle model exercises a profound hold over much safety thinking and is both cause and a symptom of blindness to major risk" (Hopkins, 2008, p.58).

9.2.1.2 A collection

The second way of reading the triangle is by regarding it as a collection of numerous separate triangles from which an average has been determined. This average teaches a lesson, namely that there in general is a few-more-many relationship between various outcomes. However, to be of *real* value for accident prevention, one needs to read the triangles separately. These separate triangles may conform to the concept, but some will not even be triangles in the broadest geometrical sense.[24]

Heinrich mostly followed this way of reading the triangle, as suggested by the various cases that he describes, and the way he mentions that it was about "unit groups" and "similar accidents." From these separate triangles, he determined an overall average (the 1:29:300 ratio) which he used as his main tool to communicate the concept.

Of course, the distinction between the first two views is not black and white or absolute. There is also the challenge of determining the right scope for these separate triangles as discussed below regarding the principle of similarity. The literature shows examples of intermediate views that do not regard the triangle as a whole, while not separating into specific scenarios either. For example, when various groups of workers or activities are compared (e.g. Powell et al., 1971), or the SIF-approach.

9.2.1.3 Bottom-up

The next two views are centred on causation and events within the triangle's scope. In the bottom-up view,[25] the events with no or minor outcomes at the bottom are seen to have the potential of developing into something serious. Taking this to extremes, one takes the "small stuff" at the bottom as the starting point and assumes that all this develops into one of the serious outcomes at the top. Following this assumption, working on events with minor or no outcomes makes a lot of sense, because preventing them is bound to prevent major accidents from happening.

The main problem with this view is that reasonably not all events at the bottom of the triangle have the potential to develop into something serious. Although even a paper-cut in the copy room can develop into a fatal accident under certain extreme circumstances (e.g. involving blood poisoning), the likelihood of this outcome is extremely small and can practically be discounted. As the SIF-studies correctly concluded, some minor events do not have the potential developing into something major.

9.2.1.4 Top-down

The counterpart is reading the triangle top-down. This takes as its premise that the events with serious outcomes at the top are usually preceded by events with less severe outcomes that can serve as warnings. This takes a major outcome event as a starting point and works downwards through a series of possible precursors. This is a view that has scenarios in mind (Hale, 2002). These are always ending in the credible/plausible potential outcome and not necessarily the worst.

9.2.1.5 Retrospective

Investigation of major accidents shows frequently that they were preceded by a larger number of minor accidents and near-misses. Had the organisation acted on these, the major accident might not have happened. Initially, Heinrich himself seems to have had a retrospective look leading to the idea of the triangle. For accidents leading to injuries, they found several accidents without injury that had happened before. These gave opportunities for prevention (HWH1927c, 1928a, 1929d, e,[26] 1930a).

One can argue this being a counterfactual view that might misdirect causation, suggesting that failing to prevent the accident from happening was one of its causes. However, that is not what Heinrich suggested. He had a prevention perspective, and then counterfactuals may be beneficial since they can lead to suggestions for improvement. Still, taking this way of reading the triangle to extremes, it may lead to misguided conclusions as one safety advisor who was heard saying "we had two accidents, so that means that 1,200 unsafe acts must have preceded them."[27]

9.2.1.6 Prospective

The opposite of a retrospective view is forward looking. This emphasises the potential of an event instead of its actual outcome. Heinrich during his research quickly realised the proactive potential, "a splendid opportunity to anticipate and prevent actual injuries" (HWH1931a, p.89). Accidents with minor or no outcomes often had the potential to create something much more serious, thereby providing a valuable warning. Acting upon them might prevent them from recurrence and major outcomes from happening.

Taking this view to extremes is also misguided. For example, when one starts reading too much prediction into near-misses, seeing everything as omens of doom, possibly wasting resources that could more wisely be spent on other problems. Alternatively, being blinded by "success." A few near-misses are nothing to worry about; after all, statistics have shown that it takes hundreds to produce an accident – ignoring Heinrich's warning that "one major injury may result from the very first accident, or from the last or from any of those that intervene" (HWH1931a, p.91).

9.2.1.7 As a heuristic

A final view is seeing the triangle as a heuristic (Le Coze, 2018), as a rule of thumb reminding and informing about various aspects of the concept. Even when the numbers are wrong, and even without any numbers at all, it serves as a pedagogical and communication tool because it is illustrative and memorable. In its most simple form, the triangle illustrates the few-more-many principle that applies to the outcomes of many types of accidents. Intuitively people can relate to this simple message because of personal experience. This can be used to stress the importance of near-miss reporting and the opportunities found therein.

As a heuristic, the triangle also illustrates the concept of visibility. In line with Lateiner's iceberg, the top (serious outcome) is most visible while many near-misses are under the waterline and draw no attention at all. "It goes without saying that actual accidents have the highest visibility, but that day-to-day behavioural acts are easily overlooked" (Van der Schaaf et al., 1991, p.4). With increased visibility, comes often an increased amount of information. In his Air Safety Information Model, Ranter (2002) therefore proposed another view, drawing a typical accident triangle (which he extended downwards by adding defects and unreported occurrences) with an inverted triangle next to it. The latter illustrated the relative amount of information available about the events. There tends to be much more information available about serious accidents and only little about near-misses or defects.

9.2.2 Underlying principles

The triangle had some underlying principles. Some of these Heinrich explained, others we can infer from his examples. Papers discussing the triangle often investigate only one or some of these principles, usually without being explicit what they aim at or how they interpret or read the triangle. However, generally this does not stop the authors making claims about the *entire* concept.

9.2.2.1 Accidents as unit of analysis

Heinrich's main unit of analysis were accidents. He defined these as "an unforeseen, improper or non-planned occurrence" (HWH1931a, p.93). This included narrow escapes from injury and even to some degree unsafe conditions and practices. His triangle therefore consisted of accidents with various degrees of outcome severity. Heinrich analysed accidents, not injuries.

Post-Heinrich triangles often used different language that may have contributed to confusing the principle, for example by distinguishing between accidents and incidents, or just noting injuries and near-misses.

9.2.2.2 Separation of accident and outcome

In Heinrich's opinion, the confusion of causes, accident, and injury hampered safety work. His work stressed the need for greater clarity, e.g. through the accident sequence, "When a person is injured the sequence of events is, first, the cause, second, the accident, third, the injury" (HWH1928a, p.9). In the triangle, the accident was separated from its outcome. Mixing up injury and accident stood in the way of prevention, because if "...the chief purpose of analysis is to furnish a clue to accident prevention, it is vital to know the cause of the *accident* itself as distinguished from the cause of the *injury*" (HWH1931a, p.40).

9.2.2.3 Outcome is largely fortuitous

Heinrich proposed that an accident can have various outcomes, including none, "...an accident need not produce an injury..." (HWH1931a, p93). Beforehand, it is uncertain what the outcome will be. This depended not purely on chance, although factors as time and place are relatively random (the difference between a fatal hit and near-miss may after all depend on an inch closer or a second earlier). One may have a sense of the probability, however. Heinrich mentioned some personal factors as physical and mental condition. Energy involved was another important factor. Outcome therefore depended on both hazard (energy) and randomness.

9.2.2.4 Similarity

One of the most important principles and often forgotten by those discussing the triangle, is the principle of similarity. Heinrich applied the triangle to a "unit group" of accidents. They were "similar" and "of the same kind" (HWH1941). This principle ties in with reading the triangle as a collection of "smaller" triangles.

In the 1980 edition, Petersen and Roos offered examples of a broad interpretation of similarity, i.e. by occupation. Heinrich's original text suggested a narrower interpretation. He does not define "unit group," "similar," or "of the same kind" explicitly. However, there are clues. A quote as "...a serious injury is not an isolated happening by itself but is the logical consummation of some 329 precisely similar accidents, all of which have the same cause" (HWH1931a, p.294) takes the similarity principle to extremes. Strictly speaking no accident outside a fully controlled laboratory ever can be "precisely similar." Since Heinrich was a practical man, one assumes that he intended "precisely" to have a looser meaning. This thought is supported by what follows:

> The employee who is injured as the result of an accident has, in the average case, made the very same mistake or the same error of judgement – violated the same instruction or taken the same chance – that finally caused the injury, many, many times before without, however, having been hurt.
>
> (HWH1931a, p.294)

This suggests a similarity by type and scenario (causes), which is supported by the various cases that Heinrich mentioned illustrating the triangle. These shine a light on how Travelers engineers derived the number of non-injury accidents, and also tell how to understand the concept of similarity/ unit groups. They deal consistently with one type of event and a repeated scenario.

However, a "unit group" is not only determined by "type of accident" (e.g. fall) or "the agencies or means by which accidents occur" (HWH1931a, p.260). Also "circumstances" help to differentiate it from other unit groups. Besides, these circumstances may affect the outcome. Heinrich gave examples of falling on a soft rug-covered floor at home versus falling from a steel erector on a skyscraper (HWH1929a, p.4).

9.2.2.5 Common causes

To be a proactive tool, the triangle needs to be able giving clues about the causes of the accidents one wants to prevent. Therefore, common causality[28] is another important principle. It is probably also the most contended. Wright and Van der Schaaf (2004) concluded that little proper research is done to test the common cause hypothesis, either because studies confounded matters, or because the data was insufficient.

Heinrich's writing on the subject leaves room for several interpretations, sometimes clarifying the subject, sometimes contributing to confusion. In addition, its applicability depends much on how one reads the triangle and whether one observes the other underlying principles. Some ways to interpret common causality are the following:

DIRECT CAUSES

Heinrich's research contained limited causal analysis. During their research, Travelers engineers assigned only *one direct* cause (code) to each of the cases (HWH1929a). For actual accident prevention, merely acting on this direct cause would be too limiting. After all, disregarding rules (cause code: "Poor discipline") in one situation requires a different approach than in another situation, something Heinrich was aware of, as discussed in an earlier chapter.

Also the description of Heinrich's cases, however brief, was richer than a mere direct cause (code). Therefore, for common causality to work, one needs to move beyond simple direct causes.

ROOT CAUSE VIEW

The thinking that actions on underlying causes can have a further reaching effect than actions on direct causes is prevalent in modern safety management, "Eliminating the causes of one organisational problem, will eliminate the causes of others" (Petersen, 2001a, p.11). However, it is found relatively seldom in sources discussing the triangle. In most cases when these papers mention "root" causes, they mistakenly claim that Heinrich indicated human behaviour as the root cause. As discussed before, Heinrich spoke mainly of direct causes.

When DeBlois (1926a) mentioned common causality, unlike Heinrich's work, the type of accident did not need to be the same.[29] He was thinking in the "modern" line of "root" causes as underlying factors in an accident. A typical contemporary example of this interpretation tells us, "The reasoning behind this scheme is simple: many accidents share common root causes, and therefore, addressing more commonplace accidents that cause no injuries can prevent accidents that cause injuries" (Guzzetti, 2013, p.8).

The root cause view sounds sensible but is not without problems and likely stretches the triangle's concept and the principle of similarity.

ALL ACCIDENTS HAVE COMMON CAUSES

A belief that all accidents share common causes is either a complete misunderstanding of the principle, or an extreme view of the above-mentioned root cause view, which both can be easily debunked by "common sense." After all, refinery explosions do have very different causes from paper cuts.

However, Mauro et al. (2018) suggest an interesting perspective, proposing that workplace accidents are an example of self-organised criticality. Other phenomena governed by self-organised criticality, as landslides, earthquakes, and forest fires, have common underlying causes. This suggests to the authors that also workplace accidents share a common underlying cause – providing scientific support for Heinrich's empirical concept.

> Our results provide scientific support for the Heinrich accident triangle, with the practical implication that suppressing the rate of severe accidents requires changing the attitude toward workplace safety in general. By creating a culture that values safety, empowers individuals, and

strives to continuously improve, accident rates can be suppressed across the full range of severities.

(Mauro et al., 2018, p.284)

While this idea is interesting, proposing "culture" as an abstract blanket cause, capturing almost everything, this may mean that it thereby misses its value (Silbey, 2009). Also, it seems to be a cause of another order when compared to breaks along fault lines for earthquakes, or perturbations by falling grains of sand that trigger a landslide. It is probably more useful to apply the thinking within narrower scenarios or hazard types.

CONFOUNDED VIEWS

Those who reject common causality often apply one of the following arguments:

- increases or decreases of serious and minor injuries do not correlate, or their correlation is even inverse,
- causes for frequency are different from causes for severity.

Davies et al. argue that "...the observation that a reduction in frequency does not lead to a similar reduction in severity" (2003, p.48) confounds the matter, and also ignores some essential points, notably similarity.

Petersen (2001) and others have suggested that non-identical causes under-lie accidents which give rise to injuries of varying severity. Examples offered to disprove the triangle, as the Texas City refinery explosion, the Challenger disaster, or Deepwater Horizon, often confound scenarios. Typically, these cases are presented with an argument that occupational hazards or minor injuries were not predictive of these accidents. However, studying the cases one learns about various precursors, such as structurally overfilling and a maintenance backlog, earlier problems with O-rings, and problems with the Macondo well. So, while the mass of accidents deals with things unrelated to the major accidents, often precursors can be found within the disaster's scenario.

The other argument is based on studies of accident data (e.g. Saloniemi & Oksanen, 1998; Barnett & Wang, 2000). These tell us, "The situation where the two are not increasing and declining simultaneously can be interpreted as evidence against identical causation and vice versa" (Saloniemi & Oksanen, 1998, p.60). However, as Wright (2002) argues, this is not a valid test for common causality because it does not look at causes. Furthermore, these studies hardly observe the principle of similarity because they look within a country or a sector (Salminen et al., 1992).

Another way of confounding common causality is by arguing that minor injuries and major injuries have different causes. This often involves mixing various types of accidents or ignores Heinrich distinguishing be-tween causes for accidents and causes for injuries. Conklin (2017b) has ex-pressed this critique in the one-liner "What hurts you does not kill you," using ankle sprains and electrocutions as examples. However, Conklin compares apples to oranges, or rather outcomes (ankle sprains) to events (electrocution). Comparing small to big apples (a fall resulting in sprained ankle versus a broken neck) would provide another, more correct, picture. Another issue is that also the things that have the potential to kill you do not necessarily kill you. Factors to consider include probability and vulnerabil-ity. This may indeed suggest that working on high voltage contains higher risk than a slippery floor, although the number of exposed employees may suggest another prioritisation.

WITHIN SCENARIOS

The most useful approach to common causality is observing the principle of similarity. A group of accidents, of the same kind or type, are likely to have similarities in their scenarios. The outcomes can be different, after all there may be separate causes or factors that determine them, but the scenarios leading up to the accident share at least some commonalities. Therefore, Hale (2002) argued it was important to analyse scenarios, or even parts of scenarios and observe the potential within these scenarios, "We should not think in terms of comparing major and minor injuries, but of understanding accident scenarios. We should compare completed and uncompleted acci-dent sequences" (Hale, 2002, p.39). This was the proactive power of this approach.

> ...the problem is not with the pyramid representation, but rather with its use without criteria. To bring about a useful analysis, one must dis-criminate between causal factors; this sounds quite right even if one is dealing with occupational accidents alone.
>
> (Jacinto & Soares, 2008, p.633)

While the principle sounds logical and appeals to common sense – every reader will probably have experienced events that did not result in bad outcomes but might have, given minor differences in circumstances – there is little scientific research that has studied the concept properly. Some research to support the approach was done in UK railways (Wright, 2002), sleepy driving (Powell et al., 2007), and recent research of thou-sands of occupational accidents in the Netherlands. The latter suggested, "Smaller severity more frequent accidents can provide information about the direct and underlying causes of bigger severity more catastrophic ac-cidents but only if looking within the same hazard category" (Bellamy, 2015, p.93).

9.2.2.6 Opportunity

The final underlying principle of the triangle deals with potential. It is about reacting on weak signals, like near-misses, and seizing them as opportunities for prevention. This was an important step into the direction of more proactive accident prevention. Acting before something bad happened: "Accidents – Not injuries – the Point of Attack" (HWH1941, p.26).

Thinking about potential brings along the problem is of where to draw the line. After all, most normal operation does have the potential to develop into something bad. However, hardly anyone will take the earlier mentioned example of a paper-cut developing into a fatality seriously. Besides, for most organisations it is impossible to react on all the near-misses and other minor events.

It is therefore important to select situations with a perceived greater potential than others, situations that are a major deviation from the "normal" situation, or those prioritised in some way by the organisation (e.g. as part of a campaign). This may be very arbitrary, making it hard or impossible for scientific validation of the principle. Possibly some preventive actions may be misdirected; however, the principle is quite useful for practical improvement. We will reflect some more on this at the end of the chapter.

9.3 Interpretations and critique

The various ways to read the triangle can be useful to understand how people have interpreted and adopted the triangle. Mixing some of the above views and principles – willingly or unwillingly – may result in confounded views and misapplications. If you for example take the triangle as a whole and use the principle of common causality together, it provides the most important source of confusion, namely that all accidents and incidents have the same causes. Therefore, let us now turn towards various interpretations and adaptations along with some critique offered over the years. Space allows only reviewing some of the most frequent.

These interpretations may not necessarily be in line with what Heinrich intended. They may even be contradictory. As James Reason at one point said regarding critique of the Swiss Cheese Model, "...they may be accusing it for failing to achieve something it never intended" (Reason et al., 2006, p.21). However, these interpretations are a reality for many practitioners and organisations because they have become the prevailing belief in their circles.

What caused the misunderstandings and interpretations? There are surely many contributing factors. These include shallow implementations (e.g. as part of BBS-programs), using the triangle as a metric (leading to the inevitable mix-up of goal and means), mixing terms and definitions, difficult access to original sources, and even wilful misconstruction of the message. While insufficient understanding of the concept may have led people in believing that reacting on minor events would contribute to preventing major accidents, surely no-one made an explicit claim that preventing ankle sprains would reduce explosions. However, such statements are offered as reasons why the triangle cannot be right.

As mentioned before, attempts to confirm or disprove the triangle rarely explain what part or interpretation of the triangle they try to investigate. Likewise, only few studies[30] do go back to what Heinrich wrote and check his premises. This section aims at correcting this situation for some interpretations.

9.3.1 Behaviour

Some say that the triangle is mainly about behaviour. This need not be so; above we have seen applications in many other domains. Anand (2015) gives a technical example with a maintenance background, Hopkins (2008) proposes a process safety triangle, and Heinrich frequently discussed conditions and proposed "engineering revision" as the first choice of action. On the other side, Heinrich attributed most accidents to "unsafe acts" and occasionally discussed the triangle with emphasis on the behavioural element: "A person suffers serious injury, in the average case, only on the 330th *violation* of safe-practice procedure..." (HWH1943a, p.13, emphasis added).

While Heinrich's message thus was more balanced, it is easy to see how it can be interpreted in a behavioural way and that some critique connects the triangle to blaming workers. "Heinrich's work encourages people to look strictly at procedures and training instead of rethinking system design" (Johnson, 2011) because "such claims let upper management off the hook." Much of this critique has most likely nothing to do with the triangle as such but is mostly due to its use in some safety programs, notably some BBS adaptations. "We need to 'work' the principle at the behavioural level and let the pointy end of the triangle take care of itself" (Marsh, 2017, p.35).

Surely, often there will be some behaviour underlying, but not always. One main problem is of course, what exactly is "unsafe behaviour" and whether one is able to spot it. Also, it is debatable what "working at the behavioural level" means. Many may interpret that as changing behaviours, approaching things from a systems perspective may be more successful.

Hopkins (2008) describes how the belief that thousands of unsafe acts were underlying the Texas City triangle model misdirected efforts and resources from major accident hazard related problems to a behavioural safety programme focussing on easily observed "unsafe behaviours" as not wearing PPE, while much more critical behaviour was "systematically missed" because it was infrequent or not obvious to those who might have observed it. A lesson learned from several major disasters is that a balanced approach is necessary. Instead of singling out one element, one should treat near-misses and errors as information about the "health" of the *system* and try to learn from this.

9.3.2 Fixed ratios

One of the most persistent beliefs around the triangle, is that Heinrich proposed a fixed or stable ratio (e.g. Long, 2014b; Dekker, 2018; Marsden, n.d.). "The triangle model assumes that these ratios are fixed in some way..." (Hopkins, 2008, p.56). Some even speak about "Heinrich's Law" when discussing the 1:29:300 ratio (e.g. Conklin, 2007; Ward, 2012; Anand, 2015; Gesinger, 2018). This has led to several studies trying to replicate or test the ratio or determine a stable ratio for a certain sector or activity (Gallivan et al., 2008; Marshall et al., 2018).

Attributing the belief in a fixed ratio to Heinrich is not correct, however. Sure enough, Heinrich proposed a "specific ratio" (Yorio & Moore, 2018, p.40) and he invariably mentioned the 1:29:300 ratio when he spoke about the triangle. His choice of wording was sometimes quite definite. Taking statements like "...it requires 330 accidents to produce only 1 major injury and 29 minor injuries" (HWH1931a, p.90) in isolation may suggest a mechanistic view and a fixed ratio. However, one has to read this in its context where he listed many examples of different ratios (HWH1929a, b, 1931a, 1941), and he emphasised regularly that these were averages and estimates: "...there are *at least* ten others" (HWH1929a, p.5, emphasis added), and "...we are able to estimate..." (HWH1929b, p.9), he wrote. Also,

> ...for every mishap resulting in an injury, there are many other accidents in industry which cause no injuries whatsoever. The investigation has enabled us to establish the conservative ratio of 10 to 1, between no-injury or potential-injury accidents and those causing injuries.
>
> (HWH1929a, p.1)

About the lowest level of the triangle, Heinrich said, "it probably never will be known exactly" (HWH1931a, p.89) and he suggested that some accidents did not have a ratio at all: "...it may be the result of an exceptional isolated accident-type that might never occur again" (HWH1929b, p.10).

When reflecting on Heinrich's continued mentioning of the 1:29:300 ratio and why this may have made sense to him, one strong suggestion is that the ratio is highly recognisable and memorable thanks to the convenient numbers and provides thereby a good way to "anchor" the idea. Most likely, after a while it stopped being a mere ratio and instead it became a shorthand, or heuristic for the idea.

One wonders about the persistence of the belief of a stable ratio and its universal applicability while any ratio is merely a snapshot – specific per industry, company, department, period of time, etc. Many authors after Heinrich pointed this out. In recent years Hale (2002) and Hollnagel (2014) discussed the subject. Research has demonstrated that there are different triangles for different hazards and activities (e.g. Bellamy et al., 2008). However, this should have been known for many years:

> ...data indicate that Heinrich's data is probably conservative, but also show wide variations as among the various industries, different establishments within the same industry, and in comparing different occupations. Many factors are involved.
>
> (Blake, 1945, p.103)

> Studies made over a long period in a wide variety of plants have indicated that for manufacturing as a whole about 29 nondisabling injuries occur, on the average, for every disabling injury. This generality has received wide acceptance as a basis for making broad comparisons. Its author, however, has pointed out that this ratio cannot be considered as representative of conditions in any specific industry and that it is to be expected that there will be wide variations in the experience of different industries...
>
> (McElroy & McCormack, 1945, p.1155)

Misunderstandings prompted Heinrich in the 1959 book to emphasise that the ratio was an average. Petersen and Roos put things even clearer:

> ...there are different ratios for different accident types, for different jobs, for different people, etc. The triangle for the accident type 'electricity' is a different looking triangle than the one for 'handling materials'. Common sense dictates totally different relationships in different types of work.
>
> (HWH1980, p.64)

Also Bird explicitly cautioned of this,

> It is worth emphasizing that the ratio study was of a certain large group of organisations at a given point in time. It does not necessarily follow that the ratio will be identical for any particular occupational group or organisation. That is not the intent. The significant point is that major injuries are rare events and that many opportunities are afforded by the more frequent, less serious events to take actions to prevent the major losses from occurring.
>
> (Bird & Germain, 1992, p.21)

Repeating Heinrich's main message, the triangle is not about the ratio or a "law." It is about opportunities for prevention and improvement.

9.3.3 Proportional reduction

Another persistent belief, found in many sources, is that "Changes in the frequency of minor injuries cause a proportional change in harm" (Seward & Kestle, 2014, p.362). Often, this is dressed in talk of "...eroding the base of the iceberg..." (McKinnon, 2012, p.27), and pictured as a triangle which is reduced in size by "cutting" off from the side, usually suggesting that serious accidents at the top will be prevented.[31]

The root for some of the misunderstanding was Heinrich himself. Passages as "...the frequency of major injuries varies directly with the frequency of no-injury or potential injury accidents" (HWH1929a, p.5) read in isolation seem to suggest proportionality. It appeals to common sense that "...by driving down the number of precursor events, the number of injuries, and in particular, serious injuries can be reduced" (Hopkins, 2008, p.56). However, this can only be true if one observes common causality and the similarity principle; if one sticks within the same scenarios one may prevent similar accidents from happening.

As an example, take a work situation where a product needs to be grinded. In the absence of protective screens and goggles there will be a certain proportion of major injuries (loss of sight in one eye), minor injuries (needing medical attention and leading to absence, like a walk to the first aid post and some rest) and no-injury accidents. When a protective screen is applied and used, we can assume that all other factors being equal the hazard is reduced – there will be fewer metal chips flying towards the eyes. It is not unreasonable to assume that this reduction can have a roughly proportional effect on all outcomes from this specific scenario (Figure 9.6).

Still, testing for proportional reduction is a common "test" of validity of the triangle. Since studies tend to only look at outcome categories and not within scenarios, this is neither a useful nor valid test for the triangle. Besides, even if one would test within scenarios, one will encounter practical problems because one needs large data sets in order to be able to rule out chance effects.

Interestingly, if one would apply safety goggles as safety measure in the above case, one influences mostly the top levels of the triangle and the ratio of no-injury accidents to minor and major increases. In this case, corrective actions widen the base of the triangle and one would see something contrary to proportional reduction – stressing the problem of using proportionality to test the triangle (Figure 9.7).

9.3.4 Prediction

As persistent as the belief in proportional reduction is believing in its predictive value. Many sources speak of prediction in relation to the triangle, but few specify what this precisely means and why or how the triangle's

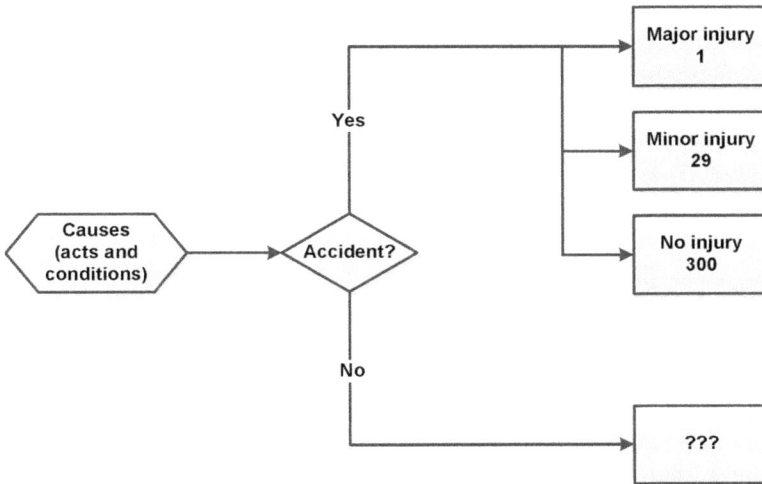

Figure 9.6 A simple event tree scenario – reducing the input might theoretically reduce outcomes proportionally.

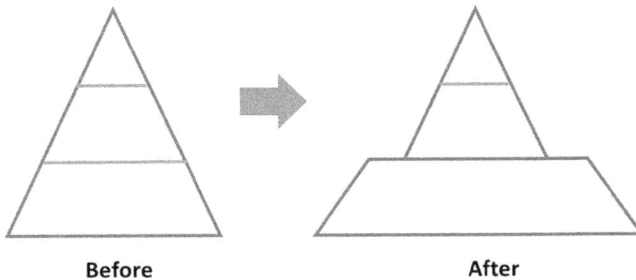

Before   After

Figure 9.7 A counter-intuitive effect.

prediction works.[32] Neither is the term found in Heinrich's writings; an "opportunity to *anticipate* and prevent" (HWH1931a, p.89) comes closest.

One important factor is how one defines prediction. There are various ways to think about prediction and what it would require. When we think of prediction, we often think in terms of very specific prediction. What will happen. How it will happen. When it will happen. What will outcomes be. It needs little discussion that the triangle cannot do any of this.

Turning to a dictionary[33] for help what prediction might mean, it leads to related terms as forecast or foresight and tells us of indicating in advance, of foretelling on the basis of observation, experience, or scientific reason, of the ability to see what will or might happen in the future and also of preparation for the future.

These are useful clues as how we can interpret, and maybe even use, the triangle's predictive force. Safety management deals often with foresight and prediction aiming to prevent bad outcomes. When implementing a mitigating measure, one is in the process of prediction, expecting to change future outcomes. Within safety management there are roughly three approaches to this.

First in a quantitative, probabilistic sense. Although some seem to suggest so, "A ratio such as this is truly invaluable, since it makes prediction possible" (McKinnon, 2012, p.26), the triangle is unlikely to provide the information necessary to make quantitative assessments. The ratios are too much prone to variations, influenced by many variables, underreporting and varying definitions and are at best valid for specific situations and scenarios, mostly limited to highly technical, relatively linear systems, based upon massive amounts of data.

Second, there is the semi-quantitative approach where one has some statistical information, giving an indication, although additional information is necessary. In this approach there may be a notion of "the more near-misses, the more likely an accident will happen," "As soon as too many near-miss incidents are reported in a specific area, it is a clear sign to the management and safety staff that a more serious accident is about to occur, and they can take preventive action" (McKinnon, 2012, p.26).

There is some sense in this. If comparing near-miss accidents to casts of a die, the more often one casts, the higher the likelihood that it will at some point land on a six. Heinrich's argument was therefore to address the event, and thereby reduce the likelihood for both the event *and* injury to happen. Heinrich said, "Statistics provide a valuable clue to the causes of future accidents" (HWH1931a, p.263). At the same time he cautioned,

> The term 'clue' is used deliberately because it is unwise to apply a remedy for the occurrence of accidents based wholly on the lessons of the past. But with the conclusions drawn from statistics as a guide, a study of present conditions may be profitably made and a final decision as to the method of existing known accident causes more readily reached.
>
> (HWH1931a, p.264)

This leads to the third form of safety management prediction: being aware of "early warnings" and precursors and using these opportunities for prevention. Even though Heinrich spoke much of statistics, his main focus in the predictive sense was that hazards and accidents with minor or no outcomes had the potential to develop into something worse.

A useful way of thinking about prediction is to "systematically imagine alternative futures" (Fuerth, 2009). Thinking about what may happen in order to be able to think about what must be done. Here, the triangle can help. Looking at possible accident scenarios (or being informed through near-misses of possible scenarios), enabling identifying precursor events, conditions, or patterns that can serve as a warning of *similar* serious accidents, we can see them as "rumble strips" (Conklin, 2007), signalling the risk of driving off the road and thereby giving the opportunity to react and correct. However, these precursors are not a warning of other accidents, like crashing because a tyre blows out.

Finally, Burin (2013) proposed a continuum in safety management that goes from reactive via proactive to predictive. The reactive approach waits until an accident happens and only then addresses the risks. In a proactive approach, one tries to do something before an accident happens by utilising a variety of sources. Prediction tries to see potential risks and do something about those. In this perspective, the triangle at minimum helped bring safety into the proactive spectrum, and possibly a bit beyond.

9.3.5 Metric

Some organisations use the metaphor, posting it on the wall, tracking the numbers of accidents and reported incidents in an iceberg or triangle shape. Heinrich was a strong advocate of improving statistics and making better use of them for accident prevention (e.g. 1929c, 1930c, 1931a, 1932c, 1935). However, nowhere does he present the triangle as a "way of tracking occupational illnesses and injuries" (Porter, 2016). He did not regard it as a metric, but as a communicative tool to explain and promote a more proactive approach to safety management.

Later authors have discussed some applications as a metric. Hale (2002) mentions uses for trending and Cooper suggested that it might "…be useful to determine the ratio between actual and potential SIFs experienced, to facilitate benchmark comparisons with others" (Cooper, 2019, p.174).

Constructing a triangle for an organisation can indeed give an indication about the "riskiness" and the state of (un)safety of that organisation in comparison to others or a national average. However, accidents and incidents are only a limited measure of safety, and the measurement is fully dependent on the information gathered (Jacinto & Soares, 2008) and the level of reporting (Jones et al., 1999; Lessin & McQuiston, 2013; Nascimento et al., 2013).

Many organisations do not stop there however, connecting goals or KPIs to these statistics. Safety literature even tells to do so. McKinnon argues that a near-miss management system "…must have a set of standards and targets. Expectations must be set for the number of near-miss incidents to be reported per month or year…" (2012, p.54). The number of near-miss reports can surely be used as a metric for the degree of reporting. Still, it is only a weak indicator, because it is unknown what a "good" number would be, and that it says nothing about the usefulness of the reports or of the actual learning.

On the surface, a requirement to "report a minimum of four near-miss events per month" (Andersen, 2018, p.96) is expected to have a positive effect. The more reported near-misses, the more opportunities after all. However, there will also be a greater chance for noise through useless reports made with the sole aim of reaching the goal. Then there is the question whether you harvest the learning and improvement.

Andersen found in his study a firm belief, "...the more near-miss reporting the safer the operation of vessels" (2018, p.10). This prompted him to ask, if the number of near-miss reports doubles, does that mean that it is twice as safe? In his opinion for most organisations the near-miss-KPI was merely "a ritual" (Andersen, 2018, p.96). On the other hand, he reflected, "...without KPIs would any near-miss events be reported and what effect would it have on learning?" (Andersen, 2018, p.70).

Turning the triangle into a "numbers game" (Dekker, 2014, p.176–177) has nothing to do with the principle, but rather with how safety is "managed" (in various meanings of the word) in some organisations. The triangle was originally an argument *against* "suppressing such learning opportunities" (Dekker, 2014, p.176) and rather an invitation to *use* these opportunities. Heinrich regarded the triangle in the first place as an opportunity. No need to wait until the harm is done, one can intervene and try to prevent accidents from happening before someone is injured.

> ...in prevention work, the importance of any individual accident lies in its *potentiality* of creating injury and *not* in the fact that it actually does, or does not, so result. When lost-time or so-called 'major' accidents are selected for study, therefore, as a basis for records and for guidance in prevention work, efforts are often misdirected, valuable data are ignored, and statistical exposure is unnecessarily limited.
>
> (HWH1931a, p.88)

This should appeal to common sense (by lack of a better term), and yet still today, many investigations start mainly after major consequences. Heinrich wrote, "...in general, attention is centred upon these more spectacular occurrences to the exclusion, in part at least, of adequate consideration of minor accidents" (HWH1929a, p.2). His quote rings true today "...present day accident prevention work is misdirected when it is based largely upon the analysis of major injuries" (HWH1941, p.30).

9.3.6 Zero

Many people argue that the only real goal for safety can be "Zero." Recently, a world-wide "Vision Zero"[34] campaign was launched. "Zero" does have a certain appeal. It has symbolic and aspirational value. However, absence of accidents is not the same as presence of safety and "zero accidents" does not necessarily equate "safe."

Several authors (Conklin, 2017a; Long, 2017a; Dekker, 2018) connect Heinrich to "zero harm," seeing the triangle as an expression of striving for zero harm, "...with ideas like Heinrich's pyramid, the notion every accident is preventable" (Conklin, 2017a, p.51; similar statements are found in e.g. Long, 2017b; Dekker, 2018). Connecting "zero" to Heinrich may seem obvious, after all Heinrich stated that nearly all accidents (98%) were preventable (HWH1928a, 1931a, 1941). Even the remaining 2% were in his view mainly because "in practice the cure for an accident cannot always be applied effectively" (HWH1931a, p.45) rather than the often-cited "acts of God."

That, however, is only a superficial interpretation. Studying his work more closely, one finds that Heinrich, while stating that nearly all accidents are preventable, was far from a zealous "Zero Harm" perspective. He was a realist who wrote, "In practice it is difficult to so perfect an industrial organisation that all of these accidents can be eliminated" (HWH1928a, p.11). Most likely, Heinrich saw the preventability of nearly all accidents merely as a hypothetical possibility and something that is judged in hindsight: "We know that we can never attain perfection – that we can never achieve the ideal; and we know therefore, that accidents will continue to occur in sufficient numbers to keep the wheels of insurance turning" (HWH1929c, p.173). Also, he was clear about balancing safety and production, "Safety at any cost, safety at all costs, is not to be desired" (HWH1930b, p.151). He already used the "reasonably" criterion, stating that employers had the responsibility of "maintaining reasonably safe operating conditions" (HWH1931a, p.271).

In other places, he nuanced the "98%"-statement drastically with practical realism:

> Not more than 2 per cent of all accidents are of the so-called 'act of God' or wholly unpreventable type. Of the remaining 98 per cent, 50 per cent are practically preventable and 48 per cent could be prevented[35] if it were not for practical consideration of cost or interference with production and profit. All preventable accidents are controllable by management.
>
> (HWH1941, p.106)

It is also interesting to note that Heinrich was not adamant that 98% are preventable, period. Instead he called them of a "preventable type" (HWH1931a, p.50) or "preventable kind" (HWH1941, p.20), meaning that they could be prevented given certain circumstances.

Still, there is some ambiguity. The text accompanying the 1959 version of the triangle, "Moral 1: Prevent the accidents and there can be no injuries" (HWH1959, p.27), can be interpreted as aiming at zero. Also, Heinrich was generally positive about the effect of "no-accident contests." However, he offered also critique, mentioning that these "contests" usually were aimed at

LTIs (as they are today). He explained with examples how LTIs were a weak measure of safety: "The occurrence of lost-time accidents or serious injuries, in itself, does not always provide an accurate measure of the conditions out of which they arise" (HWH1931a, p.123). He also recognised that these "contests" are only useable as temporary actions because they wear off.

The link between zero and the triangle is mostly based on post-Heinrich interpretations of the triangle that a reduction of minor events will prevent major ones, and eventually make it possible to reach a "goal" of zero accidents.

9.4 Some reflections

This chapter concludes with some reflections about advantages and limitations and the triangle's relevance for today's safety work. Summing up, the triangle told,

1   Paying attention to "small things" increases the knowledge base (frequency may be more important than severity).
2   Potential is more important than actual outcome (a risk-based approach).
3   One may prevent serious outcomes from happening by reacting on events and conditions with no or minor consequences (a more proactive approach).
4   For the purpose of accident prevention, all events within a unit group are regarded equally important, regardless their outcome.

9.4.1 Opportunity

The triangle's main message was one of opportunity. A proactive approach to safety by reacting on events with minor or no consequences. McKinnon describes these events as "a gift" (2012, p.21), enabling action before something bad happens. Bird also echoed Heinrich's original message,

> The 1–10–30–600 relationships in the ratio indicate quite clearly how foolish it is to direct our major effort at the relatively few events resulting in serious or disabling injuries when there are so many significant opportunities that provide a much larger basis for more effective control of total accident losses.
>
> (Bird & Germain, 1992, p.21)

Reacting on weak signals (Weick & Sutcliffe, 2001) brings many advantages. It is more proactive instead of reactive measures after serious outcomes have occurred. Typically, they happen more frequently, so there are more opportunities to act. They are also cheaper and less traumatic because no or little damage has been done. They may be easier to discuss because nothing bad has happened yet (Van der Schaaf et al., 1991).

Assuming management and decision makers see the value of reacting on weak signals and are willing to make an effort, the idea is good. However, there are many misunderstandings and interpretations that may lead astray. And even if one avoids these there are limitations.

9.4.2 Challenges and limitations

9.4.2.1 Recognition and identification

The first problem is how to recognise weak signals. What precursors are a warning of potential accidents? What are signals? What is noise? What is a deviation? What is normal work? And what if normal work is a predecessor? As the term indicates, many precursors are *weak* signals and because they are weak, their connection to risk and potential harm may be unclear.

> Some of these precursors are a natural, even necessary, part of normal operations, yet others are unchangeable within current working constraints and still others may play no significant role in any future scenario.
>
> (Reason, 1991, p.21)

Even when there may be evidence towards failure it is not always an indication that a significant adverse event may happen. Weak signals are often missed. Analysis of major accidents and historical events (Wohlstetter, 1965; Chan, 1979) has taught that they are much easier to detect in hindsight. Often, only then one can see the importance of a certain clue which may have been obscured at the time.

While there is often talk of "safe" and "unsafe," events may be not clearly black and white but highly subjective and depend on how one feels about things. Recognising weak signals depends on knowledge, experience, assumptions, and interpretation. What looks risky to one may not look risky to others. Wohlstetter notes that depending upon "…the background of expectation, it is only natural to ignore small cues that might, in review of the whole or on the simple count, add up to something significant" (1965, p.699).

Vaughan (2002) discussed several challenges regarding the interpretation of signals and action based on them. There are problems with both how signals are transmitted, and how they are interpreted. Are signals taken seriously or are they denied or reinterpreted? When the signals reach someone, it depends on many factors how they are interpreted and whether the recipient decides to act. Relevant factors include knowledge and experience (some signals require no particular skill to recognise them for what they are, others may be much more subtle and depend on tacit knowledge of operators to be taken seriously), complexity of the situation (are causal relationships clear, can the signal be understood), social and cultural factors (shared beliefs, assumptions, histories, priorities), frames of reference, influences from others (peers or superiors), the context they are received in, patterns of information, previous decisions (e.g. about the acceptance of certain risks or conditions, or decisions on action or inaction), and institutional rules and logics (this is the way we handle stuff around here). To complicate matters further, often signals are mixed, first indicating that something was wrong, then suggesting all was well – the latter leading to a view that no action is necessary.

Interestingly, near-misses may not always carry information about danger, but sometimes the opposite. Instead of providing a warning, these events can be interpreted as confirmation of decisions and actions, since people take them as evidence of success, not failure. Studies by Dillon and Tinsley suggested that people often missed the "warnings" attached to near-misses. Instead of regarding them as events that almost resulted in bad outcomes, they saw them rather as successful outcomes that avoided harm (Dillon & Tinsley, 2008, 2015).

9.4.2.2 Prioritisation

When weak signals are recognised and identified it is a challenge which should be selected for action and what to do with them. It may be tempting to prioritise "low hanging fruit." These may be attractive for intervention by managers or consultants. On the other hand, there will be issues that have the potential to create major issues. Metrics that seemingly tell that everything is okay may misguide this decision as illustrated by cases like Texas City. Turner (1978) discussed the so-called "decoy effect": events that drew away the attention of the things that (at least in hindsight) really mattered.[36]

Factors as differing agendas and conflicting objectives play a role in different interpretation and construction of "facts" between stakeholders. Vaughan (1996) coined the term "normalisation of deviance," a process by which signals that are initially regarded as an indication of increasing risk upon analysis and assessment are reclassified as "acceptable."

Many organisations have goals to increase the reporting of events which may raise the issue of volume. Handling reports and learning costs time and resources. Macrae (2015) is critical of the focus on quantity ("report everything") and argues that learning from incidents is misunderstood and misapplied. One should not only focus on the number of reports but primarily on how to deal with them, increase the quality, and use them to learn critical lessons instead of merely skimming a mass of minor things. He states that too many reports are "swamping important signals with noise" (Macrae, 2015, p.72), drawing attention away from the things that really matter. This is a real problem for many organisations. McKinnon (2012) emphasises the need to determine which of the large number of minor cases will be useful for improvement. Several authors have countered Macrae's concerns. "Ultimately in any system there will be reports that are of no use, but this is less of a problem than no reports at all" (Hopkins & Maslen, 2015, p.121). In general, safety literature tends to see underreporting as a greater problem than its opposite, overreporting,

> The reporting of critical incidents is therefore an essential part of the learning process and the underreporting of incidents is likely to have a negative effect on the probability and severity of future incidents, subsequent organizational learning as well as the performance of the system.
>
> (Van der Westhuizen & Stanz, 2017, p.200)

9.4.2.3 Effectiveness

Reacting on precursor events with the aim to prevent recurrence of similar events or to mitigate hazards is generally perceived as a positive thing. That applies also to the assumed effects of actions taken. However, these effects may not be as clear as imagined. Often things may seem obvious, and there may be little doubt about what to do, other situations are not that clear-cut. If one does *not* act on the observation and something goes wrong, one can be blamed for the adverse outcome – maybe even though everything was normal at face value, and applicable rules were followed. If one *does* act, there is uncertainty about the success of the intervention, and there may be complaints about "unnecessary" delays, costs, or hassle. After all, it is notoriously hard proving that an action actually prevented bad outcomes. One can make this plausible, but never prove. In some cases, attempts to do so may be comparable to rain dancing.

Even worse, interventions may lead to unsafe situations, or adverse outcomes. Examples are increased bureaucracy and administration or negative effects on production because of safety interventions. Actions to mitigate hazards can also create new risks. More barriers and control can make situations less transparent and more complex (Perrow, 1984) or create "fragility" (Taleb, 2012). There may also be the effect of risk homeostasis (Wilde, 2014). People's behaviour often alters because of preventive measures. This may diminish the positive effect of preventive interventions.

Besides, if one reacts to a certain precursor or hazard, one may eliminate or reduce the likelihood for one scenario. This means that the probability that an accident (or injury) happens in that specific way is lower, but this does not rule out the possibility in others. And, as it is possible to underestimate the potential of a precursor event, one may also overreact. One never knows for sure whether the incident one reacted upon really would develop its full potential and one may be dealing with a decoy, preventing a minor issue while leaving a more important risk unaddressed.

9.4.2.4 Applicability

Some authors stress the uniqueness of accidents, casting doubt on common causality. Of course, the details of each accident are unique, and there are some challenges determining similarity. The safety profession has struggled with this for decades:

> The great bulk of accidents, however, are not very homogenous in these regards, and one of the great difficulties in accident research has been to find enough factors common to enough hazardous situations for specific remediable actions to prevent sizable numbers of accidents.
>
> (McFarland, 1963, p.687)

However, that does not mean that this is a hopeless task. In practice, the principle has proven to work. Also, as Hopkins (2001) suggests regarding Three Miles Island, it is not necessary to predict the entire scenario to avoid such an accident. Often it is enough to identify the risk or hazard one is addressing, and a partial scenario may be enough.

On the other side, there will be scenarios for which the few-more-many concept does not apply. Even Heinrich indicated the possibility of unique scenarios that only happened once and never again.[37] Also, it has been suggested that the principle works best for (semi)-linear systems (Hollnagel, 2014a) because scenarios and precursors are much clearer for these systems than for complex systems. However, even in complex systems there may be "common patterns" that can serve as precursors or warning signals. While the system cannot be predicted and controlled as a linear system can, patterns within the complexity can be understood and influenced and the controls may be rather emergent than engineered (Carrillo, 2019).

9.4.3 Conclusion

Despite all these problems, the triangle indicates opportunities. If one does not make the effort to pay attention to small things, one denies oneself the choice of opportunity. Strip away all the numbers and forget later interpretations. Essentially, the triangle was a call for a proactive approach to safety, which should be a good thing. Putting argument in other words:

> The same incident causes all the different levels of injury, but the increasing severity bears no relation to the activity undertaken. The degree of loss of control is the same in each case, but the outcomes are different. Yet, we are likely to respond differently based on the outcome rather than the loss of control that prompted the accident.
>
> (Marriott, 2018, p.53)

Marriott does not add a ratio of course, preventing being distracted by numbers.

Notes

1 Heinrich only did a 2D shape triangle; therefore, one finds mainly the term triangle throughout this book. From the sources, it is most likely Bird who first came with the 3D pyramid shape.

2 In fact, the first hints are already present in the earlier papers on hidden cost (HWH1926, 1927a, b).

3 The causes of fatalities are much the same as the causes that predominate in the minor injuries – which conclude that if we prevent minor accidents, the fatalities will to a large extent take care of themselves. (HWH1927c, p.231)

4 I wonder whether this should have said "case" instead of "cause." This would give more meaning in the context and "minor cause" is not discussed elsewhere in Heinrich's work.

5 This quote shows how Heinrich may have contributed to later misinterpretations of the triangle, as it, taken out of context, can be interpreted as if action on any minor injury accident will improve any major. However, the quote is preceded by an example that illustrates the principle within one common scenario that has various outcomes.

6 One may wonder what Heinrich meant by the word "foundation" as he only uses it for the paper's title and the caption of the triangle chart. The text suggests that it means what one is basing one's prevention work on (HWH1929a, p.2). This interpretation is supported by the description found in Heinrich's 1943 and 1944 management books which also speak of "low consequence events" as the "background" of serious injuries. McKinnon suggested many decades later: "Near-miss incidents are truly the foundation of major injuries, the building blocks of accidents, and warning signs that loss is imminent" (2012, p.xvii). I like the final part. The middle is nonsense; they are not building blocks (several near-misses do not combine to make an accident) but rather unfinished versions. The foundation is more in the metaphor of the triangle.

7 The table shows very clearly that the "no-injury" column is filled with numbers that are based on the 1:10 estimate.

8 Despite the fact that Heinrich presented them as "cause" in the table and text, these are actually causal categories ("labels") and not causes as such. Accidents for which the same cause codes are chosen (by abstract labels) may not necessarily share the same more specific causes (Note that I choose to say *more* specific, because no accident will likely ever share the *exact* specific causes with another accident).

It is a rather striking coincidence that the cause code "mechanical hazards" in the tables ends up with 10% – exactly the same as in the *Origin of Accidents* ratio of 88:10:2…

9 The ratios per causal code vary between 1:28.3:293 and 1:29.6:306.

10 The text changes from the early papers (HWH1929a, b) to the 1931 book. It was absent in the 1940s reprint.

11 In his 1927 article on accidental costs, Heinrich gave another argument to concentrate on minor events, in this case because of the money involved: "…accident costs in the aggregate, as measured by payments for compensation and medical aid, is made up chiefly of a great volume of minor injuries" (HWH1927a, p.19).

12 One exception: "The causes of fatalities are much the same as the causes that predominate in the minor injuries – from which we conclude that if we prevent minor accidents the fatalities will to a large extent take care of themselves" (HWH1927c, p.231).

13 In the fourth edition of *Industrial Accident Prevention* (HWH1959, p.31), Heinrich mentioned a study of over 5,000 cases to determine the no-injury frequency.

This number was only mentioned once (although repeated HWH1980, p.60), and one wonders whether this was a mistake and the 50,000 mentioned earlier (HWH1929a, b), and in the 1931 table, also repeated in the 1954 Travelers brochure) is meant here. This mistake proved to be a cause of confusion and critique (Manuele, 2002; Dekker, 2019) since it came out of the blue for people who are unfamiliar with the 1929 papers.

14 N.B. Heinrich speaks of "one-on-one" where "the first accident produces an injury" (HWH1929a, p.5). More correctly, this would be a "one-on-zero" ratio.

15 While one may critique the research, one must admit that it sounds more convincing than merely assuming at least one order of magnitude which would have resulted in a similar estimate.

16 We now know that learning from weak signals is not always as easy as it seems. See some reflections at the end of this chapter. Richardson seems to have realised this to a certain degree: "The general tendency is to pass over the occurrence as fortunately past, and to take no steps to prevent a similar recurrence."

17 Emphasis added here. It is interesting that Richardson would mention "bottom" here without a graphic representation, as the triangle, to refer to.

18 These omissions may be a main reason for some critique on his research as later authors only used the books and hardly looked at Heinrich's other work.

19 A critic might say that if the lines are meant to represent "unit groups," for a better representation each section should have had a different size, of course, but it is merely a model, not a correct representation of reality. Also, it probably looks better this way.

Others have suggested that the new representation may have given the impression that causal lines run from the bottom to the top, while the separated blocks from the 1929/1931 ziggurat do not give that impression. This suggestion should be rejected, however. The text accompanying the early graphic representation makes clear causal statements.

20 So far, I have been unable to be locate the original source. It is mentioned in various sources (Hubbard & Neil, 1986; HSE, 1997; McKinnon, 2012; Rausand, 2013).

21 So far, I have been unable to locate the original research. It is mentioned in among others (Masimore, 2007; Freibott, 2012).

22 Ironically, even the term "fatalities and serious injuries" was used by Heinrich on several occasions.

23 Oktem's "risk pyramid" was extended downwards, "Foreshadowing Events and Observations" adding a dimension of risk perception to it.

24 As an example, one might take the accident scenarios of falling from great height, or a track worker being hit by a train. In these cases, there may be many near-misses and some accidents with serious outcome, but it is unlikely that there will be many minor/first aid injuries. Instead of a triangle shape, one rather expects a deformed hourglass shape.

25 There are several possibilities to explain bottom-up and top-down views. For the distinction between them, the book follows the thought of Swuste et al. (2019).

26 The use in the *Message to the Foremen* papers is quite interesting:

> Keep this ratio of accidents to injuries in mind when next you hear the assertion 'This is the first time an accident of this kind ever happened.' It may be the first time a serious injury occurred, but hundreds of similar accidents undoubtedly occurred which, fortunately, produced no injury, and these may be observed and checked by the very man who should hold himself responsible for them – the foreman.
>
> (HWH1929e, p.51, 52)

27 This is based on personal experience. I cannot recall where, when, or by whom, but professional courtesy would prevent me from disclosing the name anyway.

28 This term should not be confused with the notion of common and special causes as defined by statisticians Shewhart and Deming. Common (chance) cause for them is variation that is "naturally" present in the system and can be predicted while special (assignable) cause comes as a surprise (Deming, 1986).

29 DeBlois wrote,

> While the detection of all the remediable causes of an accident is most important, it does not follow that either the proximate or the principal remediable cause is necessarily the logical point for applying prevention. Careful analysis of a group of accidents may disclose the constant appearance of the same contributory cause in all of them, in which case it may well be that the greater progress can be made through concentrating on its elimination than upon the prevention of separate and widely different proximate causes.
>
> (HWH1926a, p.52)

30 Exceptions include Hale (2002), Wright (2002), Ward (2012), Hollnagel (2014), and Rebbitt (2014).

31 The critical reader may note that this depiction is not a proportional reduction at all. Proportional reduction would mean that if the number of near-misses at the bottom is reduced by 25%, the top level would also be reduced by 25%. The line therefore should not be drawn parallel to the right side of the triangle, but rather come from the top (a bit as in Heinrich's 1941 triangle). Besides, with the iceberg metaphor in mind – even if you melt half of the iceberg, around 10% of it will be above the waterline!

32 Some clear misconception exists, notably taking a mechanical view at prediction that I will not discuss further. Examples are,

> The Safety Pyramid shows that a multitude of minor incidents are *required* for one major incident to occur and even more near-misses should occur for some minor incidents.
>
> (Awolusi & Marks, 2015, emphasis added)

> One of the most important principles of Heinrich's original pyramid is that fatalities cannot occur without a foundation of less severe incidents.
>
> (Seward & Kestle, 2014, p.363)

33 Start, for example, with https://www.merriam-webster.com/dictionary/predict, and follow up with some of the words that you find.

34 http://visionzero.global/

35 Heinrich does not explain where he takes the 50% and 48% from. Most likely, he just made an educated guess, or was influenced by Gilmour (1913). Interestingly, he lends a slight majority to the "preventable" side instead of simply splitting evenly.

36 Williams already suggested this effect: "...it may mean that educational activity has cut down the minor accidents while at the same time hazardous plant conditions or lack of careful analysis and supervision of hazardous occupations has permitted the serious and costly accidents to increase" (Williams, 1927, p.80).

37 And of course, you only know what you know – precursors to unknown and unimaginable events are obviously hard to detect. Often these may be a combination of normal situations that cumulate into something unwanted. So-called "Black Swans" are hard to forecast, until after the event.

Chapter 10

Other main themes

This chapter collects the remaining three main themes from Heinrich's work. The shortness of this chapter must not be misunderstood. One of these main themes, practical remedy, takes a substantial number of pages in Heinrich's work. All three must have been important to him, seeing as how they return throughout his work, and it is important for the reader to get a more complete picture of his work and thinking.

10.1 Professionalisation of safety

When Heinrich started working with safety, this was not recognised by the scientific and engineering societies as an engineering discipline. In the *Origin of Accidents* papers, he wrote, "The prevention of accidents is a science, but it is not so recognized, nor is it treated scientifically today" (HWH1928a, p.9). Heinrich contributed to changing this situation, but already during the early years of the safety movement, professionals started discussing qualifications, education, and role.

Chester Rausch (1914) did a lengthy paper discussing the use and misuse of the term "engineer," its meaning and scope, and the challenges of combining the term with safety because safety concerned itself not only with technical issues and laws of nature but also with uncertain issues as human behaviour. Also, the wide scope of safety issues and possible preventive measures conflicted with an engineer's tendency to specialisation. After reviewing Carnegie Institution's six principal qualifications for engineers (character, judgement, efficiency, understanding, knowledge, and technique), and acknowledging the practical as well as theoretical sides of engineering, Rausch agreed to broadly apply the title of safety engineer.

> Any person who, through adequate training and appropriate experience, understands and can interpret the laws of that control the materials and forces of nature and their effect upon the human body, and is capable of controlling or using these forces in a manner that prevents their harmful effect upon human life, may be said to be a safety engineer.
> (Rausch, 1914, p.302)

In an address before the Third NSC congress the same year, Connelly (1914) argued for the need that safety would enter training of all engineers, as well as the need for specialised safety engineers[1] of which he specified some "qualities" including varied technical, scientific, and economic knowledge, and analytical and social skills.

DeBlois (1926a) also had a chapter on *Safety Engineers* in his book, much about what kind of person they should be, the importance of their enthusiasm for safety and even attention for their salary and possibilities for promotion. Some years earlier, DeBlois delivered a speech on the subject before the ASME in which he described safety engineering as "true engineering" to achieve "a fundamental requisite in the design of structures and machines" (DeBlois, 1918, p.22). He described the safety engineer's many facets in design, analysis, organisation, supervision, inspection, engineering, promoting, and knowledge of basic psychology. However, few schools and universities taught safety, leaving it to graduates to find out for themselves.

Keefer (1926) confirmed this – despite various initiatives such as the *Safety Fundamentals* lectures (Safety Institute of America, 1920). A questionnaire among colleges showed that over half paid little or no attention to safety. Only 6 out of 72 colleges had safety principles and practices in selected courses. Those who should lead the elimination of accident hazards because of their theoretical and practical knowledge – engineers – hardly recognised this responsibility. According to Keefer, this could be improved by teaching engineers about the importance of safety during their education.

Just as Heinrich tried to further the professionalisation of management, he was very interested in the professionalisation of safety engineers and advancing safety as a recognised profession. While not dealing explicitly with the subject, much of his early work implicitly deals with a professionalisation through understanding of terms, better causal analysis, providing models and principles, suggesting a more proactive approach and improving statistical information. One example of the latter was the call for safety engineers and life insurance underwriters to work together and reap the mutual benefits (HWH1932c).

Heinrich's contribution during the 31st National Safety Congress in October 1942 was the first explicit speech and paper (HWH1943b) on professionalism in safety. Pieces of *What Makes a Safety Engineer?* even appeared in the financial column of the *Hartford Courant* the day after.[2]

Stressing the importance of accident prevention, safety engineers were portrayed as patriotic heroes, "He is the man of the hour in many ways." Because there was a shortage, "More qualified men are urgently needed and can be obtained only by the education and training of men who have basic qualifications" (HWH1943b, p.46). Heinrich listed seven characteristics:

1   The right spirit and attitude,
2   Being practical, logical, and have a sense of proportion with knowledge of general physics and mechanics,

3   Practical knowledge of human nature,
4   Able to express himself intelligently both orally and on paper,
5   Analytical ability and be able to draw conclusions and make decisions from observations, reports, and oral descriptions,
6   Familiar with the mechanical operations and processes of the plant,
7   Having confidence in his own ability, and strength of character, reasonable force, and considerable tact (HWH1943b, p.47,48).

However, in line with his emphasis on integrating safety in other business activities, Heinrich's closing remarks suggested "...safety engineering, will perhaps not be a separate profession but be so merged with all other activities as to be regarded as something inherent in every man's natural life and activities" (NSC, 1942b, p.123). Also, while Heinrich argued for educated safety engineers, he was no advocate of credentialism: "However, it is the ability to do the job that counts, no matter where or how this ability is acquired" (HWH1943b, p.48).

Heinrich's very last paper, written a few months after he went into retirement, also dealt with the subject that had been close to his heart for decades and can be interpreted as an appeal to continue his work. In *Recognition of Safety as a Profession: A Challenge* (HWH1956), he argued for the need for safety professionals to guide the work, based on common-sense reasoning that safety work benefited from being done by qualified persons.[3] He called for a special curriculum for the education of safety professionals. This must be varied and broad in its subjects and experience was necessary, suggesting that there was both "art and science of safety" (HWH1956, p.39).

10.2 Social engagement

Throughout his career, Heinrich showed social engagement. Chapter 5 mentioned how he proposed a work plan during the Depression and on many occasions, he combined the economic benefits of accident prevention with humanitarian motives or moral responsibility for employers to take care of workers.

There is also a humanitarian side to the question. Forgetting the dollars and cents, and keeping in mind the fact that the conservation of human life is at stake – that the happiness and well-being of your employees is involved – it would seem to be true that the time and effort spent in the prevention of accidents is a work well your while, for its own sake alone.
(HWH1925, p.260)

He spoke often of the economic benefits of accident prevention as a "further incentive" (HWH1931a, p.16) in addition to the humanitarian motives.

The basis of his safety management ladder (HWH1950a) spoke of a "Desire to serve humanity, industry and country."

> In safety, the greatest possible accident-preventing condition is that which is constituted by the regard of the individual (1) for his own well-being, (2) for the dependents who may suffer if he is incapacitated for work, and (3) for the losses suffered by his employer, by industry in general, and by the community and society, if through injury to himself he becomes incapable of accepting his share of work and responsibilities.
>
> (HWH1930b, p.149)

Heinrich was a patriot, which showed already before the War, "The country depends on you – the key man of industry, the foreman – to maintain that leadership among nations, of which we are so justly proud, in the safe, peace-time development of industry" (HWH1929e, p.52). The War, however, brought this even more to the front. It also brought out some of his most lyrical writing, as can be seen in the *Motion Study* papers,

> Marching men, rolling caissons, sailing and flying ships, plunging tanks, and reverberating riveters – motions and more motions – motions for fighting and motions for war production – all must be kept going – all must be directed usefully for victory.
>
> (HWH1942a, p.5, 1942b, p.175)

Among Heinrich's writings, *It's so Easy to be Patriotic* is one with the most interesting titles, expressing Heinrich's patriotism and a wish to "make secure the American way of life" (HWH1943a, p.12). The article dealt with how those "in the apparent safety of our homes" could contribute to the war efforts. "Each person must *take an action and helpful part* in it" (HWH1943a, p.12) he stressed individual responsibility, giving suggestions as not wasting resources, investing in war bonds, and by working safely and not wasting manpower and resources through accidents and their direct and hidden costs. He ended optimistic, "Safety no longer needs to be sold. Industry and government now demand it... It is universally accepted as a well worthwhile and an inherent part of good business" (HWH1943a, p.14). Then, however, Heinrich quoted the President saying, "It is not only our enemies who kill valuable Americans... ALL OF US CAN PARTICIPATE IN THE SAVING OF OUR MANPOWER" (HWH1943a, p.14). This might be constructed into the worst form of blame possible as is found in much war-time safety propaganda, framing accidents and lost time in terms of sabotage. Heinrich adopted this in the *Motion Study* papers, closing with a tremendous patriotic appeal: "Stop this senseless slaughter and maiming of workmen, stop this unintentional sabotage of your production effort, and thus help in a smashing big way to WIN THE WAR!" (HWH1942b, p.178).

Another interesting example of Heinrich's social engagement was found in an open letter, to the Hartford Courant during the 1948 Berlin Blockade.[4] This was the first major Cold War crisis. The Soviets had blocked all land and water access to Berlin. The Allies organised an airlift that lasted over a year bringing supplies into the isolated city. "Stalin has found the soft spot in the armor of Uncle Sam into Russia is viciously and with Satanic cleverness, boring in," he opened his letter and proposed "The method, simply stated, is nothing but kindly, serious and positive action." Then he detailed a seven-step plan that Heinrich thought would call Russia's bluff, but undoubtedly have escalated the situation.

Another form of social engagement in safety, was expanding from the workplace to other areas. Heinrich was chairman of the Traffic and Transportation Committee of the Hartford Chamber of Commerce after the war.[5] Many other authors from the safety movement spent efforts on home safety. Heinrich was rather late in his career to pick up on that subject with a speech at the 1951 National Safety Congress.

In his opinion, safety engineers had an opportunity and therefore also responsibility (a connection he would make regularly) to help improve home safety. To harness this, he suggested a "practical community program" where the knowledge and abilities of safety engineers were used for improvement. These abilities included dealing with obstacles and lending structure to the action programme. He also saw a community responsibility for industrial management and suggested safety education. He closed with a typical Heinrich appeal: "Being practical people they think the way to begin is to begin. There is no better time to begin that now, and it seems that we have already had enough talk. Action is the key now" (HWH1951, p.8).

10.3 Practical remedy

Heinrich was a practical man. On various occasions he wrote that knowledge or criticism in itself was not useful. These must come with practicable suggestions for improvement. Knowledge of causes was necessary. Not for the sake of knowledge, but for effective actions (e.g. 1931d, p.97). Therefore, it is not surprising that his work was littered with discussions of practical solutions and recognisable examples that often tie in with the other main themes.

There is relatively little attention for these practical solutions in this book, but they were a substantial part of Heinrich's work. In his papers, and especially his books, one finds major parts dedicated to the presentation of practical measures and technical solutions which are outdated, while others stand even today. These are as varied as ways to teach better ways of lifting weight (HWH1944), approaches for supervision (HWH1943d, 1944, 1949), and the guarding of machines.

For the latter, he proposed a prototypical hierarchy of control of elimina-
tions at source, guarding, revising, and relocating (HWH1944, p.57). Documentation of his teaching the principles of machine safeguarding to students
(HWH1952b) is left. Also, engineering revision, the adjusting and improving
of the workplace was to be the first consideration. In *Industrial Accident
Prevention* the chapter on the guarding of machines took up much space
even though this decreased in size from 101 (HWH1931a) to 65 (HWH1959).
The reduced number of pages was probably because of technical progress
where guarding had become more of an integrated part of design, and the
availability of other literature on the subject, such as standards. However,
later editions of *Industrial Accident Prevention* did not give less attention to
practical remedies. "Old" subjects as illumination and first aid, were supplemented by new subjects, including personal protection devices, motor fleet,
and radiation.

Not only needed solutions to be practical, Heinrich emphasised the need
for specific solutions. Comparing the 1937 approach of safety to that of
20 years earlier, he stated that one should concentrate on the things that
mattered instead of a general blanket-approach. Safety actions should be
specific and tailored to the company and its needs, "Having selected outstanding hazards of a specific nature, subsequent action is adapted to the
facts of the individual case" (HWH1937a, p.104). This was essentially what
one today calls a risk-based approach: "In making a selection, consideration
must be given to the frequency and severity of past losses, exposure, probability of future frequent or severe losses, ..." (HWH1937a, p.104).

Finally, Heinrich dismissed "one size fits all" approaches. He did several
sector specific papers, adjusting his general message to the specific challenges of sectors like construction (HWH1925, 1938b), gas manufacturing
(HWH1931d), and small plants (HWH1932a, 1959). This was also in line
with his emphasis of finding "causes" and "reasons" because these would
tell about the specific needs for the circumstances.

Notes

1 Interestingly, Connelly spoke of the "science of Safety First."
2 *Hartford Courant*, 28 October 1942.
3 Heinrich's thought is shared today, "...the professionalization of the safety role
   is widely considered necessary for advancing the quality of safety professional
   practice and improving the regard for safety professionals" (Provan et al., 2017,
   p.100).
4 *Hartford Courant*, 11 July 1948.
5 *Hartford Courant*, 9 August 1946.

Chapter 11

Heinrich in the 21st century

In the preface of his book, DeBlois wrote "Industrial safety is still in its adolescence – the late formative period – undergoing expansion, rationalization and the preliminary stages of standardization" (DeBlois, 1926a, p.vii). As we have seen, Heinrich had a huge impact on safety theory and practice. Did Heinrich, following a few years after and building on DeBlois's work, help bringing safety to maturity? Did Heinrich help to evolve from the "movement phase" and advance to a "permanent status" as DeBlois says?

If we would have asked Heinrich, he would surely reply positively, judging by the *Safety Wins a Place in the Sun* (HWH1934) and *Industrial-Accident Prevention in 1937* (HWH1937) papers that sing praises on the advances in safety with reference to his own contributions. Many of today's scholars will probably disagree. Seeing Heinrichian models and metaphors in use, they may acclaim that we are still stuck in safety's infancy. Partly, this may be attributed to how new generations perceive their predecessors. At the same time, reflection upon previous methods and teaching suggests some maturity, "Part of professional development consists of cultivating an appreciation of historical roots of current issues, questions, and concepts" (Weick, 1995, p.64).

So, let us reflect for a moment.[1] What does this mean for the future? Questions to consider, some of which discussed previously in this book, include: Heinrich's ideas were ground-breaking at the time, but what is Heinrich's relevance today? Should we still teach (about) Heinrich? Why is it important to learn from the past? Is there value to his metaphors, models, and methods today? Uses and misuses? Can we learn anything from Heinrich?

The book opened with the question whether one should bother about "ancient" safety, answering positively. It is important to study where we come from, where things went wrong, what went right. As the previous chapters showed, Heinrich's work continues to create discussions[2] and that is one reason to engage with his thoughts in a critical, open-minded way. For some time in the foreseeable future, his work will be used as a yardstick/showcase of traditional safety management, which new movements will compare themselves against. Then, it will be important to understand the origins and be able to nuance claims some of which may be cartoon-like. Some things we may need to improve, others discard, but in some cases, we may also need to restore some of the basics and explain them properly, or with their context and limitations instead of just simply repeating or dismissing them.

One of the simplest and most frequent forms of judgement and subtle discrediting of Heinrich's work is authors making a point out of Heinrich's work being "dated" or just being "old," playing on the instinctive human preference of "new" above "old":

> It is indicative of the development of safety thinking that a model developed almost a century ago remains an active, and indeed heavily relied upon, part of our current approach to safety management.
>
> (Marriott, 2018, p.23)

Similar arguments are found in Gesinger (2018), others speak of "tired and outdated" (Long, 2018c) and point out that "thinking on which it is based is getting on in age – soon a century" (Dekker, 2014, p.123). The point is not elaborated upon, however, just as if age were a self-explanatory disqualifier. Being "old" in itself should not be a relevant argument, however. Newton's laws and Pythagoras' theorem are even older. Is that a reason to question them?

Whether one likes it or not, Heinrich's work is here to stay – at least for a while longer. Taleb's "Lindy Effect" suggests time as the only effective judge of non-perishable things, "an idea will fail if it is not useful." Things that survive have shown to possess some kind of resilience. After all, they have been exposed to potentially harmful factors, but are around, in some form or another. Taleb even links to Popper's thoughts, "The longer an idea has been around without being falsified, the longer its future life expectancy" (Taleb, 2018, p.149).

Obviously, major differences exist between situations from the 1920s and 1930s and today. Complexity, scale, and interconnectedness have increased (Hollnagel, 2014a), and, with them, uncertainty, unpredictability, and possibly a need for approaches that better suit these circumstances than Heinrich's metaphors and approaches. Yet, while the times we are living in are quite different from Heinrich's, many of the problems he encountered or noticed are still around today. And have humans actually changed that much? (Hovden et al., 2010). James Nyce commented,

> We suffer often from a romantic notion that the past is somehow fundamentally different than the present. Perhaps nowhere is more evident than when we turn to technology and society. It is admittedly often difficult to get a handle on what the pace and effect of technology change was like in the past... but still do we have to fall into some kind of position where the present is always "ahead" of the past?[3]

With that in mind, it is interesting to look at some Heinrich quotes that would fit well in contemporary safety literature. For example, "No intelligent employee will deliberately expose himself unnecessarily to danger" (HWH1929b, p.11) might be regarded as a predecessor to the more positive sounding "People do not come to work to do a bad job" (Dekker, 2014, p.6). "Almost equally ineffective are many of the attempts today to check the rising frequency of industrial accidents with the aid of unsuitable tools and methods" (HWH1931b, p.5) would probably fit well into one of the many papers related to SIF-reduction. And, "...accident analysis in general today is improper and too limited in range" (HWH1928, p.9) sounds like contemporary critique of investigations often only starting after major outcomes, not investigating deeply enough or investigating only certain elements (e.g. Roed-Larsen & Stoop, 2012).

There are some clear parallels between Heinrich regularly describing how the introduction of new production processes introduced more and new risks, the very same argument that we also find with contemporary authors as Hollnagel and Dekker. Another insight, valid today, is how insincere, lip-service safety programmes create scepticism and cynicism among workers:

> ...many an employee listens to such an appeal with his 'tongue in cheek', that he questions to a certain extent the sincerity and the motives of his company executives who are apparently so solicitous of his safety welfare...
>
> (HWH1931c, p.50)

It is striking that Heinrich said things that are found with "new view" and contemporary safety thinkers many years later in almost exact same form. As the Bible taught over 2,500 years ago, there is "Nothing New Under the Sun."[4]

Throughout the book the relevance, advantages, and downsides of most of Heinrich's approaches have been discussed. One thread in this discussion is that, although the world has become increasingly complex and coupled, there still is a need for practical solutions. Another conclusion is that many of the subjects Heinrich spoke about still resonate with what safety practitioners experience on a day-to-day basis, e.g.:

- The tension between economical aspects and safety and health is at least as relevant today as it was in (and before) Heinrich's days. Safety measures cost money and resources, but also come with benefits.
- The role of management and responsibility.
- Dealing with the human element – both as an asset and as a problem.
- The integration of production and safety as the recommended way of managing safety (and leading in safety).[5]
- Learning from weak signals and being more proactive, instead of waiting for accidents with major outcomes to happen.

For problems in daily practice as these, he tried to provide practical solutions. One can wonder whether his ideas are not entirely obsolete. Some of his general principles might still be suitable today, providing a structured and practical approach to problems. His metaphors have proven their power and maybe they still can be used as heuristics and images to explain and remember them.

All models are wrong, Box (1976) taught us, but some are useful. Reviewing critique, one may wonder how much "wrongness" one can accept, and how to define "usefulness." Any model comes with limits of applicability. However, many people tend to ignore or forget these. Perhaps because this suggests uncertainty or makes things more complicated. It is important to be aware that application depends on situation. Tools need to be taught and considered with their limitations and drawbacks (Hollnagel et al., 2006, p.245). "The traditional approaches may be good enough; suited to some workplaces but not to others and suited to understanding some accidents but not others" (Hovden et al., 2010, p.954).

Having concluded that his work can be useful to discuss and that we need to regard context and limitations rather than reject it outright, what else can we learn from Heinrich? There are some things that characterised his approach. He was really good at selling his message. He knew how to communicate to various audiences, including professionals and managers, and seeing the lasting power of his thoughts, several generations. He also had a gift for turning things into practical approaches and easy to remember tools. He was good at illustrating his points with examples and stories. And, even though the scientific quality of his work may be questioned today, he was interested in research and facts. One could phrase this as in him having an inquisitive mind. Adopting these characteristics may aid safety work.

Besides his strong emphasis for practical solutions there is another element that characterises him and his work: his optimism. He saw many things as opportunities and talked about them this way. He would probably have applauded the new approaches in safety that look at what goes right instead of only concentrating on what goes wrong. Heinrich would probably have had no problem extending the triangle downwards, including normal work. In fact, phrasing the triangle differently, one can say that things go mostly right and that the 300 no-injury events are an indication of (mostly) failing in a safe way.

Opportunities speak of proactivity. "We are often misdirecting our efforts and ignoring valuable data" (HWH1929b, p.9) argued at the time for using accidents with minor or no outcomes as a basis for safety work. Hollnagel made the very similar argument many years later. Heinrich argued that one should not only look at accidents with major consequences, or lost-time, but also at minor accidents, or near-misses (i.e. other things that go wrong). Hollnagel (2014) argued that we should not only look at what goes wrong (1% or 0.1% of the actions) but also learn from the 99% or 99.9% of actions that go right. Ironically, most organisations still focus on lost-time injuries and other major accidents.

Hollnagel's recent work even presents conditions for learning that can be interpreted as the triangle in a new guise. Opportunities to learn must be frequent enough for learning to happen and they need to be similar so that lessons can be generalised. Hollnagel adds possibilities to verify as another condition which can be seen as outcome. Plotting accidents, incidents, and everyday work on this scale of frequency and similarity even creates a figure that looks like a tilted triangle (Hollnagel, 2018, p.40–42).

Heinrich most likely would have approved of the term "opportunities to learn." His work, with all its flaws and positive sides, gives us opportunities to learn, to reflect upon, and shows us some of the origins of safety theory and practice. As argued in this book, engaging with these sources with an open mind, putting them within their context, trying to understand, and reflecting critically upon them can be worthwhile, and most likely contributes to better understanding of contemporary approaches and assessing the value and claims these make. I hope this book has contributed something to this, and I would like to challenge the reader to continue on this path.

11.1 Famous last words

Closing this book with the words of my friend, Jos Villevoye:

> Heinrich made a significant and memorable contribution to the evolution of safety management. Wrong, right or partly agreeable. I'd like to use his legacy to create a flexible mindset and not to create a rigid divide.[6]

Amen to that.

Notes

1 Likewise, Heinrich reflected much on the safety work of his time.
2 Jesse Bird (1976) wrote in his biography that "In death as in life Heinrich stimulated discussion."
3 Personal mail, 21 February 2018.
4 Ecclesiastes 1:9.
5 Dan Petersen wrote,

> …this expressed some of Heinrich's best thinking, but for some reason, this is the one principle presented by Heinrich that safety people have not seemed to live by in the past.
>
> (Petersen, 1971, p.15)

6 https://www.linkedin.com/feed/update/urn:li:activity:6435573289950814208?commentUrn=urn%3Ali%3Acomment%3A%28activity%3A6435573289950814208%2C6435743770133831680%29&replyUrn=urn%3Ali%3Acomment%3A%28activity%3A6435573289950814208%2C6435748391736741888%29

Appendix I

The Heinrich bibliography

Heinrich, H.W. (1923a) Welded Joints in Unfired Pressure Vessels. *The Travelers Standard*, XI (7): 153–154.

Heinrich, H.W. (1923b) Bedford Stone Club Safety Talk, November 6, 1923. [speaker notes].

Heinrich, H.W. (1925) The Contractors' and Builders' Safety Problem. *The Travelers Standard*, XIII (12): 250–260.

Heinrich, H.W. (1926) The Incidental Cost of Accidents to the Employer. *The Travelers Standard*, XIV (12): 244–257.

Heinrich, H.W. (1927a) The "Incidental" Cost of Accidents. *National Safety News*, February 1927: 18–20.

Heinrich, H.W. (1927b) Incidental Cost of Accidents to the Employer. *Monthly Labor Review*, 25 (2): 46–50.

Heinrich, H.W. (1927c) The Incidental Cost of Accidents to the Employer. *The Travelers Standard*, XV (11): 221–231.

Heinrich, H.W. (1928a) The Origin of Accidents. *National Safety News*, 18 (1): 9–12.

Heinrich, H.W. (1928b) The Origin of Accidents. *The Travelers Standard*, XVI (6): 121–137.

Heinrich, H.W. (1928c) The Origin of Accidents. *Transactions of the National Safety Council, 17th Annual Safety Congress*, New York, October 1–5, 1929: 127–131.

Heinrich, H.W. (1929a) The Foundation of a Major Injury. *The Travelers Standard*, XVII (1): 1–10.

Heinrich, H.W. (1929b) The Foundation of a Major Injury. *National Safety News*, 19 (1): 9–11.

Heinrich, H.W. (1929c) Relation of Accident Statistics to Industrial Accident Prevention. *Proceedings of the Casualty Actuarial Society 1929–1930*: 170–174.

Heinrich, H.W. (1929d) A Message to Foremen. *The Travelers Standard*, XVII (12): 247–252.

Heinrich, H.W. (1929e) A Message to Foremen. *National Safety News*, 20 (6): 23–24, 51–52.

Heinrich, H.W. (1929f) Remarks at Manager's Conference, January 9, 1929. [speaker notes].

Heinrich, H.W. (1930a) The Foreman's Responsibility for Accidents. *Monthly Labor Review*, 30 (2): 83–88.

Heinrich, H.W. (1930b) The Reward of Merit. *The Travelers Standard*, XVIII (7): 141–152.

Heinrich, H.W. (1930c) Cost of Industrial Accidents to the State, the Employer, and the Man. *Monthly Labor Review*, 31 (5): 72–87.

Heinrich, H.W. (1931a) *Industrial Accident Prevention—A Scientific Approach*. New York: McGraw-Hill.

Heinrich, H.W. (1931b) Unsafe Practices and Conditions. *The Travelers Standard*, XIX (1): 5–12.

Heinrich, H.W. (1931c) Creating Interest in Safety. *The Travelers Standard*, XIX (3): 48–56.

Heinrich, H.W. (1931d) Safety in Gas Manufacturing. *The Travelers Standard*, XIX (5): 93–102.

Heinrich, H.W. (1932a) Safety in the Small Plant. *The Travelers Standard*, XX (2): 28–34.

Heinrich, H.W. (1932b) Practical Safety Methods Applied to the Sand and Gravel Industry. *The Travelers Standard*, XX (5): 106–102.

Heinrich, H.W. (1932c) The Safety Engineer Aids the Life Underwriter. *National Safety News*, August 1932: 21–22.

Heinrich, H.W. (1932d) Planning the Day's Work. *The Travelers Standard*, XX (4): 72–77.

Heinrich, H.W. (1933a) Mastery of Machine. *The Travelers Standard*, XXI (1): 9–18.

Heinrich, H.W. (1933b) Conservation and Essential in Industrial Recovery. *The Travelers Standard*, XXI (3): 52–59.

Heinrich, H.W. (1933c) The Cost of Industrial Accidents. *The Hercules Mixer*, January 1933: 12.

Heinrich, H.W. (1933d) Safety and Production. *The Hercules Mixer*, February 1933: 31.

Heinrich, H.W. (1933e) Accident Cause-Analysis. *The Hercules Mixer*, March 1933: 58.

Heinrich, H.W. (1933f) Accident Causes and Their Analyses: Classification of Causes. *The Hercules Mixer*, April 1933: 78.

Heinrich, H.W. (1933g) Physical and Supervisory Causes of Industrial Accident Prevention. *The Hercules Mixer*, May 1933: 98.

Heinrich, H.W. (1933h) Determining Supervisory Causes of Industrial Accidents. *The Hercules Mixer*, June 1933: 118.

Heinrich, H.W. (1933i) Application of Cause-Analysis to an Entire Industry. *The Hercules Mixer*, July/August 1933: 138.

Heinrich, H.W. (1934) Safety Wins a Place in the Sun. *The Travelers Standard*, XXII (5): 112–118.

Heinrich, H.W. (1935) Use of Accident Records in the Prevention of Machine Accident. *Transactions of the National Safety Council, 24th Annual Safety Congress*, Louisville, KY, October 14–18, 1935: 30–32.

Heinrich, H.W. (1937a) Industrial-Accident Prevention in 1937. *The Travelers Standard*, XXV (6): 101–106.

Heinrich, H.W. (1937b) A Break in The Vicious Cycle. *The Travelers Standard*, XXV (10): 185.

Heinrich, H.W. (1938a) It's Up to the Foreman! *The Industrial Supervisor*, 4 (7): 4–5, 14.

Heinrich, H.W. (1938b) Accident Cost in the Construction Industry. *Transactions of the National Safety Council, Silver Jubilee Safety Congress*, Chicago, IL, October 10–14, 1938: 374–377.

Heinrich, H.W. (1940) The Unsafe Habits of Men. *The Travelers Standard*, XXVIII (6): 112–118.

Heinrich, H.W. (1941) *Industrial Accident Prevention* (second edition). New York: McGraw-Hill.

Heinrich, H.W. (1942a) Men in Motion. *The Industrial Supervisor*, 10 (8): 4–5, 11.

Heinrich, H.W. (1942b) Keep 'em Moving (War-Time Motion Study and Foremanship). *The Travelers Standard*, XXX (9): 174–178.

Heinrich, H.W. (1942c) The Foreman's Place in the Safety Program. *Transactions of the 31st National Safety Congress*, Chicago, IL, October 27–28, 1942: 191–194.

Heinrich, H.W. (1943a) It's so Easy to be Patriotic. *The Travelers Standard*, XXXI (1): 12–14.

Heinrich, H.W. (1943b) What Makes a Safety Engineer? *The Travelers Standard*, XXXI (3): 46–48.

Heinrich, H.W. (1943c) The Warehouseman's Safety Problem. *The Travelers Standard*, XXXI (7): 135–140.

Heinrich, H.W. (1943d) *The Supervisor's Safety Manual.* New York: Safety Engineering Magazine and Alfred M. Best Company.

Heinrich, H.W. (1944) *Basics of Supervision.* New York: Alfred M. Best Company.

Heinrich, H.W. (1945a) Key Men of Industry (part I). *The Industrial Supervisor*, January 1945, 13 (1): 4–6.

Heinrich, H.W. (1945b) Key Men of Industry (part II). *The Industrial Supervisor*, February 1945, 13 (2): 12–14.

Heinrich, H.W. (1945c) Key Men of Industry (part III). *The Industrial Supervisor*, March 1945, 13 (3): 12–14.

Heinrich, H.W. (1945d) What Makes a Safety Engineer. *The Casualty and Surety Journal*, March 1945: 11–14.

Heinrich, H.W. (1945e) European Aftermath. *The Travelers Beacon*, July 1945: 5–7.

Heinrich, H.W. (194?) *The Foundation of a Major Injury* (reprint). Hartford, CT: Travelers Insurance Company.

Heinrich, H.W. (1948) Digest of Engineer's Meeting—March 10, 1948. [personal notes].

Heinrich, H.W. (1949) *Formula for Supervision: Outlining the Application of Supervisory Control to Secure, Safe, Efficient Work Performance.* New York: National Foremen's Institute, Inc.

Heinrich, H.W. (1950a) *Industrial Accident Prevention* (third edition). New York: McGraw-Hill.

Heinrich, H.W. (1950b) The Human Element in the Cause and Control of Accidents. *Transactions of the 1950 National Safety Congress.* Chicago, IL: National Safety Council: 7–10.

Heinrich, H.W. (1951) The Safety Engineer and Home Safety. *Transactions of the 39st National Safety Congress.* Chicago, IL: National Safety Council: 6–8.

Heinrich, H.W. (1952a) Safety Organization—A Function of Management. *15 May 1952, Eighteenth Annual Meeting, Virginia State-Wide Safety Conference*, Richmond, VA. [speaker notes].

Heinrich, H.W. (1952b) Principles of Machine Safeguarding. 27 October 1952, University of Toronto. [speaker notes].

Heinrich, H.W. (1954) Effective Work in Accident Prevention: Express Accident Cost in Plant Language (Engineering and Loss Control Division, Handbook Supplement H-15). November 1954. [draft notes].

Heinrich, H.W. (1956) Recognition of Safety as a Profession: A Challenge. *Transactions of the 44th National Safety Congress.* Chicago, IL: National Safety Council: 37–40.

Heinrich, H.W. & Blake, R.P. (1956) The Accident Cause Ratio 88:10:2. *National Safety News*, May 1956: 18–22.

Heinrich, H.W. & Granniss, E. (1959) *Industrial Accident Prevention* (fourth edition). New York: McGraw-Hill.

Lateiner, A.R. & Heinrich, H.W. (1969) *Management and Controlling Employee Performance*. New York: Lateiner Publishing.

Heinrich, H.W., Petersen, D. & Roos, N. (1980) *Industrial Accident Prevention—A Safety Management Approach*. New York: McGraw-Hill Book Company.

References

Adams, E. (1976) Accident Causation and the Management System. *Professional Safety*, October 1976.

Adams, J. (1995) *Risk*. Oxford: Routledge.

Ainsworth, C., et al. (1944) *Men, Minutes and Victory. Three Year Progress Report. National Committee for the Conversation of Manpower in War Industries*. Washington, DC: United States Department of Labor, Division of Labor Standards.

Aldrich, M. (1997) *Safety First: Technology, Labor, and Business in the Building of American Work Safety, 1870–1939*. Baltimore, MD: Johns Hopkins University Press.

Alexander, M.W. (1926) Need of Safety from the Employer's Point of View. *The Annals of the American Academy of Political and Social Science*, 123: 6–8.

Allison, R.W., Hon, C.K.H. & Xia, B. (2019) Construction Accidents in Australia: Evaluating the True Costs. *Safety Science*, 120: 886–896.

Amalberti, R. (2001) The Paradoxes of Almost Totally Safe Transportation Systems. *Safety Science*, 37 (2): 109–126.

American Engineering Council (1928) *Safety and Production: An Engineering and Statistical Study of the Relationship between Industrial Safety and Production: A Report*. New York: Harper.

American Standards Association (1941a) *American Recommended Practice for Compiling Industrial Accident Causes. Part I—Selection of Accident Factors*. New York: American Standards Association.

American Standards Association (1941b) *American Recommended Practice for Compiling Industrial Accident Causes. Part II—Detailed Classification of Accident Factors*. New York: American Standards Association.

American Statistical Association (1937) Accident Statistics. *Journal of the American Statistical Association*, 32 (200): 715–716.

Anand, N. (2015) Caught in Numbers, Lost in Focus. *Seaways*, August 2015: 5–8.

Anand, N. (2018) Near Miss Reporting: A (Mis)leading Indicator of Safety? *Standard Safety*, July 2018: 4–5.

Andersen, M.G. (2018) *A Field Study in Shipping: Near-Miss, a Mantra with Dubious Effect on Safety* (Master's Thesis, Lund University).

Anderson, P. (2003) *Managing Safety at Sea* (PhD Thesis, Middlesex University).

Anonymous (1914) Safety Axioms. *Safety Engineering*, 28 (3): 146.

Anonymous (May 1956) Bill Heinrich Retires. *Risk Control*: 7.

Ashe, S.W. (1917) *Organization in Accident Prevention*. New York: McGraw-Hill.

ASME (1947) *Form for Use in Self-Appraisal of Industrial Plants*. New York: The American Society of Mechanical Engineers.

Awolusi, I. & Marks, E. (2015) Near Miss Reporting to Enhance Safety in the Steel Industry. *AISTech 2015 Proceedings*.

Barnett, A. & Wang, A. (2000) Passenger Mortality Risk Estimates Provide Perspectives about Flight Safety. *Flight Safety Digest*, 19 (4): 1–12.

Baron, J. & Hersey, J.C. (1988) Outcome Bias in Decision Evaluation. *Journal of Personality and Social Psychology*, 54 (4): 569–579.

Beagle, L.A. (1917) Industrial Accident Prevention. *Safety Engineering*, 33 (3): 136–140.

Bellamy, L.J. (2015) Exploring the Relationship between Major Hazard, Fatal and Non-fatal Accidents Through Outcomes and Causes. *Safety Science*, 71 (Part B): 93–103.

Bellamy, L.J., Ale, B.J.M., Whiston, J.Y., Mud, M.L., Baksteen, H. Hale, A.R., Papazoglou, I.A., Bloemhoff, A., Damen, M. & Oh, J.I.H. (2008) The Software Tool Storybuilder and the Analysis of the Horrible Stories of Occupational Accidents. *Safety Science*, 46 (2): 186–197.

Besnard, D. & Hollnagel, E. (2012) I Want to Believe: Some Myths about the Management of Industrial Safety. *Cognition, Technology & Work*, 16 (1): 13–23.

Beyer, D.S. (1916) *Industrial Accident Prevention*. Boston, MA: Houghton Mifflin Company.

Beyer, D.S. (1917) Accident Prevention. *The Annals of the American Academy of Political and Social Science*, 70, Modern Insurance Problems: 238–243.

Bhattacharya, S. (2009) *The Impact of the ISM Code on the Management of Occupational Health and Safety in the Maritime Industry* (PhD Thesis, Cardiff University).

Bird Jr., F.E. & Davies, R.J. (1996) *Safety and the Bottom Line*. Loganville, GA: Institute Publishing Inc.

Bird Jr., F.E. & Germain, G.L. (1966) *Damage Control: A New Horizon in Accident Prevention and Cost Improvement*. New York: American Management Association.

Bird Jr., F.E. & Germain, G.L. (1992) *Practical Loss Control Leadership* (second revised edition). Loganville: DNV and International Loss Control Institute, Inc.

Bird Jr., F.E. & Loftus, R.G. (1976) *Loss Control Management*. Loganville: Institute Press, Inc.

Bird Jr., F.E. & Schlesinger, L.E. (1970) Safe-Behavior Reinforcement. *American Society of Safety Engineers Journal*, 15, June 1970: 16–24.

Bird, J.M. (1976) *History Heinrich: The Work Life and Education of the Travelers Best Known Safety Pioneer*. Hartford, CT: Manuscript.

Blake, R.P. (1945) *Industrial Safety*. New York: Prentice-Hall, Inc.

Blauw, S. (2018) *Het Bestverkochte Boek Ooit (met deze Titel): Hoe Cijfers ons Leiden, Verleiden en Misleiden*. Amsterdam: De Correspondent.

Blokland, P.J. & Reniers, G. (2019) Measuring (Un)safety. A Broad Understanding and Definition of Safety, Allowing for Instant Measuring of Unsafety. *Chemical Engineering Transactions*, 77: 253–258.

Boston Elevated Railway Company (1930) *Safety on the "El"*. Boston, MA: Boston Elevated Railway Company.

Bowie, G.W. (1916) Epigrams and Aphorisms, Old and New on 'Safety First'. *Safety Engineering*, 32 (4): 238–240.

Box, G.E.P. (1976) Science and Statistics. *Journal of the American Statistical Association*, 71 (356): 791–799.

Brown, R.L. (1978) Adapting Token Economy Systems in Occupational Safety. *Accident Analysis & Prevention*, 10 (1): 51–60.

Bureau of Labor Statistics (1920) *Standardization of Industrial Accident Statistics. Reports of the Committee on Statistics and Compensation Insurance Cost of the International Association of Industrial Accident Boards and Commission, 1915–1919: Bulletin of the United States Bureau of Labor Statistics, No. 276*. Washington, DC: United States Government Printing Office.

Bureau of Labor Statistics (1930) *Proceedings of the Sixteenth Annual Meeting of the International Association of Industrial Accident Boards and Commissions, Buffalo, NY, October 8–11, 1929. Bulletin of the United States Bureau of Labor Statistics, No. 511*. Washington, DC: United States Government Printing Office.

Bureau of Labor Statistics (1931) *Proceedings of the Seventeenth Annual Meeting of the International Association of Industrial Accident Boards and Commissions, Wilmington, Del, September 22–26, 1930. Bulletin of the United States Bureau of Labor Statistics, No. 536*. Washington, DC: United States Government Printing Office.

Bureau of Labor Statistics (1934) *Discussions of Industrial Accidents and Diseases: At the 1933 Meeting of the International Association of Industrial Accident Boards and Commissions, Chicago, Ill: Bulletin of the United States Bureau of Labor Statistics, No. 602*. Washington, DC: United States Government Printing Office.

Burin, J. (2013) Being Predictive in a Reactive World. *ISASI Forum*, January–March 2013: 23–24.

Busch, C. (2019a) Brave New World: Can Positive Developments in Safety Science and Practice also have Negative Sides? *MATEC Web Conference*, 273: 01003.

Busch, C. (2019b) *Heinrich's Local Rationality: Shouldn't 'New View' Thinkers Ask Why Things Made Sense to Him?* (Master's Thesis, Lund University).

Cameron, C. (1973) Accident Proneness. *Accident Analysis & Prevention*, 7: 49–53.

Carnegie, D. (1936) *How to Win Friends and Influence People*. New York: Simon & Schuster.

Carrillo, R.A. (2019) *The Relationship Factor in Safety Leadership. Achieving Success through Employee Engagement*. Milton Park: Routledge.

Casey, T.W. & Griffin, M. (2020) *LEAD Safety. A Practical Handbook for Frontline Supervisors and Safety Practitioners*. Boca Rotan, FL: CRC Press.

Centers for Disease Control and Prevention (2014) *World Polio Day Infographic*. Retrieved 25 December 2019 from https://www.cdc.gov/globalhealth/ immunization/infographic/world-polio-day-2014.htm

Chan, S. (1979) The Intelligence of Stupidity: Understanding Failures in Strategic Warning. *The American Political Science Review*, 73 (1): 171–180.

Chaney, L.W. (1926) The Need for More Definite Analysis of Accident Causes. *The Annals of the American Academy of Political and Social Science*, 123, Industrial Safety: 41–45.

Chase, S. (1929) *Men and Machines*. New York: The Macmillan Company.

Chase, S. (1931) *Waste and the Machine Age* (revised edition). New York: League for Industrial Democracy.

Chhokar, J.S. & Wallin, J.A. (1984) Improving Safety Through Applied Behavior Analysis. *Journal of Safety Research*, 15 (4): 141–151.

Cialdini, R.B. (1984) *Influence. The Psychology of Persuasion*. New York: Harper Collins.

Collins, R. (2010) Heinrich and Beyond. *Process Safety Progress*, 30 (1): 2–5.

Collins, R. (2011) Heinrich's Fourth Dimension. *Open Journal of Safety Science and Technology*, 1 (1): 19–29.

Conklin, T. (2007) Preventing Serious Accidents with the Human Performance Philosophy. *Nuclear Weapons Journal*, 2007 (1): 17–18.

Conklin, T. (2012) *Pre-accident Investigations: An Introduction to Organizational Safety.* Boca Raton, FL: CRC Press.

Conklin, T. (2016) *Pre-Accident Investigations: Better Questions.* Boca Raton, FL: CRC Press.

Conklin, T. (2017a) *Workplace Fatalities: Failure to Predict.* Santa Fe, NM: PreAccident Media.

Conklin, T. (2017b) *PAPod 143- Heinrich Was Wrong—Admit It—and Move On...* Retrieved 16 October 2017 from http://preaccidentpodcast.podbean.com/e/papod-143-heinrich-was-wrong-admit-it-and-move-on/

Connelly, C.B. (1914) Safety as a Part of the Education of an Engineer. *Safety Engineering*, 28 (6): 471–476.

Cooper, M.D. (2019) The Efficacy of Industrial Safety Science Constructs for Addressing Serious Injuries & Fatalities (SIFs). *Safety Science*, 120: 164–178.

Cousins, J.A. (1918) Health Plus Safety Equals Efficiency. *Safety Engineering*, 35 (4): 215–216.

Cowee, G.A. (1916) *Practical Safety Methods and Devices. Manufacturing and Engineering.* New York: D. Van Nostrand Company.

Davies, J., Ross, A., Wallace, B. & Wright, L. (2003) *Safety Management: A Qualitative Systems Approach.* London: Taylor & Francis.

Dean, C.C. (1997) Primer of Scientific Management by Frank B. Gilbreth: A Response to Publication of Taylor's Principles in The American Magazine. *Journal of Management History*, 3 (1): 31–41.

DeBlois, L.A. (1918) The Safety Engineer. In: National Safety Council (Ed.) *The Teaching of Safety in Technical Schools and Universities.* Chicago, IL: National Safety Council: 22–26.

DeBlois, L.A. (1919) Supervision as a Factor in Accident Prevention. *Safety Engineering*, 38 (4): 199–202.

DeBlois, L.A. (1926a) *Industrial Safety Organization for Executive and Engineer.* New York: McGraw-Hill.

DeBlois, L.A. (1926b) Progress in Accident Prevention. *Monthly Labor Review*, 22 (3): 1–3.

Dekker, S.W.A. (2002) Reconstructing Human Contributions to Accidents: The New View on Error and Performance. *Journal of Safety Research*, 33 (3): 371–385.

Dekker, S.W.A. (2011) *Drift into Failure: From Hunting Broken Components to Understanding Complex Systems.* Farnham: Ashgate.

Dekker, S.W.A. (2014) *The Field Guide to Understanding 'Human Error'.* Farnham: Ashgate.

Dekker, S.W.A. (2017) Rasmussen's Legacy and the Long Arm of Rational Choice. *Applied Ergonomics*, 59: 554–557.

Dekker, S.W.A. (2018) *The Safety Anarchist. Relying on Human Expertise and Innovation, Reducing Bureaucracy and Compliance.* Milton Park: Routledge.

Dekker, S.W.A. (2019) *Foundations of Safety Science: A Century of Understanding Accidents and Disasters.* Boca Raton, FL: CRC Press.

Dekker, S.W.A. & Hollnagel, E. (2004) Human Factors and Folk Models. *Cognition, Technology & Work*, 6: 79–86.

Deming, W.E. (1986) *Out of the Crisis.* Cambridge, MA: MIT Press.

Difford, P.A. (2011) *Redressing the Balance: A Commonsense Approach to Causation.* Bridgwater: Accidental Books.

Dillon, R.L. & Tinsley, C.H. (2008) How Near-Misses Influence Decision Making under Risk: A Missed Opportunity for Learning. *Management Science*, 54(8): 1425–1440.

Dillon, R.L. & Tinsley, C.H. (2015) Near-Miss Events, Risk Messages, and Decision Making. *Environment Systems and Decisions*, 36 (1): 34–44.

DNV (1992) *Safety and Quality Management Guidelines.* Høvik: Det Norske Veritas Classification AS.

Donkin, R. (2001) *Blood Sweat & Tears. The Evolution of Work.* New York: Texere.

Dow, M.A. (1928) *Stay Alive! In Which Jim the Truckman Gently Kicks the Drivin' Fools and Walkin' Yaps.* New York: Marcus Dow Publishers.

Downey, E.H. (1916) Relative Importance of Accident Causes in Industries. *Safety Engineering*, 32 (5): 266–268.

Dula, C.S. & Geller, E.S. (2007) *Creating a Total Safety Traffic Culture.* AAA Foundation for Traffic Safety. Retrieved 19 July 2017 from https://www.aaafoundation.org/sites/default/files/DulaGeller.pdf

Dunlap, E.S., Basford, B. & Smith, M. (2019) Remodelling Heinrich: An Application for Modern Management. *Professional Safety*, May 2019: 44–52.

Dwyer, T. (1992) Industrial Safety Engineering—Challenges of the Future. *Accident Analysis & Prevention*, 24 (3): 265–273.

Eastman, C. (1910) *Work-Accidents and the Law. The Pittsburgh Survey.* New York: Russell Sage Foundation Publications.

Environmental Protection Agency (n.d.) *How BenMAP-CE Estimates the Health and Economic Effects of Air Pollution.* Retrieved 16 December 2019 from https://www.epa.gov/benmap/how-benmap-ce-estimates-health-and-economic-effects-air-pollution

Federal Aviation Administration (2012) *Avoid the Dirty Dozen: 12 Common Causes of Human Factors Errors.* Washington, DC: Federal Aviation Administration.

Fisher, E.B. (1922) *Mental Causes of Accidents.* Boston, MA: Houghton Mifflin Company.

Fletcher, J.A. (1972) *The Industrial Environment. Total Loss Control. A Guide for Managers and Supervisors.* Willowdale, ON: National Profile Ltd.

Folkehelseinstituttet (2020) *COVID-19-Epidemien: Risikovurdering og Respons i Norge, Versjon 3.* Oslo: Folkehelseinstituttet.

Freibott, B. (2012) Sustainable Safety Management: Incident Management as a Cornerstone for a Successful Safety Culture. *WIT Transactions on the Built Environment*, 134: 257–270.

Fuerth, L. (2009) *What's the Difference between Prediction and Foresight?* Retrieved 3 September 2019 from https://www.youtube.com/watch?v=Ejh6wKmG1QU

Fulwiler, R.D. (2002) The New Safety Pyramid. *Occupational Hazards*, 64(1): 50–54.

Gallivan, S., Taxis, K., Franklin, B.D. & Barber, N. (2008) Is the Principle of a Stable Heinrich Ratio a Myth? A Multimethod Analysis. *Drug Safety*, 31 (8): 637–642.

Geller, E.S. (1996) *The Psychology of Safety: How to Improve Behaviors And Attitudes.* Boca Raton, FL: CRC Press.

Geller, E.S., Berry, T.D., Ludwig, T.D, Evans, R.E., Gilmore, M.R. & Clarke, S.W. (1990) A Conceptual Framework for Developing and Evaluating Behavior Change Interventions for Injury Control. *Health Education Research: Theory and Practice*, 5 (2): 125–137.

Germain, G.L. & Clark, M.D. (2007) *A Tribute to Frank E. Bird Jr. (1921–2007).* Retrieved 28 January 2020 from https://www.dnvgl.com/oilgas/international-sustainability-rating-system-isrs/tribute-to-frank-bird.html

Gesinger, S. (2018) *The Fearless World of Professional Safety in the 21st Century.* Milton Park: Routledge.

Gilbreth, F.B. (1911) *Motion Study: A Method for Increasing the Efficiency of the Workman.* New York: D. Van Nostrand Company.

Gilbreth, F.B. (1912) *Primer of Scientific Management.* New York: D. Van Nostrand Company.

Gilmour, G. (1913) Safety Engineering. *The Travelers Standard*, I (8): 141–160.

Glenn, D.D. (2011) Job Safety Analysis. Its Role Today. *Professional Safety*, March 2011: 48–57.

Groeneweg, J. (1992) *Controlling the Controllable: The Management of Safety.* Leiden: DSWO Press.

Groot, G. de (2005) *Basisboek Veiligheid.* Zeist: Kerkebosch.

Guarnieri, M. (1992) Landmarks in the History of Safety. *Journal of Safety Research*, 23 (3): 151–158.

Guldenmund, F.W. (2010) *Understanding and Exploring Safety Culture.* Oisterwijk: BOXPress.

Guzzetti, J. (2013) The Agony and the Ecstasy of Utilizing Safety Data for Modern Accident Prevention and Investigation. *International Society of Air Safety Investigators. Submission for the August 2013 ISASI Seminar*, Vancouver, Canada.

Hale, A.R. (2002) Conditions of Occurrence of Major and Minor Accidents: Urban Myths, Deviations and Accident Scenarios. *Tijdschrift voor Toegepaste Arbowetenschappen*, 15 (3): 34–41.

Hamann, C.J. & Peek-Asa, C. (2017) Examination of Adult and Child Bicyclist Safety-Relevant Events Using Naturalistic Bicycling Methodology. *Accident Analysis & Prevention*, 102: 1–11.

Harari, Y.N. (2011) *Sapiens: A Brief History of Humankind.* London: Vintage and Penguin Random House.

Hawkins, R.D. & Fuller, C.W. (1996) Risk Assessment in Professional Football: An Examination of Accidents and Incidents in the 1994 World Cup Finals. *British Journal of Sports Medicine*, 30: 165–170.

Hawkins, R.D. & Fuller, C.W. (1998) An Examination of the Frequency and Severity of Injuries and Incidents at Three Levels of Professional Football. *British Journal of Sports Medicine*, 32: 326–332.

Hayhurst, E.R. (1932) Industrial Accident Prevention: A Scientific Approach. *American Journal of Public Health and the Nation's Health*, 22(1): 119–120.

Heath, C. & Heath, D. (2007) *Made to Stick: Why Some Ideas Survive and Others Die.* New York: Random House.

Heraghty, D., Dekker, S.W.A. & Rae, A. (2018) Accident Report Interpretation. *Safety*, 46 (4): 1–25.

Heyne, H.P. (1918) Reduction in Accidents Means Increased Efficiency. *Safety Engineering*, 35 (4): 213–214.

Hill, N. (1937) *Think and Grow Rich*. Meriden, CT: The Ralston Society.

Hollnagel, E. (2000) Analysis and Prediction of Failures in Complex Systems: Models and Methods. In: P. Elzer (Ed.) *Lecture Notes in Control and Information Sciences*. London: Springer: 39–41.

Hollnagel, E. (2004) *Barriers and Accident Prevention*. Aldershot: Ashgate.

Hollnagel, E. (2009) *The ETTO Principle: Efficiency-Thoroughness Trade-Off— Why Things That Go Right Sometimes Go Wrong*. Farnham: Ashgate.

Hollnagel, E. (2014a) *Safety I and Safety II: The Past and Future of Safety Management*. Farnham: Ashgate.

Hollnagel, E. (2014b) Human Factors/Ergonomics as a Systems Discipline? "The Human Use of Human Beings" Revisited. *Applied Ergonomics*, 45: 40–44.

Hollnagel, E. (2018) *Safety-II in Practice. Developing the Resilience Potentials*. Milton Park: Routledge.

Hollnagel, E., Pariès, J., Woods, D.D. & Wreathall, J. (2011) *Resilience Engineering in Practice*. Aldershot: Ashgate Publishing, Ltd.

Hollnagel, E., Woods, D.D. & Leveson, N. (2006) *Resilience Engineering: Concepts and Precepts*. Aldershot: Ashgate Publishing, Ltd.

Hopkins, A. (1994) The Limits of Lost Time Injury Frequency Rates. In: *Positive Performance Indicators for OHS: Beyond Lost Time Injuries*. Canberra: Australian Government Publishing Service, Commonwealth of Australia. Retrieved 25 January 2002 from http://www.nohsc.gov.au/OHSInformation/NOHSCPublications/fulltext/docs/h2/ppio/PPIO1-06.htm

Hopkins, A. (2001) Was Three Miles Island a 'Normal Accident'? *Journal of Contingencies and Crisis Management*, 9(2): 65–72.

Hopkins, A. (2006) What Are We to Make of Safe Behaviour Programs? *Safety Science*, 44: 583–597.

Hopkins, A. (2008) *Failure to Learn. The BP Texas City Refinery Disaster*. Sydney: CCH.

Hopkins, A. (2014) Issues in Safety Science. *Safety Science*, 67: 6–14.

Hopkins, A. & Maslen, S. (2015) *Risky Rewards: How Company Bonuses Affect Safety*. Farnham: Ashgate.

Hovden, J., Albrechtsen, E. & Herrera, I.A. (2010) Is There a Need for New Theories, Models and Approaches to Occupational Accident Prevention? *Safety Science*, 48 (8): 950–956.

Howard, C.H. (1917) An Accident Is an Inefficiency. *Safety Engineering*, 33 (2): 65.

Howarth, D. (2000) *Discourse*. Buckingham: Open University Press.

HSE (1997) *Successful Health and Safety Management. HSG65* (2nd edition). Norwich: HSE Books.

HSE (1999) *The Cost to Britain of Workplace Accidents and Work-related Ill Health in 1995/96 (HSG101)*. Sudbury: HSE Books.

Hubbard, R.K.B. & Neil, J.T. (1986) Major-Minor Accident Ratios in the Construction Industry. *Journal of Occupational Accidents*, 7 (4): 225–237.

Hubbard, S.D. (1921) Why Should We Not Prevent Accidents? *Safety Engineering*, 42 (1): 24–26.

Hudson, P. (2007) Implementing a Safety Culture in a Major Multi-National. *Safety Science*, 45 (6): 697–722.

Hummerdal, D. (2014) *What's in a Name?* Retrieved 4 June 2018 from http://www.safetydifferently.com/whats-in-a-name/

Hutchins, B.L. & Harrison, A. (1911) *A History of Factory Legislation* (2nd edition). London: P.S. King & Son.

Hydén, C. (1987) *The Development of a Method for Traffic Safety Evaluation: The Swedish Traffic Conflicts Technique.* Lund: Lunds Universitet, Institutionen For Trafikteknik.

ICS (2013) Implementing an Effective Safety Culture. Basic Advice for Shipping Companies and Seafarers. *IMO Symposium on the Future of Ship Safety*, 2013.

Immel, H.D. (1942) Factory Inspection in the War Effort. *Transactions of the 31st National Safety Congress*, Chicago, IL. October 27–28, 1942: 127–130.

Inouye, J. (2018) *Serious Injury and Fatality Prevention: Perspectives and Practices.* Itasca, IL: Campbell Institute and National Safety Council.

Inspectie SZW (2017) *Staat van Ernstige Ongevallen. Weer Veilig Thuis uit je Werk.* Den Haag: Inspectie SZW.

Jacinto, C. & Soares, G. (2008) The Added Value of the New ESAW/Eurostat Variables in Accident Analysis in the Mining and Quarrying Industry. *Journal of Safety Research*, 39 (6): 631–644.

Jacobs, A.M. & Grainger, J. (1994) Models of Visual Word Recognition—Sampling the State of the Art. *Journal of Experimental Psychology: Human Perception and Performance*, 29 (6): 1311–1334.

Jacobs, H.H. (1961) Research Problems in Accident Prevention. *Social Problems*, 8 (4): 329–341.

Johnson, A. (2011) *Examining the Foundation: Were Heinrich's Theories Valid, and Do They Still Matter?* Retrieved 28 April 2019 from http://www.safetyandhealth-magazine.com/articles/print/6368-examining-the-foundation

Johnson, C.W. (2003) *Failure in Safety-Critical Systems: A Handbook of Incident and Accident Reporting.* Glasgow: Glasgow University Press.

Johnson, L. (2011) Is Safety Pyramid a Myth? Study Suggests New Approach to Injury Prevention. *Canadian Occupational Safety.* Retrieved 28 April 2019 from http://www.cos-mag.com/occupational-hygiene/30549-is-safetypyramid-a-myth-study-suggests-new-approach-to-injury-prevention/

Jones, S., Kirchsteiger, C. & Bjerke, W. (1999) The Importance of Near Miss Reporting to Further Improve Safety Performance. *Journal of Loss Prevention in the Process Industries*, 12: 59–67.

Kanigel, R. (1997) *The One Best Way: Frederick Winslow Taylor and the Enigma of Efficiency.* London: Little, Brown & Company.

Karol, P.J. (2019) *Selling Safety: Lessons from a Former Front-Line Supervisor.* Boca Rotan, FL: CRC Press.

Keefer, W.D. (1926) Training Engineers in Safety. In: G.M. Whipple (Ed.) *The Twenty-Fifth Yearbook of the National Society for the Study of Education. Part I: The Present Status of Safety Education.* Bloomington, IN: Public School Publishing Company: 319–325.

Kennedy, J.J. (1920) Safety and Efficiency. *Safety Engineering*, 40 (5): 216.

Kennedy, R.D. (1914) Efficiency in Safety Work. *Safety Engineering*, 28 (6): 477–480.

Khanzode, V., Maiti, J. & Ray, P.K. (2012) Occupational Injury and Accident Research: A Comprehensive Review. *Safety Science*, 50 (5): 1355–1367.

Kjellén, U. (2000) *Prevention of Accidents Through Experience Feedback.* London: Taylor & Francis.

Koenig, F. (1916) The ABC of "Safety First": Always be Careful. *Safety Engineering*, 32 (1): 20–23.

Komaki, J., Barwick, K.D. & Scott, L.R. (1978) A Behavioral Approach to Occupational Safety: Pinpointing and Reinforcing Safe Performance in a Food Manufacturing Plant. *Journal of Applied Psychology*, 63 (4): 434–445.

Komaki, J., Collins, R.L. & Penn, P. (1982) The Role of Performance Antecedents and Consequences in Work Motivation. *Journal of Applied Psychology*, 67 (3): 334–340.

Kossoris, M.D. (1939) A Statistical Approach to Accident Prevention. *Journal of the American Statistical Association*, 34 (207): 524–532.

Kossoris, M.D. (1944) *Accident-Record Manual for Industrial Plants: Bulletin of the United States Bureau of Labor Statistics, No. 772*. Washington, DC: United States Government Printing Office.

Krause, T.R., Hidley, J.H. & Hodson, S.J. (1990) *The Behavior-based Process. Managing Involvement for an Injury-Free Culture*. New York: Van Nostrand-Reinhold.

Krause, T.R. & Murray, G. (2012) On the Prevention of Serious Injuries and Fatalities. *Proceedings of the ASSE Professional Development Conference and Exposition*, Denver, CO. June 3–6, 2012.

Kuhn, T. (1962) *The Structure of Scientific Revolutions*. Chicago, IL: University of Chicago Press.

La Duke, P. (2014) *The Power of Pyramids: How Using Outmoded Thinking about Hazards can be Deadly*. Retrieved 23 April 2018 from https://philladuke.wordpress.com/2014/08/30/the-power-of-pyramids-how-using-outmoded-thinking-about-hazards-can-be-deadly/

La Duke, P. (2018a) *Never Let Facts Screw up a Good Opinion*. Retrieved 15 April 2020 from https://philladuke.wordpress.com/2018/03/04/never-let-facts-screw-up-a-good-opinion/

La Duke, P. (2018b) *80% of Safety Practitioners Are Idiots*. Retrieved 15 April 2020 from https://philladuke.wordpress.com/2018/12/02/80-of-safety-practitioners-are-idiots/

La Duke, P. (2019) Safety can Never be a Science. Retrieved 25 June 2019 from https://philladuke.wordpress.com/2019/06/25/safety-can-never-be-a-science/

Lange, F.G. (1926) *Handbook of Safety and Accident Prevention*. New York: The Engineering Magazine Company.

Lansburgh, R.H. (1928) *Industrial Management* (2nd edition). New York: John Wiley & Sons Inc.

Lansburgh, R.H. & Spriegel, W.R. (1940) *Industrial Management* (3rd edition). New York: John Wiley & Sons Inc.

Larouzee, J. & Le Coze, J.-C. (2020) Good and Bad Reasons: The Swiss Cheese Model and its Critics. *Safety Science*, 126: 104660.

Lateiner, A.R. (1954) *The Techniques of Supervision*. New York: National Foreman's Institute.

Lateiner, A.R. (1958) If We're to Stop Accidents Preventing Injuries Is Not Enough. *Industrial Supervisor*, 26 (11): 3–5, 14.

Lateiner, A.R. (1960) What's Wrong with the Safety Engineer? *Transactions of the 1960 National Safety Council*, 19: 6–10.

Lateiner, A.R. (1961) Revitalize Your Safety Program. *Petroleum Refiner*, May 1961: 259–262.

Lateiner, A.R. (1969) *The Lateiner Method of Accident Control.* Palm Beach, FL: Lateiner.

Lateiner, A.R. (1974) *Modern Techniques of Supervision* (12th revised printing). Santa Barbara, CA: Lateiner Publishing.

Lateiner, A.R. (1988) *Modern Techniques of Supervision* (16th revised printing). Santa Barbara, CA: Lateiner Publishing.

Le Coze, J.-C. (2012) Towards a Constructivist Program in Safety. *Safety Science,* 50: 1873–1887.

Le Coze, J.-C. (2013) New Models for New Times. An Anti-Dualist Move. *Safety Science,* 59: 200–218.

Le Coze, J.-C. (2018) Safety, Model, Culture. In: C. Gilbert, B. Journé, H. Laroche & C. Bieder (Eds.) *Safety Cultures, Safety Models: Taking Stock and Moving Forward.* Cham: Springer: 81–92.

Lessin, N. & McQuiston, T.H. (2013) An Inverse Relationship between Injuries and Fatalities: What Is Surprising—And What Is Not. *American Journal of Industrial Medicine,* 56(5): 505–508.

Leveson, N. (2011) Applying Systems Thinking to Analyze and Learn from Events. *Safety Science,* 49 (1): 55–64.

Li, J. & Hale, A. (2014) Identification of, and Knowledge Communication among Core Safety Science Journals. *Safety Science,* 74: 70–78.

Lischeid, W.E. & Bird, J.M. (2008) *H.W. Heinrich—Up Close and Personal.* Hartford, CT: Manuscript.

Long, I. (2018) *Simplicity in Safety Investigations: A Practitioner's Guide to Applying Safety Science.* Milton Park: Routledge.

Long, R. (2012) *For the Love of Zero: Human Fallibility and Risk.* Kambah: Scotoma Press.

Long, R. (2013) *A Comparison of Safety Paradigms.* Retrieved 3 June 2018 from https://safetyrisk.net/a-comparison-of-safety-paradigms/

Long, R. (2014a) *The Seduction of Measurement in Risk and Safety.* Retrieved 3 June 2018 from https://safetyrisk.net/the-seduction-of-measurement-in-risk-and-safety/

Long, R. (2014b) *Nonsense Curves and Pyramids.* Retrieved 3 June 2018 from https://safetyrisk.net/nonsense-curves-and-pyramids/

Long, R. (2015) *Do You Believe in Good and Bad Luck?* Retrieved 3 June 2018 from https://safetyrisk.net/do-you-believe-in-good-and-bad-luck/

Long, R. (2017a) *Safety Curves and Pyramids.* Retrieved 3 June 2018 from https://safetyrisk.net/safety-curves-and-pyramids/

Long, R. (2017b) *The Real Story of Zero.* Retrieved 3 June 2018 from https://safetyrisk.net/the-real-story-of-zero/

Long, R. (2017c) *Curves and Pyramids.* Retrieved 3 June 2018 from https://vimeo.com/124273239

Long, R. (2017d) *Zero Accident Vision Non-Sense.* Retrieved 3 June 2018 from https://safetyrisk.net/zero-accident-vision-non-sense/

Long, R. (2018a) *The Fear of Freedom in Safety.* Retrieved 13 December 2018 from https://safetyrisk.net/the-fear-of-freedom-in-safety

Long, R. (2018b) *I am a Spreadsheet King.* Retrieved 3 June 2018 from https://safetyrisk.net/i-am-a-spreadsheet-king/

Long, R. (2018c) *Fallibility and Risk. Living with Uncertainty.* Kambah: Scotoma Press.

Lundberg, J., Rollenhagen, C. & Hollnagel, E. (2010) What You Find Is Not Always What You Fix—How Other Aspects than Causes of Accidents Decide Recommendations for Remedial Actions. *Accident Analysis & Prevention*, 42 (6): 2132–2139.

Macrae, C. (2015) The Problem with Incident Reporting. *BMJ Quality and Safety*, 25: 71–75.

Maine Department of Labor and Industry (1932) *Industrial Safety Bulletin, Nov. 1932*. Augusta, GA: Department of Labor and Industry.

Malcolm-Smith, G. (1964) *The Travelers. 100 Years*. Hartford, CT: Travelers Insurance Company.

Manuele, F.A. (2002) *Heinrich Revisited: Truisms or Myths*. Itasca, IL: National Safety Council.

Manuele, F.A. (2008) Serious Injuries & Fatalities: A Call for a New Focus on their Prevention. *Professional Safety*, December 2008: 32–39.

Manuele, F.A. (2011a) Accident Costs: Rethinking Ratios of Indirect to Direct Costs. *Professional Safety*, January 2011: 39–47.

Manuele, F.A. (2011b) Reviewing Heinrich: Dislodging Two Myths from the Practice of Safety. *Professional Safety*, October 2011: 52–61.

Manuele, F.A. (2013) *On the Practice of Safety* (fourth edition). Hoboken, NJ: Wiley.

Manuele, F.A. (2014a) *Heinrich Revisited: Truisms or Myths* (second edition). Itasca, IL: National Safety Council.

Manuele, F.A. (2014b) Incident Investigation. Our Methods Are Flawed. *Professional Safety*, October 2014: 34–43.

Marriott, C. (2018) *Challenging the Safety Quo*. Milton Park: Routledge.

Marsden, E. (2017) *Heinrich's Domino Model of Accident Causation*. Retrieved 15 April 2020 from https://risk-engineering.org/concept/Heinrich-dominos

Marsden, E. (n.d.) *The Heinrich/Bird Pyramid: Pioneering Research Has Become a Safety Myth*. Retrieved 25 December 2019 from https://risk-engineering.org/concept/Heinrich-Bird-accident-pyramid

Marsh, T. (2017) *A Definitive Guide to Behavioural Safety*. Milton Park: Routledge.

Marshall, P., Hirmas, A. & Singer, M. (2018) Heinrich's Pyramid and Occupational Safety: A Statistical Validation Methodology. *Safety Science*, 101: 180–189.

Martin, D.K. & Black, A.A. (2015) Preventing Serious Injuries & Fatalities. *Professional Safety*, September 2015: 35–43.

Masimore, L. (2007) *Proving the Value of Safety*. Milwaukee, WI: Rockwell Automation.

Mathews, R. & Wacker, W. (2007) *What's Your Story? Storytelling to Move Markets, Audiences, People, and Brands*. Upper Saddle River, NJ: FT Press.

Mauro, J.C., Diehl, B., Marcellin, R.F. & Vaughn, D.J. (2018) Workplace Accidents and Self-organized Criticality. *Physica A: Statistical Mechanics and its Applications*, 506: 284–289.

McElroy, F.S. & McCormack, G.R. (1945) Causes of Accidents in Slaughtering and Meat Packing, 1943. *Monthly Labor Review*, 61 (6): 1153–1168.

McFarland, R.A. (1963) A Critique of Accident Research. *Research Methodology and Potential in Community Health and Preventive Medicine*, 107 (2): 686–695.

McKinnon, R.C. (2012) *Safety Management. Near Miss Identification, Recognition, Investigation*. Boca Raton, FL: CRC Press.

McSween, T. & Moran, D.J. (2017) Assessing and Preventing Serious Incidents with Behavioral Science: Enhancing Heinrich's Triangle for the 21st Century. *Journal of Organizational Behavior Management*, 37 (3–4): 283–300.

Merriam Webster: Scientific (n.d.) Retrieved 26 November 2018 from https://www. merriam-webster.com/dictionary/scientific

Metzgar, C. (2002) Heinrich Revisited: Truisms or Myths. *Professional Safety*, 47 (6): 27–28.

Mix, M.W. (1914) Safety as an Asset to Manufacturers. *Safety Engineering*, 28 (4): 295–299.

Mowery, H.W. (1915) "Carelessness", or Incorrect Accident Classification? *Safety Engineering*, 30(5): 365–367.

Muller, J.Z. (2018) *The Tyranny of Metrics*. Princeton, NJ: Princeton University Press.

Münsterberg, H. (1918) *Business Psychology*. Chicago, IL: La Salle Extension University.

Muntz, E.E. (1932) Industrial Accidents and Safety Work. *The Journal of Educational Sociology*, 5 (7): 397–412.

Nascimento, F.A.C., Majumdar, A. & Ochieng, W.Y. (2013) Investigating the Truth of Heinrich's Pyramid in Offshore Helicopter Transportation. *Transportation Research Record: Journal of the Transportation Research Board*, 2336: 105–116.

National Conservation Bureau (1942) *Handbook of Industrial Safety Standards* (Revised 1942 Edition, Including Supplement on Wartime Protection). New York: National Conservation Bureau.

National Safety Council (1942a) Fundamental Causes of Accidents. *Transactions of the 31st National Safety Congress*, Chicago, IL. October 27–28, 1942: 79–94.

National Safety Council (1942b) Government Safety Service in Industry. *Transactions of the 31st National Safety Congress*, Chicago, IL. October 27–28, 1942: 103–123.

National Safety Council (1961) *Supervisors Safety Manual. Better Production Through Accident Prevention* (2nd edition). Chicago, IL: National Safety Council.

Nazeri, N. & Donohue, G. (2008) Analyzing Relationships between Aircraft Accidents and Incidents: A Data Mining Approach. *Third International Conference on Research in Air Transportation*, Fairfax, VA. June 2008.

Niskanen, T. & Saarsalmi, O. (1983) Accident Analysis in the Construction of Buildings. *Journal of Occupational Accidents*, 5: 89–98.

Oktem, U.G. (2002) Near-Miss: A Tool for Integrated Safety, Health, Environmental and Security Management. Prepared for Presentation at the 37th Annual AIChE Loss Prevention Symposium.

Oktem, U.G., Wong, R. & Oktem, C. (2010) Near-Miss Management: Managing the Bottom of the Risk Pyramid. *Risk & Regulation*, July 2010: 12–13.

Orth, G.A. (1926) Does Accident Prevention Pay? *The Annals of the American Academy of Political and Social Science*, 123, Industrial Safety: 20–26.

Perrow, C. (1984) *Normal Accidents—Living with High-Risk Technologies*. New York: Basic Books.

Peters, J. & Pauw, J. (2004) *Intensieve Menshouderij—Hoe Kwaliteit Oplost In Rationaliteit*. Schiedam: Scriptum.

Petersen, D. (1971) *Techniques of Safety Management*. New York: McGraw Hill.

Petersen, D. (1996) *Human Error Reduction and Safety Management* (3rd edition). New York: Wiley and Van Nostrand Reinhold.

Petersen, D. (2000) The Behavioral Approach to Safety Management. *Professional Safety*, March 2000: 37–39.

Petersen, D. (2001a) *Safety Management: A Human Approach* (3rd edition). Des Plaines, IL: American Society of Safety Engineers.

Petersen, D. (2001b) *Authentic Involvement*. Chicago, IL: National Safety Council and NSC Press.

Petersen, P.B. (1990) Early Beginnings: Occupational Safety Management 1925–1935. *Journal of Managerial Issues*, 2 (4): 382–405.

Phimister, J., Oktem, U.G., Kleindorfer, P. & Kunreuther, H. (2003) Near-Miss Incident Management in the Chemical Process Industry. *Risk Analysis*, 23: 445–459.

Pink, D.H. (2009) *Drive: The Surprising Truth about What Motivates Us*. New York: Riverhead Books.

Pink, D.H. (2012) *To Sell Is Human: The Surprising Truth about Persuading, Convincing and Influencing Others*. New York: Riverhead Books.

Pollina, V. (1962) Safety Sampling. *Journal of the ASSE*, 7, August 1962: 19–22.

Porter, N. (2016) *Forgetting the Heinrich Pyramid with Incident Management Software*. Retrieved 18 April 2019 from https://www.gensuite.com/forgetting-the-heinrich-pyramid-with-incident-management-software/

Powell, N.B., Schechtman, K.B., Riley, R.W., Guilleminault, C., Chiang, R.P. & Weaver, E.M. (2007) Sleepy Driver Near Misses May Predict Accident Risks. *Sleep*, 30 (3): 331–342.

Powell, P.I., Hale, M., Martin, J. & Simon, S. (1971) *2.000 Accidents. A Shop Floor Study of their Causes based on 42 Months' Continuous Observation*. London: National Institute of Industrial Psychology.

Prem, K.P., Ng, D. & Mannan, S.S. (2010) Harnessing Database Resources for Understanding the Profile of Chemical Process Industry Incidents. *Journal of Loss Prevention in the Process Industries*, 23: 549–560.

Provan, D.J., Dekker, S.W.A. & Rae, A.J. (2017) Bureaucracy, Influence and Beliefs: A Literature Review of the Factors Shaping the Role of a Safety Professional. *Safety Science*, 98: 98–112.

Pryor, P. (2019) Developing the Core Body of Knowledge for the Generalist OHS Professional. *Safety Science*, 115: 19–27.

Pupulidy, I. (2015) *The Transformation of Accident Investigation: From Finding Cause to Sensemaking* (PhD Thesis, Tilburg University).

Qureshi, Z.H. (2008) *A Review of Accident Modelling Approaches for Complex Critical Sociotechnical Systems*. Edinburgh South Australia: Command, Control, Communication and Intelligence Division.

Ranter, H. (2002) Access to Air Safety Information. *2nd Annual CIS & Eastern Europe Airline Engineering & Maintenance Conference*, Budapest. October 9–10, 2002.

Rasmussen, J. (1990) Human Error and the Problem of Causality in Analysis of Accidents. *Philosophical Transactions of the Royal Society of London. Series B, Biological Sciences*, 327: 449–462.

Rasmussen, J. (1997) Risk Management in a Dynamic Society: A Modelling Problem. *Safety Science*, 27 (2/3): 183–213.

Rasmussen, J. & Lind, M. (1981) *Coping with Complexity*. Risø-M-2293, Electronics Department, Risø National Laboratory, Roskilde, Denmark.

Rausand, M. (2013) *Risk Assessment: Theory, Methods, and Applications*. Hoboken, NJ: Wiley.

Rausch, C.C. (1914) The Safety Engineer—His Qualifications and Duties. *Safety Engineering*, 38 (5): 261–262; 297–302.

Reason, J. (1990) *Human Error*. Cambridge: Cambridge University Press.

Reason, J. (1991) Too Little and Too Late: A Commentary on Accident and Incident Reporting Systems. In: T.W Van der Schaaf, D. Lucas & A.R. Hale (Eds.) *Near Miss Reporting as a Safety Tool*. Oxford: Butterworth-Heinemann Ltd: 9–26.

Reason, J. (1997) *Managing the Risks of Organisational Accidents*. Farnham: Ashgate.

Reason, J. (2008) *The Human Contribution*. Farnham: Ashgate.

Reason, J., Hollnagel, E. & Paries, J. (2006) *Revisiting the "Swiss Cheese" Model of Accidents*. EEC Note No. 13/06, Eurocontrol.

Rebbitt, D. (2014) Pyramid Power. A New View of the Great Safety Pyramid. *Professional Safety*, September 2014: 30–34.

Richardson, A.S. (1916) Minor and "Near" Accidents. *Safety Engineering*, 32 (4): 189.

Rieder, R. & Bepperling, S. (2011) Heinrich Triangle for Ground Operation. *Journal of System Safety*, September–October, 2011: 23–28.

Rivers, P.E. (2006) Loss Management Information System. Teaching Students How to Make Better Decisions. *Professional Safety*, May 2006: 42–50.

Roed-Larsen, S. & Stoop, J. (2012) Modern Accident Investigation—Four Major Challenges. *Safety Science*, 50 (6): 1392–1397.

Roels, J. (1992) *Ongevalsonderzoek en -Rapportage. (S137)*. Den Haag: Arbeidsinspectie.

Sabet, P.G.P., Adal, H., Jamshidi, M.H.M. & Rad, K.G. (2013) Application of Domino Theory to Justify and Prevent Accident Occurance in Construction Sites. *IOSR Journal of Mechanical and Civil Engineering*, 6 (2): 72–76.

Safety Differently (n.d.) Retrieved 9 April 2017 from http://www.safetydifferently.com/about/

Safety Institute of America (1920) *Safety Fundamentals*. New York: Safety Institute of America.

Salminen, S., Saari, J., Saarela, K.L. & Räsänen, T. (1992) Fatal and Non-fatal Occupational Accidents: Identical Versus Differential Causation. *Safety Science*, 15 (2): 109–118.

Saloniemi, A.E. & Oksanen, H. (1998) Accidents and Fatal Accidents: Some Paradoxes. *Safety Science*, 29 (1): 59–66.

Schreiber, H.V. (1916) Studying the Causes of Accidents. *Safety Engineering*, 32 (5): 230–231.

Schwartz, B. (2015) *Why We Work*. London: TED Books and Simon & Schuster.

Seward, M. & Kestle, L. (2014) Health and Safety Practices on Christchurch's Post-Earthquake Rebuild Projects: How Relevant is Heinrich's Safety Pyramid? In: A.B. Raiden & E. Aboagye-Nimo (Eds.) *Proceedings of the 30th Annual AR-COM Conference*, September 1–3, 2014, Portsmouth: Association of Researchers in Construction Management: 361–370.

Silbey, S. (2009) Taming Prometheus: Talk about Safety and Culture. *Annual Review of Sociology*, 35: 341–369.

Simon, H.A. (1969) *The Sciences of the Artificial*. Cambridge, MA: MIT Press.

Simonds, R.H. & Grimaldi, J.V. (1956) *Safety Management: Accident Cost and Control*. Homewood, IL: Richard D. Irwin, Inc.

Skinner, B.F. (1953) *Science and Behaviour*. New York: The Macmillan Company.

Sklet, S. (2002) *Methods for Accident Investigation*. Trondheim: ROSS and NTNU.

Slocombe, C.S. & Bingham, W.V. (1927) Men Who Have Accidents: Individual Differences among Motormen and Bus Operators. *Personnel Journal*, 6: 251–257.

Slovic, P. (1987) Perception of Risk. *Science*, 236: 280–285.

Smith, T.A. (2011) Safety Management: A Personal Development Strategy. *Professional Safety*, March 2011: 58–68.

Stabile, D.R. (1987) The DuPont Experiments in Scientific Management: Efficiency and Safety, 1911–1919. *The Business History Review*, 61 (3): 365–386.

Stack, H.J. (1953) *Safety for Greater Adventures: The Contributions of Albert Wurts Whitney*. New York: Center for Safety Education.

Stack, H.J., Siebrecht, E.B. & Elkow, J.D. (1949) *Education for Safe Living*. Englewood Cliffs, NJ: Prentice-Hall.

Starbuck, W.H. & Milliken, F.J. (1988) Challenger: Fine-Tuning the Odds Until Something Breaks. *Journal of Management Studies*, 25 (4): 319–340.

State of Queensland: Department of Natural Resources, Mines and Energy (2018) *Resources Safety and Health: Compliance Policy*. MIN/2018/4325. Version 1.02.

Stewart, M. (2009) *The Management Myth: Debunking Modern Business Philosophy*. New York: Norton.

Stone, R.W. (1931) Reviewed Work: Industrial Accident Prevention by H.W. Heinrich. *Social Service Review*, 5 (2): 323–324.

Svensson, Å. & Hydén, C. (2006) Estimating the Severity of Safety Related Behaviour. *Accident Analysis & Prevention*, 38 (2): 379–385.

Swuste, P.H.J.J. (2016) Is Big Data Risk Assessment a Novelty? *Safety and Reliability*, 36 (3): 134–152.

Swuste, P.H.J.J., Gulijk, C. van & Zwaard, W. (2010) Safety Metaphors and Theories: A Review of the Occupational Safety Literature of the US, UK and The Netherlands, Till the First Part of the 20th Century. *Safety Science,* 48: 1000–1018.

Swuste, P.H.J.J., Gulijk, C. van & Zwaard, W. (2013) Safety Management According to Heinrich. *ORP 2013: XI Congreso Internacional de Prevención de Riesgos Laborales*, Santiago de Chile, April 3–5, 2013.

Swuste, P.H.J.J., Gulijk, C. van, Zwaard, W., Lemkowitz, S., Oostendorp, Y. & Groeneweg, J. (2016) Developments in the Safety Science Domain, in the Fields of General and Safety Management Between 1970 and 1979, the Year of the Near Disaster on Three Mile Island: A Literature Review. *Safety Science*, 86: 10–26.

Swuste, P.H.J.J., Gulijk, C. van, Zwaard, W., Lemkowitz, S., Oostendorp, Y. & Groeneweg, J. (2019) *Van Veiligheid Naar Veiligheidskunde*. Alphen a/d Rijn: Vakmedianet.

Swuste, P.H.J.J., Gulijk, C. van, Zwaard, W. & Oostendorp, Y. (2014) Occupational Safety Theories, Models and Metaphors in the Three Decades since World War II, in the United States, Britain and the Netherlands: A Literature Review. *Safety Science*, 62: 16–27.

Swuste, P.H.J.J. & Sillem, S. (2018) The Quality of the Post Academic Course 'Management of Safety, Health and Environment (MoSHE) of Delft University of Technology. *Safety Science*, 102: 26–37.

Taleb, N.N. (2012) *Antifragile: Things that Gain from Disorder*. New York: Random House.

Taleb, N.N. (2018) *Skin in the Game: Hidden Asymmetries in Daily Life*. New York: Random House.

Taxis, K., Gallivan, S., Barber, N. & Dean Franklin, B. (2005) *Can the Heinrich Ratio be Used to Predict Harm from Medication Errors?* Report to the Patient

Safety Research Programme (Policy Research Programme of the Department of Health).

Taylor, F.W. (1911) *The Principles of Scientific Management*. New York: Harper.

Tetlock, P.E. & Gardner, D. (2015) *Superforecasting: The Art & Science of Prediction*. New York: Crown Publishers.

Toft, Y., Dell, G., Klockner, K.K. & Hutton, A. (2012) Models of Causation: Safety. In: HaSPA (Health and Safety Professionals Alliance) (Ed.) *Core Body of Knowledge for the Generalist OHS Professional*. Tulamarine: Safety Institute of Australia: 1–35.

Tolman, W.H. & Kendall, L.B. (1913) *Safety: Methods for Preventing Occupational and Other Accidents and Disease*. New York: Harper & Brothers Publishers.

Townsend, A.S. (2013) *History of Health and Safety*. Retrieved 22 May 2018 from http://www.safetydifferently.com/history-of-health-and-safety/

Townsend, A.S. (2014) *Safety Can't be Measured: An Evidence-based Approach to Improving Risk Reduction*. Farnham: Gower Publishing.

Travelers Insurance Company (1913a) The Engineering and Inspection Service of The Travelers. *The Travelers Standard*, I (8): 161–175.

Travelers Insurance Company (1913b) Industrial Accidents and the Travelers Inspection Service. *The Travelers Standard*, I (15): 297–314.

Travelers Insurance Company (1914) *The Employee and Accident Prevention*. Hartford, CT: Travelers Insurance Company.

Travelers Insurance Company (1915) Accident Prevention Service in 1941. *Safety Engineering*, 29 (6): 533–536.

Travelers Insurance Company (1918) *Foremen and Accident Prevention* (third, revised edition). Hartford, CT: Travelers Insurance Company.

Travelers Insurance Company (1920) *Safety in the Machine Shop*. Hartford, CT: Travelers Insurance Company.

Travelers Insurance Company (1925) Three Others Who Were Promoted. *Protection*, March 1925.

Travelers Insurance Company (1927) Supervising and Senior Engineers Plan an Intensive 1927 Campaign. *Protection*, 2 March 1927: 8–9.

Travelers Insurance Company (1928) *The Travelers Yearbook for 1928*. Hartford, CT: Travelers Insurance Company.

Travelers Insurance Company (1929) *The Travelers Yearbook for 1929*. Hartford, CT: Travelers Insurance Company.

Travelers Insurance Company (1934) *The Accident Sequence*. Hartford, CT: Travelers Insurance Company.

Travelers Insurance Company (1938) *Chart 4235: Relation of Engineering Service to Underwriting Profit Sequence of Engineering Functions*. Hartford, CT: Travelers Insurance Company.

Travelers Insurance Company (1945) *The Hazard Thru Track*. Hartford, CT: Travelers Insurance Company.

Travelers Insurance Company (1949) *The Five Factors in the Accident Sequence*. Hartford, CT: Travelers Insurance Company.

Travelers Insurance Company (1954) *The Origin of Accidents*. Hartford, CT: Travelers Insurance Company.

Turner, B.A. (1978) *Man Made Disaster*. London: Wykeham Publications.

Turner, B.A. (1994) Causes of Disaster: Sloppy Management. *British Journal of Management*, 5: 215–219.

Van Alphen, W.J.T., Gort, J., Stavast, K.I.J. & Zwaard, A.W. (2008) *Leren van Ongevallen: Een Overzicht van Analysemethodieken*. Den Haag: SDU Uitgevers.

Van der Schaaf, T.W., Lucas, D. & Hale, A.R. (Eds.) (1991) *Near Miss Reporting as a Safety Tool*. Oxford: Butterworth-Heinemann Ltd.

Van der Westhuizen, J. & Stanz, K. (2017) Critical Incident Reporting Systems: A Necessary Multilevel Understanding. *Safety Science*, 96: 198–208.

Van Gulijk, C., Swuste, P.H.J.J., Ale, B. & Zwaard, W. (2009) Ontwikkeling van Veiligheidskunde in het Interbellum en de Bijdrage van Heinrich. *Tijdschrift voor Toegepaste Arbowetenschap*, 20 (3): 80–95.

Van Gulijk, C. & Swuste, P.H.J.J. (2011) Lessen uit het Verleden, de 88:10:2 Regel voor Ongevalsoorzaken: (N)iemand Verantwoordelijk voor Veiligheid!? Retrieved 22 July 2018 from https://www.veiligheidskunde.nl/xu/document/cms/streambin.asp?requestid=9F84CDE9-0517-4688-9568-CBFCD78E6764

Van Schaack, D. (1910) *Safeguards for the Prevention of Industrial Accidents*. Hartford, CT: Ætna Life Insurance Co.

Vaughan, D. (1996) *The Challenger Launch Decision*. Chicago, IL: The University of Chicago Press.

Vaughan, D. (2002) Signals and Interpretive Work: The Role of Culture in a Theory of Practical Action. In: K.A. Cerulo (Ed.) *Culture in Mind: Toward a Sociology of Culture and Cognition*. New York: Routledge: 28–54.

Vernon, H.M. (1935) *Accidents and their Prevention*. London: Cambridge University Press.

Vincoli, J.W. (1994) *Basic Guide to Accident Investigation and Loss Control*. New York: Van Nostrand Reinhold.

Visser, E., Pijl, Y.J., Stolk, R.P., Neeleman, J. & Rosmalen, J.G.M. (2007) Accident Proneness, Does it Exist? A Review and Meta-Analysis. *Accident Analysis and Prevention*, 39, 556–564.

Viteles, M.S. (1932) *Industrial Psychology*. New York: W.W. Norton and Company, Inc.

Walker, G. (2017) Redefining the Incidents to Learn From: Safety Science Insights Acquired on the Journey from Black Boxes to Flight Data Monitoring. *Safety Science*, 99: 14–22.

Ward, R.B. (2012) Revisiting Heinrich's Law. *Chemeca 2012: Quality of Life Through Chemical Engineering*, Wellington, New Zealand. September 23–26, 2012. Barton: Engineers Australia: 1179–1187.

Watson, J.B. (1913) Psychology as the Behaviorist Views It. *Psychological Review*, 20: 158–177.

Weaver, D. (1971) Symptoms of Operational Error. *Professional Safety*, October 1971.

Webster: Science (1930a) Retrieved 27 April 2019 from https://archive.org/details/webstersnewinter00webs/page/1894

Webster: Scientific (1930b) Retrieved 27 April 2019 from https://archive.org/details/webstersnewinter00webs/page/1895

Weick, K.E. (1995) *Sensemaking in Organizations*. London: Sage.

Weick, K.E. & Sutcliffe, K.M. (2001) *Managing the Unexpected: Assuring High Performance in an Age of Complexity*. San Francisco, CA: Jossey-Bass.

Wilde, G.J.S. (2014) *Target Risk 3: Risk Homeostasis in Everyday Life*. Retrieved 28 August 2017 from http://riskhomeostasis.org/

Will, M.L. (1979) *The Loss Control Approach to Industrial Safety* (MSc Thesis, University of Cape Town). Retrieved 31 December 2019 from https://open.uct.ac.za/bitstream/item/12921/thesis_ebe_1979_will_ml.pdf

Williams, J. & Roberts, S. (2018) Integrating the Best of HOP and BBS: A Holistic Approach to Improving Safety Performance. *Professional Safety*, October 2018: 40–48.

Williams, S.J. (1922) *How Much Do Industrial Accidents Cost: Bulletin of the United States Bureau of Labor Statistics, No. 304: 58–62*. Washington, DC: United States Government Printing Office.

Williams, S.J. (1926) Development of the Safety Movement. In: G.M. Whipple (Ed.) *The Twenty-Fifth Yearbook of the National Society for the Study of Education. Part I: The Present Status of Safety Education*. Bloomington, IN: Public School Publishing Company: 5–14.

Williams, S.J. (1927) *The Manual of Industrial Safety*. Chicago, IL and New York: A.W. Shaw Company.

Williams, S.J. & Hillegas, M.B. (1926) Realization of the Educational Aspect of the Problem. In: G.M. Whipple (Ed.) *The Twenty-Fifth Yearbook of the National Society for the Study of Education. Part I: The Present Status of Safety Education*. Bloomington, IN: Public School Publishing Company: 15–23.

Wilson, R. (2017) The Museum of Safety: Responsibility, Awareness and Modernity in New York, 1908–1923. *Journal of American Studies*, 51(3): 915–938.

Winter, J.C.F. de & Dodou, D. (2011) Why the Fitts List has Persisted Throughout the History of Function Allocation. *Cognition, Technology & Work*, 16: 1–11.

Wohlstetter, R. (1965) Cuba and Pearl Harbour: Hindsight and Foresight. *Foreign Affairs*, 43: 691–707.

Woods, D.D., Dekker, S.W.A., Cook, R., Johannesen, L. & Sarter, N. (2010) *Behind Human Error* (second edition). Farnham: Ashgate.

Woods, D.D., Johannesen, L.J., Cook, R.I. & Sarter, N.B. (1994) *Behind Human Error: Cognitive Systems, Computers and Hindsight*. Wright-Patterson AFB, OH: CSERIAC.

World Health Organisation (2010) *Injuries and Violence: The Facts*. Geneva: World Health Organisation.

Wright, L. (2002) *The Analysis of UK Railway Accidents and Incidents: A Comparison of their Causal Patterns* (PhD Thesis, University of Strathclyde).

Wright, L. & Van der Schaaf, T.W. (2004) Accident Versus Near Miss Causation: A Critical Review of the Literature: An Empirical Test in the UK Railway Domain, and Their Implications for Other Sectors. *Journal of Hazardous Materials*, 111(1–3): 105–110.

Yates, F. & Mather, K. (1963) Ronald Aylmer Fisher, 1890–1962. *Biographical Memoirs of Fellows of the Royal Society*, 9: 91–129.

Yorio, P.L. & Moore, S.M. (2018) Examining Factors that Influence the Existence of Heinrich's Safety Triangle Using Site-Specific H&S Data from More than 25.000 Establishments. *Risk Analysis*, 38 (4): 839–852.

Index

Note: Page numbers followed by "n" denote endnotes.

ABC model 63
accident 109, 128, 133, 205–206, 209,
  215, 224–225; facts 37, 40, 77–80, 110,
  128, 130, 137, 141, 146, 184, 188, 194,
  199; model 35, 61, 111, 129, 132–133,
  138, 140–145; outcome 102, 128, 139,
  201, 206–207, 218, 220–229, 234–244;
  no-injury 19, 44, 68, 89, 145, 206–218,
  223–226, 231, 234, 236–238, 242,
  257; practical remedies (*see* action);
  preventable 18, 49, 110, 116–120,
  129, 159, 195–196, 239; proneness 35,
  40, 50, 53, 143–144, 159–163, 166,
  174n17; repeater 162–163; sequence
  (*see* dominos); statistics 22, 38, 40,
  80, 110, 113–114, 120, 201–202, 236;
  type 22, 114, 128, 207, 218, 224,
  226–229, 233
Ackoff, R. 106
action 79, 143, 252–253; based on
  analysis 84, 109, 156–158, 189, 194,
  200; based on weak signals 240–243;
  behavioural 123–124; ineffective 105,
  138, 230; preventive and corrective 36,
  40–41, 101, 114–115, 182–183, 214; *see
  also* hierarchy
Acts of God 115, 139, 239
Adams, J. 168
American Engineering Council 96–98
American Standards Association 27,
  120, 122
ancestry 66, 129–130, 144, 161, 163–165,
  175n22
ANSI Z16 *see* cause code
Ariely, D. 1

Ashe, S.W. 48, 49, 78, 153
ASME Standard for Self-evaluation 27,
  34, 42, 198
ASSE 4, 30, 65, 198
authority *see* management
Axioms 25, 35, 38, 42, 66, 78, 176, 189,
  192, 194–197, 201

balance 106
Basics of Supervision 26, 34, 130, 142,
  180–181, 197, 213
Bedford Stone Club talk 16, 42, 78, 96,
  164, 176, 179, 187, 189, 192, 194
behaviourism 41, 62–64, 155–156
Behaviour Based Safety (BBS) 62–65, 66,
  72n26, 122, 219, 231; critique 64, 231
best way 72n28, 84–87, 97, 153
Beyer, D.S. 20, 46, 48, 55, 78, 83, 99,
  111, 167
bias 5, 7, 8, 142, 148n15, 173n6, 179
biased data 74, 93n16, 125–127, 138
Bingham, W.V. 52–53, 162
Bird jr., F.E. 4, 6, 59–60, 63, 103, 139,
  185, 198, 206, 218, 221, 233, 240
Bird, J. 12, 16, 20, 22–23, 28, 31n3, 33n43
Blake, R.E. 24–25, 54–55, 68, 103, 119,
  123–125, 127, 139, 168
blame 13, 94, 123–124, 126, 142, 167,
  168–169
Boston Elevated Railway Company
  52–53, 162
Box, G.E.P. 257
Burbank, J.A. 21, 26, 29
Bureau of Labor Statistics 22, 185
Butler, L.F. 19–20

carelessness 47, 52, 72n33, 110–111, 113, 122, 151, 166–168, 169, 176, 210
Carnegie, D. 83
cause code 22, 34, 37, 113–114, 120–122, 127, 137, 185; boiler 27, 122
causes 40, 46–47, 109–111, 151, 156–158; analysis 78–80, 109–110, 133, 138, 168, 183–185, 200; attribution 18, 35–37, 40, 48, 64, 70, 109, 112–113, 123, 125–127, 140, 146, 166, 168, 231; common 206, 211, 226–229, 244, 247n28; definition 110, 135–136, 146n1; direct 18, 89, 124–125, 130, 134–138, 146, 154, 156–158, 164, 211, 226; dualism 115–127; mechanical 116–119, 152, 154, 195; mental 50, 52, 125, 174n13; physical (see mechanical causes); proximate (see direct causes); real 110, 112–115, 168, 208; reasons 26, 64, 79, 110, 122–123, 127, 128, 130, 141–143, 151, 157, 174n12, 195–196; reversal of 114; root 61, 64, 66, 110, 112, 132, 135–136, 145, 146, 219, 227; secondary (see reasons); so-called 48, 70, 113–114; subcause (see reasons); subreason (see underlying); supervisory 32n18, 35–36, 116–117, 127, 154, 155, 179; true (see real causes); underlying 60, 110, 128, 137, 138, 140, 143–145, 151, 159, 164, 174n12
causality: multi 51, 61, 112, 119, 126, 132, 139, 140–141; single 112, 126, 140
Center for Safety Education 24, 54–55
chain of supervision 181–182
Chase, S. 51, 71n12, 118
Cold War 252
common sense 3, 16, 57, 81, 90, 101, 114, 133, 155, 179, 202, 227, 229, 233, 238, 250
complexity 67, 134, 135, 148n27, 149n31, 158, 242, 244, 255
Conklin, T. 124, 229
constructivism 89, 112–114, 126–127, 146
corrective action see action
costs of accidents 14, 28, 40, 49, 81, 90, 106–107, 150, 195–196; critique 66, 88–90, 103–104; hidden 16–17, 35, 41, 44, 51, 56, 60, 86, 99–103; indirect (see hidden); see also ratio indirect costs
Cowee, G.A. 43, 48–49, 78

creating interest see motivation
Crosby, R.J. 21, 24, 47, 54
Cutter, W.A. 24, 54
Civil Works Administration (CWA) 21, 105

DeBlois, L.A. 32n36, 43, 49–50, 54, 56, 71n15, 76, 95, 111–112, 123, 128, 136, 165, 178, 186, 192–193, 211, 249, 254
decoy effect 242, 243
Deepwater Horizon 66, 220, 228
Dekker, S.W.A. 4, 9n3, 74, 82, 197, 256
Deming, W.E. 79, 247n28
Department of Labor 13, 24, 25, 26, 37, 122, 152, 180, 197
Depression 20, 43, 56, 94, 104–105, 250
Det Norske Veritas (DNV) 60, 139, 198
discipline 49, 50, 142, 146, 156, 160, 167, 169, 182–183, 196–197, 227
dominos 22, 25, 35, 38, 40, 57, 120, 128–139, 141–145, 164, 175n22, 217–218; Bird's 60, 139, 130; critique 66, 131–138, 140–141, 156; developments 60, 61, 68, 130, 138–140; Lateiner's 130, 139
Dow, M.A. 53
Downey, E.H. 111
Dupin, C. 172n2
DuPont 49, 63, 97, 176, 186

early warning see weak signal
Eastman, C. 110–111, 122–123, 167
economics of safety 40, 94–96, 196; see also costs of accidents
education see safety education
efficiency 25, 40, 47, 48, 86, 95–99, 101, 103, 107, 179, 192
efficiency thoroughness trade-off (ETTO) 5, 7, 134
engineering revision 41, 51, 62, 123, 141–142, 146, 153, 157–158, 160, 178, 182–183, 197, 231, 253
eugenics 163–164

fatality see accident outcome
fault of person 120, 129–130, 142; see also causes: reason
Fisher, B. 43, 50, 52, 125–126, 151, 155, 159, 165
Five Why 134
Ford, H. 84
foremen see management

foresight *see* prediction
Formula for Supervision 27, 34, 38, 55, 183, 203n9
Foundation of a Major Injury 18, 20, 128, 206–209; *see also* ratio accident

Gantt, H. 189
Garner, J.R. 158
Geller, S.E. 63–64
Gilmour, G. 46–47, 111
good business 40, 96, 101, 103, 105, 107
Granniss, E. 20–22, 24–27, 29–30, 47, 54, 76, 197
Grimaldi, J. 24, 54, 68, 103
guarding 19, 42, 50–51, 62, 78, 83, 111, 118, 123, 143, 150, 152–154, 178, 187; prime remedy 41, 157–158, 173n10; principles 23, 252–253; standards 38

Hale, A.R. 65, 89, 220, 229, 232, 237
Hartford 16; intelligentsia 55, 76
hazard 43, 85, 123, 127, 222, 243; identification 79, 85, 153, 197; in accident sequence 129, 131, 134–135, 142–143, 148n19, 157–158; in Axioms 195–196; influence on outcome 225; type 228, 233–234
Hazard Thru' Track 26, 42, 197
Heinrich, H.W.: family 11, 16, 21, 23, 26, 28–29, 31n12; optimism 56, 251, 257; personal life (*see* family); symbol 69
Heinrich's work 34–44; activism 91; awards 27, 30; citations 3–4; critique 2, 36, 65–66, 68–69, 88–91, 93n16, 103–104, 123, 124–128, 131–138, 140–141, 156, 159, 184, 209, 203–239; data 20, 32n14, 65–66, 74, 88–91, 92, 93n16, 104, 115, 125, 209; derivatives 57, 68–69; development 35–38; framework 35, 38, 42, 45, 48, 57, 78, 194; importance and influence 2, 11, 56–68, 120, 124, 133, 138–140, 217–220, 254–258; longevity 56–58; metaphors 39, 41, 42, 45, 56–58, 68–69, 129–131, 154, 181, 198–199, 217, 254–255; models 39, 68–69; persuasion 45, 68, 90, 106; phases 35; research 17–18, 19, 76, 88–91, 100–102, 115–119, 166, 209–210, 215; storytelling 29, 55–58, 71n17, 83, 87, 150, 164, 187, 192; themes 39–43
Heinrich's Law *see* triangle

hierarchy: control 47, 64, 142, 156, 253; supervisory action 182, 197
Hill, N. 83
Hollnagel, E. 4, 67, 82, 112–113, 132–133, 257–258
Hoover, H. 99–100
Hopkins, A. 9n6, 132, 187, 222, 231, 244
human element 40–41, 48, 49, 50, 52–53, 118, 130, 150, 187, 256; asset 41, 169–171, 256; control 40, 150–157; error 9n2, 61–62, 67, 122, 144, 165–166, 226; fatigue 38, 40, 42, 117, 150, 158, 163; man failure 21n18, 36, 40, 41, 44, 64, 116, 118, 123–124, 128, 141, 150–152, 157–158, 165, 168; violation 52, 87, 151, 155, 167, 169, 188, 210, 213, 227, 231
human &organisational performance (HOP) 62, 67
humanitarian 10n10, 40, 43, 46, 50, 95, 97, 101, 151, 154, 186, 196, 203n11, 250

iceberg *see* triangle
incentives 38, 61, 94–95, 101, 156, 196, 250
Industrial Accident Prevention 2, 3–4, 20, 25, 27, 29, 34, 36, 37–38, 45, 55, 61, 66, 77–78, 82–83, 90, 118–119, 141, 154, 158, 161, 174n12, 190, 198, 202, 212, 214, 218, 253
initiative 87, 154, 170, 171, 179, 189–190
injury *see* accident outcome
insurance companies 10n10, 13–14, 16, 19, 37, 46, 92n8, 95, 152, 209, 249
integrating safety and production 40, 41, 51, 56, 96, 100–101, 103, 157, 176, 191–192, 199, 250, 256
International Safety Rating System 60, 198
investigation *see* cause analysis

Kinsey, A. 91
Knight, E.P. 54
Komaki, J. 63

lagging indicators 201
Lange, F.G. 49, 50–51, 54, 78, 99, 111, 126, 158, 164, 167, 184
language 5, 9n5, 10n10, 27, 45, 53, 76, 113, 146, 151, 169, 173n6, 179, 181, 193, 213

Lansburgh, R.H. 96, 177, 178, 203n2
Lateiner, A.R. 30, 34, 59, 130, 139,
    175n22, 176, 185, 217, 224
Le Coze, J-C 5, 57–58, 68–69
leadership *see* management
Lindy effect 255
linearity 67, 129, 131–134, 236, 244
Lippert, F. 131, 139, 142
local rationality 2, 9n3, 75, 157
Long, R. 11, 69, 201
loss control 59–60, 68, 103, 139,
    185, 218

man failure *see* human element
management: books 34, 180–183;
    education (*see* professionalization);
    employer 13, 19, 41, 46–47, 79, 82,
    92n11, 94–96, 103, 116–117, 153,
    156, 169, 176–177, 186–191, 239, 250;
    executive 78–79, 85–86, 98–99, 154,
    171, 176, 186–189, 190–194; foremen
    14, 21, 26, 48–51, 140, 147n9, 153,
    175n23, 177–186; leadership 41,
    98, 155, 173n6, 179, 180, 189, 198;
    misconceptions 201–202; tools 56,
    193, 197
Manuele, F.A. 4, 11, 36, 65–66, 74,
    88, 104, 124, 127, 135, 159, 184,
    209, 212
Marriott, C. 2, 11, 202, 244
Marx, K. 203n11
Mastery of the Machine 20, 37, 51, 118
Message to the Foremen 71n17, 87, 190,
    211, 246n26
metaphor 8n2, 39, 41, 42, 45, 56–58,
    68–69, 129–131, 133–134, 135, 181,
    198–199, 211, 217, 237, 254–255, 257;
    see also chain of supervision; domino;
    safety ladder; triangle
metric 92n9, 230, 237–238, 242
motion study 81, 85–87, 92n12
motivation 38, 61, 139, 154, 156,
    160, 169
Münsterberg, H. 159

National Safety Council 18, 21, 24, 47,
    55, 65, 104, 118, 185, 248
near-miss *see* no-injury accident
new employee 50, 158, 163
"new view" safety 1, 62, 67–68,
    82, 85, 132, 170–171, 185,
    220, 256
non-English speaking 163–165
Nyce, J. 255

one size fits all 88, 155, 253
opportunity 41, 90, 111, 161, 177, 184,
    187, 189, 190, 195–196, 200, 207, 211,
    213, 216, 218, 223, 224, 230, 234–235,
    237–238, 240, 244, 257–258
Origin of Accidents 18, 20, 73, 102, 109,
    110, 113, 115–117, 127, 140–141, 150,
    179, 206, 209, 248; chart 35–36, 40,
    117, 119–120, 121, 170; *see also* ratio
    direct causes

patriotism 43, 249, 251
Pavlov, I.P. 63, 155
PDCA 79, 182, 199
persuasion 45, 57, 90, 106, 142, 182–183,
    196–197
Petersen, D. 4, 6, 30, 45, 60–62, 65, 90,
    104, 125, 132, 141, 186, 219, 225,
    228, 233
Peterson, G.E. 13, 29
Pittsburgh Survey 111, 167
planning 79, 85, 111, 120, 139, 182,
    188, 198
potential 43, 67, 89, 206, 209–210, 219,
    222–223, 229–230, 236, 238, 240,
    241–243
practicality 42, 48, 49, 50, 53, 56, 57–58,
    73–74, 80–82, 89, 105, 128, 133–137,
    145, 160–161, 181, 193–194, 202, 230,
    239, 252–253, 257
precursor 90, 219, 223, 228, 234, 236,
    241, 243
prediction 234–237
preventive action *see* action
professionalisation: foremen 180–183;
    safety engineers 35, 40, 42, 46, 161,
    197, 248–250
proportionality 234–235, 247n31
psychology 40, 53–54, 61, 63, 64, 65,
    128, 130, 143, 145, 154–155, 156,
    158–161, 174n12–14
public awareness 95
pyramid *see* triangle

racism 163–164
Rasmussen, J. 9, 67, 106, 112, 168
ratio: accident (1:29:300) 18–19, 59,
    205–209, 211, 216, 218–219; critique
    65–66, 88–91, 93n16, 103–104, 123,
    124–128, 209, 203–239; direct causes
    (88:10:2) 18, 36, 64, 123, 124–128;
    fixed 232–233; indirect costs (1:4) 17,
    40, 98–104; stable (*see* fixed); *see also*
    Origin of Accidents; triangle

Rausch, C. 248
Reason, J. 107, 140, 144, 230
reductionism 149n31
remedy *see* action
resilience 67, 88, 170–172, 255
responsibility 79, 85, 123–124, 126,
    155, 189–192; allocating 168–169;
    employer 13, 41, 46–47, 176–177,
    187–188, 239, 250; executives 99;
    foremen 25, 178–181; management
    36, 44, 200, 256; safety engineer 252;
    workers 64, 94
Reward of Merit 80, 171–172
Richardson, A.S. 44, 201–211
risk 23, 26, 43, 61, 64, 67, 168, 171, 190,
    209, 237, 241; assessment 146; factors
    158; group 163; homeostasis 106, 243;
    risk perception 7, 168
Risteen, A. 14–15, 47, 76
Roos, N. 30, 61–62, 90, 218–219, 225, 233
Russell, B. 132

safety: bureaucracy 80, 107n3, 201–202,
    243; congress 17, 18, 24, 54, 95, 100,
    110, 120, 128, 131, 147n9, 170, 171,
    174n14, 184, 206, 249, 252; culture 50,
    62, 115, 130, 144, 164–165, 227–228;
    education 14, 24, 27, 42, 49, 68, 137,
    153–154, 165, 178, 184; engineer
    14, 42, 45–46, 122, 160–161, 192,
    197, 248–250; first 17, 95, 154, 166,
    178, 188, 193; handbooks 58–54, 68,
    103; home 35, 252; ladder 38, 42,
    198–199, 251; management 4, 38, 42,
    103–104, 193–200, 236; margin 106;
    organisation 38, 42, 50, 78, 170, 187,
    191–194, 198, 200; rules 49, 117, 124,
    127, 151, 153, 155–156, 166–167, 170,
    183, 213, 227, 243
Safety in the Small Plant 78, 128, 141, 202
Safety on the "El" 53, 159
Safety Wins a Place in the Sun 56, 79,
    190, 191, 254
Safety-II 67–68, 171
scenario 89, 187, 207, 212, 220, 222, 223,
    226, 228–229, 234–237, 243, 244
Schlesinger, L.E. 63
Schreiber, H.V. 111
scientific approach 40, 73–91, 188,
    197; principles 3, 36, 38, 40, 42,
    73–74, 77–79, 86, 183, 188, 193–194,
    197–198, 257
Scientific Management 13, 55, 74, 84–88,
    92n13, 96–97, 153, 189

second story 6, 10n11
sexism 10n10, 164
self-help 71n15, 83, 171
self-organised criticality 227–228
selling 70n10, 83, 154, 156, 185,
    200, 257
Serious Incidents & Fatalities (SIF) 62,
    66–67, 219–220, 222, 223, 237
Shewhart, W. 55, 71n14, 79, 247n28
similarity 109, 215, 222, 225–226, 227,
    228, 229, 234, 244, 258
simplicity 45, 58, 133–134, 202
Skinner, B.F. 61, 63, 173
Slocombe, C.S. 52–53, 162
social engagement 35, 43, 250–252
social environment 66, 129–130, 144,
    164–165, 175n22
Stack, H.J. 24, 32n23, 54, 55, 68,
    71n13
Stone, R.W. 44, 92n92
stop-rule 132, 137, 145, 149n34
storytelling *see* Heinrich's work
Sullivan, R.J. 21
supervision *see* management
supervisors *see* foremen
Supervisor's Safety Manual 26, 180, 213
statistics *see* accident statistics
Swiss Cheese Model 57, 68, 107,
    140, 230
systems 66, 124, 132, 144, 168, 231, 244

Taleb, N.N. 255
Taylor, F.W. 55, 74, 80, 83–88, 92n10
Taylorism *see* Scientific Management
Texas City disaster 66, 220, 222, 228,
    231, 242
theorem 74, 130–131
Travelers Insurance Co. 13–14, 22, 25,
    45, 56, 187
triangle 18, 25, 41, 57–59, 201, 257–258;
    applications 64, 66–67, 217–220; Bird
    triangle 59, 218; challenges 241–243;
    confounding 209, 226, 228–230;
    critique 36, 65, 68–69, 88–89, 215,
    230–240; development 36, 209–216,
    217–220, 257–258; interpretations
    68–69, 217, 221–230, 230–238,
    240–244; principles 128, 205, 221–230,
    244; reading 221–224; research 102,
    209–210

U.S. Steel 116, 176, 186, 191
Uniform Boiler & Pressure Vessel
    Society 16, 29–30, 34

unit group 212, 222, 225–226, 240
unsafe acts *see* human element
unsafe conditions *see* mechanical cause

Van Schaack, D. 48, 166
Vaughan, D. 241–242
Vernon, H.M. 68, 76, 163
Villevoye, J. 258
visibility 217, 224
Viteles, M.S. 54, 55, 159

War Department 26
Watson, J.B. 63, 155–156, 173n7

weak signal 41, 230, 236, 240–241;
challenges 241–244; opportunity
240, 244
Weaver, D. 60, 62, 139
Whitney, A.W. 32n24, 55, 71n13, 96, 98
Williams, S. 43, 49, 51, 54, 68, 99, 211
worker compensation 13–14, 17, 46,
94–95, 99–100, 122, 127, 193
World War: First 15, 21, 71n12, 87,
96; Second 20, 26, 28, 43, 55, 57, 59,
180, 197

zero 238–240; harm 154, 239

For Product Safety Concerns and Information please contact our EU
representative GPSR@taylorandfrancis.com
Taylor & Francis Verlag GmbH, Kaufingerstraße 24, 80331 München, Germany

www.ingramcontent.com/pod-product-compliance
Ingram Content Group UK Ltd.
Pitfield, Milton Keynes, MK11 3LW, UK
UKHW020718141025
8378UKWH00036B/805

* 9 780367 704568 *